Monographs on Statistics and Applied Probability 143

Statistical Learning with Sparsity
The Lasso and Generalizations

Trevor Hastie
Stanford University
USA

Robert Tibshirani
Stanford University
USA

Martin Wainwright
University of California, Berkeley
USA

CRC Press
Taylor & Francis Group
Boca Raton London New York

CRC Press is an imprint of the
Taylor & Francis Group, an **informa** business
A CHAPMAN & HALL BOOK

CRC Press
Taylor & Francis Group
6000 Broken Sound Parkway NW, Suite 300
Boca Raton, FL 33487-2742

First issued in paperback 2020

© 2015 by Taylor & Francis Group, LLC
CRC Press is an imprint of Taylor & Francis Group, an Informa business

No claim to original U.S. Government works

ISBN-13: 978-1-4987-1216-3 (hbk)
ISBN-13: 978-0-367-73833-4 (pbk)

Library of Congress Cataloging-in-Publication Data

Hastie, Trevor.
   Statistical learning with sparsity : the lasso and generalizations / Trevor Hastie, Rob Tibshirani, Martin Wainwright.
       pages cm. -- (Chapman & Hall/CRC monographs on statistics & applied
       probability ; 143)
   "A CRC title."
   Includes bibliographical references and index.
   ISBN 978-1-4987-1216-3 (hardcover : alk. paper) 1. Mathematical statistics. 2. Least
   squares. 3. Linear models (Statistics) 4. Proof theory. I. Tibshirani, Robert. II.
   Wainwright, Martin (Martin J.) III. Title.

QA275.H38 2015
519.5--dc23   2015014842

Visit the Taylor & Francis Web site at
http://www.taylorandfrancis.com

and the CRC Press Web site at
http://www.crcpress.com

MONOGRAPHS ON STATISTICS AND APPLIED PROBABILITY

General Editors

F. Bunea, V. Isham, N. Keiding, T. Louis, R. L. Smith, and H. Tong

1. Stochastic Population Models in Ecology and Epidemiology *M.S. Barlett* (1960)
2. Queues *D.R. Cox and W.L. Smith* (1961)
3. Monte Carlo Methods *J.M. Hammersley and D.C. Handscomb* (1964)
4. The Statistical Analysis of Series of Events *D.R. Cox and P.A.W. Lewis* (1966)
5. Population Genetics *W.J. Ewens* (1969)
6. Probability, Statistics and Time *M.S. Barlett* (1975)
7. Statistical Inference *S.D. Silvey* (1975)
8. The Analysis of Contingency Tables *B.S. Everitt* (1977)
9. Multivariate Analysis in Behavioural Research *A.E. Maxwell* (1977)
10. Stochastic Abundance Models *S. Engen* (1978)
11. Some Basic Theory for Statistical Inference *E.J.G. Pitman* (1979)
12. Point Processes *D.R. Cox and V. Isham* (1980)
13. Identification of Outliers *D.M. Hawkins* (1980)
14. Optimal Design *S.D. Silvey* (1980)
15. Finite Mixture Distributions *B.S. Everitt and D.J. Hand* (1981)
16. Classification *A.D. Gordon* (1981)
17. Distribution-Free Statistical Methods, 2nd edition *J.S. Maritz* (1995)
18. Residuals and Influence in Regression *R.D. Cook and S. Weisberg* (1982)
19. Applications of Queueing Theory, 2nd edition *G.F. Newell* (1982)
20. Risk Theory, 3rd edition *R.E. Beard, T. Pentikäinen and E. Pesonen* (1984)
21. Analysis of Survival Data *D.R. Cox and D. Oakes* (1984)
22. An Introduction to Latent Variable Models *B.S. Everitt* (1984)
23. Bandit Problems *D.A. Berry and B. Fristedt* (1985)
24. Stochastic Modelling and Control *M.H.A. Davis and R. Vinter* (1985)
25. The Statistical Analysis of Composition Data *J. Aitchison* (1986)
26. Density Estimation for Statistics and Data Analysis *B.W. Silverman* (1986)
27. Regression Analysis with Applications *G.B. Wetherill* (1986)
28. Sequential Methods in Statistics, 3rd edition *G.B. Wetherill and K.D. Glazebrook* (1986)
29. Tensor Methods in Statistics *P. McCullagh* (1987)
30. Transformation and Weighting in Regression *R.J. Carroll and D. Ruppert* (1988)
31. Asymptotic Techniques for Use in Statistics *O.E. Bandorff-Nielsen and D.R. Cox* (1989)
32. Analysis of Binary Data, 2nd edition *D.R. Cox and E.J. Snell* (1989)
33. Analysis of Infectious Disease Data *N.G. Becker* (1989)
34. Design and Analysis of Cross-Over Trials *B. Jones and M.G. Kenward* (1989)
35. Empirical Bayes Methods, 2nd edition *J.S. Maritz and T. Lwin* (1989)
36. Symmetric Multivariate and Related Distributions *K.T. Fang, S. Kotz and K.W. Ng* (1990)
37. Generalized Linear Models, 2nd edition *P. McCullagh and J.A. Nelder* (1989)
38. Cyclic and Computer Generated Designs, 2nd edition *J.A. John and E.R. Williams* (1995)
39. Analog Estimation Methods in Econometrics *C.F. Manski* (1988)
40. Subset Selection in Regression *A.J. Miller* (1990)
41. Analysis of Repeated Measures *M.J. Crowder and D.J. Hand* (1990)
42. Statistical Reasoning with Imprecise Probabilities *P. Walley* (1991)
43. Generalized Additive Models *T.J. Hastie and R.J. Tibshirani* (1990)
44. Inspection Errors for Attributes in Quality Control *N.L. Johnson, S. Kotz and X. Wu* (1991)
45. The Analysis of Contingency Tables, 2nd edition *B.S. Everitt* (1992)
46. The Analysis of Quantal Response Data *B.J.T. Morgan* (1992)
47. Longitudinal Data with Serial Correlation—A State-Space Approach *R.H. Jones* (1993)

To our parents:

Valerie and Patrick Hastie

Vera and Sami Tibshirani

Patricia and John Wainwright

and to our families:

Samantha, Timothy, and Lynda

Charlie, Ryan, Jess, Julie, and Cheryl

Haruko and Hana

Contents

Preface

In this monograph, we have attempted to summarize the actively developing field of statistical learning with sparsity. A sparse statistical model is one having only a small number of nonzero parameters or weights. It represents a classic case of "*less is more*": a sparse model can be much easier to estimate and interpret than a dense model. In this age of big data, the number of features measured on a person or object can be large, and might be larger than the number of observations. The sparsity assumption allows us to tackle such problems and extract useful and reproducible patterns from big datasets.

The ideas described here represent the work of an entire community of researchers in statistics and machine learning, and we thank everyone for their continuing contributions to this exciting area. We particularly thank our colleagues at Stanford, Berkeley and elsewhere; our collaborators, and our past and current students working in this area. These include Alekh Agarwal, Arash Amini, Francis Bach, Jacob Bien, Stephen Boyd, Andreas Buja, Emmanuel Candes, Alexandra Chouldechova, David Donoho, John Duchi, Brad Efron, Will Fithian, Jerome Friedman, Max G'Sell, Iain Johnstone, Michael Jordan, Ping Li, Po-Ling Loh, Michael Lim, Jason Lee, Richard Lockhart, Rahul Mazumder, Balasubramanian Narashimhan, Sahand Negahban, Guillaume Obozinski, Mee-Young Park, Junyang Qian, Garvesh Raskutti, Pradeep Ravikumar, Saharon Rosset, Prasad Santhanam, Noah Simon, Dennis Sun, Yukai Sun, Jonathan Taylor, Ryan Tibshirani,[1] Stefan Wager, Daniela Witten, Bin Yu, Yuchen Zhang, Ji Zhou, and Hui Zou. We also thank our editor John Kimmel for his advice and support.

Stanford University
and
University of California, Berkeley

Trevor Hastie
Robert Tibshirani
Martin Wainwright

---

[1]Some of the bibliographic references, for example in Chapters 4 and 6, are to Tibshirani$_2$, R.J., rather than Tibshirani, R.; the former is Ryan Tibshirani, the latter is Robert (son and father).

Chapter 1

Introduction

"I never keep a scorecard or the batting averages. I hate statistics. What I got to know, I keep in my head."

This is a quote from baseball pitcher Dizzy Dean, who played in the major leagues from 1930 to 1947.

How the world has changed in the 75 or so years since that time! Now large quantities of data are collected and mined in nearly every area of science, entertainment, business, and industry. Medical scientists study the genomes of patients to choose the best treatments, to learn the underlying causes of their disease. Online movie and book stores study customer ratings to recommend or sell them new movies or books. Social networks mine information about members and their friends to try to enhance their online experience. And yes, most major league baseball teams have statisticians who collect and analyze detailed information on batters and pitchers to help team managers and players make better decisions.

Thus the world is awash with data. But as Rutherford D. Roger (and others) has said:

"We are drowning in information and starving for knowledge."

There is a crucial need to sort through this mass of information, and pare it down to its bare essentials. For this process to be successful, we need to hope that the world is not as complex as it might be. For example, we hope that not all of the 30,000 or so genes in the human body are directly involved in the process that leads to the development of cancer. Or that the ratings by a customer on perhaps 50 or 100 different movies are enough to give us a good idea of their tastes. Or that the success of a left-handed pitcher against left-handed batters will be fairly consistent for different batters.

This points to an underlying assumption of simplicity. One form of simplicity is *sparsity*, the central theme of this book. Loosely speaking, a sparse statistical model is one in which only a relatively small number of parameters (or predictors) play an important role. In this book we study methods that exploit sparsity to help recover the underlying signal in a set of data.

The leading example is linear regression, in which we observe N observations of an outcome variable y_i and p associated predictor variables (or features) $x_i = (x_{i1}, \ldots x_{ip})^T$. The goal is to predict the outcome from the predictors, both for actual prediction with future data and also to discover which predictors play an important role. A linear regression model assumes that

$$y_i = \beta_0 + \sum_{j=1}^{p} x_{ij}\beta_j + e_i, \qquad (1.1)$$

where β_0 and $\beta = (\beta_1, \beta_2, \ldots \beta_p)$ are unknown parameters and e_i is an error term. The method of least squares provides estimates of the parameters by minimization of the least-squares objective function

$$\underset{\beta_0, \beta}{\text{minimize}} \sum_{i=1}^{N} (y_i - \beta_0 - \sum_{j=1}^{p} x_{ij}\beta_j)^2. \qquad (1.2)$$

Typically all of the least-squares estimates from (1.2) will be nonzero. This will make interpretation of the final model challenging if p is large. In fact, if $p > N$, the least-squares estimates are not unique. There is an infinite set of solutions that make the objective function equal to zero, and these solutions almost surely overfit the data as well.

Thus there is a need to constrain, or *regularize* the estimation process. In the *lasso* or ℓ_1-*regularized regression*, we estimate the parameters by solving the problem

$$\underset{\beta_0, \beta}{\text{minimize}} \sum_{i=1}^{N} (y_i - \beta_0 - \sum_{j=1}^{p} x_{ij}\beta_j)^2 \ \ \text{subject to} \ \|\beta\|_1 \leq t \qquad (1.3)$$

where $\|\beta\|_1 = \sum_{j=1}^{p} |\beta_j|$ is the ℓ_1 norm of β, and t is a user-specified parameter. We can think of t as a budget on the total ℓ_1 norm of the parameter vector, and the lasso finds the best fit within this budget.

Why do we use the ℓ_1 norm? Why not use the ℓ_2 norm or any ℓ_q norm? It turns out that the ℓ_1 norm is special. If the budget t is small enough, the lasso yields sparse solution vectors, having only some coordinates that are nonzero. This does not occur for ℓ_q norms with $q > 1$; for $q < 1$, the solutions are sparse but the problem is not convex and this makes the minimization very challenging computationally. The value $q = 1$ is the smallest value that yields a convex problem. Convexity greatly simplifies the computation, as does the sparsity assumption itself. They allow for scalable algorithms that can handle problems with even millions of parameters.

Thus the advantages of sparsity are interpretation of the fitted model and computational convenience. But a third advantage has emerged in the last few years from some deep mathematical analyses of this area. This has been termed the "bet on sparsity" principle:

Use a procedure that does well in sparse problems, since no procedure does well in dense problems.

We can think of this in terms of the amount of information N/p per parameter. If $p \gg N$ and the true model is not sparse, then the number of samples N is too small to allow for accurate estimation of the parameters. But if the true model is sparse, so that only $k < N$ parameters are actually nonzero in the true underlying model, then it turns out that we can estimate the parameters effectively, using the lasso and related methods that we discuss in this book. This may come as somewhat of a surprise, because we are able to do this even though we are not told *which* k of the p parameters are actually nonzero. Of course we cannot do as well as we could if we had that information, but it turns out that we can still do reasonably well.

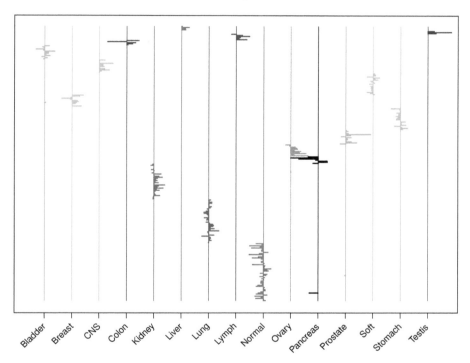

Figure 1.1 *15-class gene expression cancer data: estimated nonzero feature weights from a lasso-regularized multinomial classifier. Shown are the 254 genes (out of 4718) with at least one nonzero weight among the 15 classes. The genes (unlabelled) run from top to bottom. Line segments pointing to the right indicate positive weights, and to the left, negative weights. We see that only a handful of genes are needed to characterize each class.*

For all of these reasons, the area of sparse statistical modelling is exciting—for data analysts, computer scientists, and theorists—and practically useful. Figure 1.1 shows an example. The data consists of quantitative gene expression measurements of 4718 genes on samples from 349 cancer patients. The cancers have been categorized into 15 different types such as "Bladder," "Breast",

"CNS," etc. The goal is to build a classifier to predict cancer class based on some or all of the 4718 features. We want the classifier to have a low error rate on independent samples and would prefer that it depend only on a subset of the genes, to aid in our understanding of the underlying biology.

For this purpose we applied a lasso-regularized multinomial classifier to these data, as described in Chapter 3. This produces a set of 4718 weights or coefficients for each of the 15 classes, for discriminating each class from the rest. Because of the ℓ_1 penalty, only some of these weights may be nonzero (depending on the choice of the regularization parameter). We used cross-validation to estimate the optimal choice of regularization parameter, and display the resulting weights in Figure 1.1. Only 254 genes have at least one nonzero weight, and these are displayed in the figure. The cross-validated error rate for this classifier is about 10%, so the procedure correctly predicts the class of about 90% of the samples. By comparison, a standard support vector classifier had a slightly higher error rate (13%) using all of the features. Using sparsity, the lasso procedure has dramatically reduced the number of features without sacrificing accuracy. Sparsity has also brought computational efficiency: although there are potentially $4718 \times 15 \approx 70,000$ parameters to estimate, the entire calculation for Figure 1.1 was done on a standard laptop computer in less than a minute. For this computation we used the **glmnet** procedure described in Chapters 3 and 5.

Figure 1.2 shows another example taken from an article by Candès and Wakin (2008) in the field of *compressed sensing*. On the left is a megapixel image. In order to reduce the amount of space needed to store the image, we represent it in a wavelet basis, whose coefficients are shown in the middle panel. The largest 25,000 coefficients are then retained and the rest zeroed out, yielding the excellent reconstruction in the right image. This all works because of sparsity: although the image seems complex, in the wavelet basis it is simple and hence only a relatively small number of coefficients are nonzero. The original image can be perfectly recovered from just 96,000 incoherent measurements. Compressed sensing is a powerful tool for image analysis, and is described in Chapter 10.

In this book we have tried to summarize the hot and rapidly evolving field of sparse statistical modelling. In Chapter 2 we describe and illustrate the lasso for linear regression, and a simple coordinate descent algorithm for its computation. Chapter 3 covers the application of ℓ_1 penalties to generalized linear models such as multinomial and survival models, as well as support vector machines. Generalized penalties such as the elastic net and group lasso are discussed in Chapter 4. Chapter 5 reviews numerical methods for optimization, with an emphasis on first-order methods that are useful for the large-scale problems that are discussed in this book. In Chapter 6, we discuss methods for statistical inference for fitted (lasso) models, including the bootstrap, Bayesian methods and some more recently developed approaches. Sparse matrix decomposition is the topic of Chapter 7, and we apply these methods in the context of sparse multivariate analysis in Chapter 8. Graph-

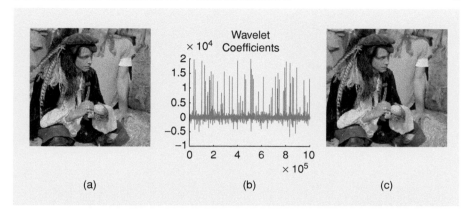

Figure 1.2 *(a) Original megapixel image with pixel values in the range* $[0, 255]$ *and (b) its wavelet transform coefficients (arranged in random order for enhanced visibility). Relatively few wavelet coefficients capture most of the signal energy; many such images are highly compressible. (c) The reconstruction obtained by zeroing out all the coefficients in the wavelet expansion but the* $25,000$ *largest (pixel values are thresholded to the range* $[0, 255]$ *). The differences from the original picture are hardly noticeable.*

ical models and their selection are discussed in Chapter 9 while compressed sensing is the topic of Chapter 10. Finally, a survey of theoretical results for the lasso is given in Chapter 11.

We note that both *supervised* and *unsupervised* learning problems are discussed in this book, the former in Chapters 2, 3, 4, and 10, and the latter in Chapters 7 and 8.

Notation

We have adopted a notation to reduce mathematical clutter. Vectors are column vectors by default; hence $\beta \in \mathbb{R}^p$ is a column vector, and its transpose β^T is a row vector. All vectors are lower case and non-bold, except N-vectors which are bold, where N is the sample size. For example $\mathbf{x}_j$ might be the N-vector of observed values for the jth variable, and $\mathbf{y}$ the response N-vector. All matrices are bold; hence $\mathbf{X}$ might represent the $N \times p$ matrix of observed predictors, and $\boldsymbol{\Theta}$ a $p \times p$ precision matrix. This allows us to use $x_i \in \mathbb{R}^p$ to represent the vector of p features for observation i (i.e., x_i^T is the ith row of $\mathbf{X}$), while $\mathbf{x}_k$ is the kth column of $\mathbf{X}$, without ambiguity.

Chapter 2

The Lasso for Linear Models

In this chapter, we introduce the lasso estimator for linear regression. We describe the basic lasso method, and outline a simple approach for its implementation. We relate the lasso to ridge regression, and also view it as a Bayesian estimator.

2.1 Introduction

In the linear regression setting, we are given N samples $\{(x_i, y_i)\}_{i=1}^N$, where each $x_i = (x_{i1}, \ldots, x_{ip})$ is a p-dimensional vector of features or predictors, and each $y_i \in \mathbb{R}$ is the associated response variable. Our goal is to approximate the response variable y_i using a linear combination of the predictors

$$\eta(x_i) = \beta_0 + \sum_{j=1}^{p} x_{ij}\beta_j. \tag{2.1}$$

The model is parametrized by the vector of regression weights $\beta = (\beta_1, \ldots, \beta_p) \in \mathbb{R}^p$ and an intercept (or "bias") term $\beta_0 \in \mathbb{R}$.

The usual "least-squares" estimator for the pair (β_0, β) is based on minimizing squared-error loss:

$$\underset{\beta_0, \beta}{\text{minimize}} \left\{ \frac{1}{2N} \sum_{i=1}^{N} \left(y_i - \beta_0 - \sum_{j=1}^{p} x_{ij}\beta_j \right)^2 \right\}. \tag{2.2}$$

There are two reasons why we might consider an alternative to the least-squares estimate. The first reason is *prediction accuracy*: the least-squares estimate often has low bias but large variance, and prediction accuracy can sometimes be improved by shrinking the values of the regression coefficients, or setting some coefficients to zero. By doing so, we introduce some bias but reduce the variance of the predicted values, and hence may improve the overall prediction accuracy (as measured in terms of the mean-squared error). The second reason is for the purposes of *interpretation*. With a large number of predictors, we often would like to identify a smaller subset of these predictors that exhibit the strongest effects.

This chapter is devoted to discussion of the *lasso*, a method that combines the least-squares loss (2.2) with an ℓ_1-constraint, or bound on the sum of the absolute values of the coefficients. Relative to the least-squares solution, this constraint has the effect of shrinking the coefficients, and even setting some to zero.[1] In this way it provides an automatic way for doing model selection in linear regression. Moreover, unlike some other criteria for model selection, the resulting optimization problem is convex, and can be solved efficiently for large problems.

2.2 The Lasso Estimator

Given a collection of N predictor-response pairs $\{(x_i, y_i)\}_{i=1}^N$, the lasso finds the solution $(\widehat{\beta}_0, \widehat{\beta})$ to the optimization problem

$$\underset{\beta_0, \beta}{\text{minimize}} \left\{ \frac{1}{2N} \sum_{i=1}^N (y_i - \beta_0 - \sum_{j=1}^p x_{ij}\beta_j)^2 \right\}$$

$$\text{subject to } \sum_{j=1}^p |\beta_j| \leq t. \tag{2.3}$$

The constraint $\sum_{j=1}^p |\beta_j| \leq t$ can be written more compactly as the ℓ_1-norm constraint $\|\beta\|_1 \leq t$. Furthermore, (2.3) is often represented using matrix-vector notation. Let $\mathbf{y} = (y_1, \ldots, y_N)$ denote the N-vector of responses, and $\mathbf{X}$ be an $N \times p$ matrix with $x_i \in \mathbb{R}^p$ in its ith row, then the optimization problem (2.3) can be re-expressed as

$$\underset{\beta_0, \beta}{\text{minimize}} \left\{ \frac{1}{2N} \|\mathbf{y} - \beta_0 \mathbf{1} - \mathbf{X}\beta\|_2^2 \right\}$$

$$\text{subject to } \|\beta\|_1 \leq t, \tag{2.4}$$

where $\mathbf{1}$ is the vector of N ones, and $\| \cdot \|_2$ denotes the usual Euclidean norm on vectors. The bound t is a kind of "budget": it limits the sum of the absolute values of the parameter estimates. Since a shrunken parameter estimate corresponds to a more heavily-constrained model, this budget limits how well we can fit the data. It must be specified by an external procedure such as cross-validation, which we discuss later in the chapter.

Typically, we first standardize the predictors $\mathbf{X}$ so that each column is centered ($\frac{1}{N}\sum_{i=1}^N x_{ij} = 0$) and has unit variance ($\frac{1}{N}\sum_{i=1}^N x_{ij}^2 = 1$). Without

---

[1]A *lasso* is a long rope with a noose at one end, used to catch horses and cattle. In a figurative sense, the method "lassos" the coefficients of the model. In the original lasso paper (Tibshirani 1996), the name "lasso" was also introduced as an acronym for "Least Absolute Selection and Shrinkage Operator."

Pronunciation: in the US "lasso" tends to be pronounced "lass-oh" (oh as in goat), while in the UK "lass-oo." In the OED (2nd edition, 1965): "lasso is pronounced lăsoo by those who use it, and by most English people too."

standardization, the lasso solutions would depend on the units (e.g., feet versus meters) used to measure the predictors. On the other hand, we typically would not standardize if the features were measured in the same units. For convenience, we also assume that the outcome values y_i have been centered, meaning that $\frac{1}{N}\sum_{i=1}^{N} y_i = 0$. These centering conditions are convenient, since they mean that we can omit the intercept term β_0 in the lasso optimization. Given an optimal lasso solution $\widehat{\beta}$ on the centered data, we can recover the optimal solutions for the uncentered data: $\widehat{\beta}$ is the same, and the intercept $\widehat{\beta}_0$ is given by

$$\widehat{\beta}_0 = \bar{y} - \sum_{j=1}^{p} \bar{x}_j \widehat{\beta}_j,$$

where $\bar{y}$ and $\{\bar{x}_j\}_1^p$ are the original means.[2] For this reason, we omit the intercept β_0 from the lasso for the remainder of this chapter.

It is often convenient to rewrite the lasso problem in the so-called Lagrangian form

$$\underset{\beta \in \mathbb{R}^p}{\text{minimize}} \left\{ \frac{1}{2N}\|\mathbf{y} - \mathbf{X}\beta\|_2^2 + \lambda\|\beta\|_1 \right\}, \tag{2.5}$$

for some $\lambda \geq 0$. By Lagrangian duality, there is a one-to-one correspondence between the constrained problem (2.3) and the Lagrangian form (2.5): for each value of t in the range where the constraint $\|\beta\|_1 \leq t$ is active, there is a corresponding value of λ that yields the same solution from the Lagrangian form (2.5). Conversely, the solution $\widehat{\beta}_\lambda$ to problem (2.5) solves the bound problem with $t = \|\widehat{\beta}_\lambda\|_1$.

We note that in many descriptions of the lasso, the factor $1/2N$ appearing in (2.3) and (2.5) is replaced by $1/2$ or simply 1. Although this makes no difference in (2.3), and corresponds to a simple reparametrization of λ in (2.5), this kind of standardization makes λ values comparable for different sample sizes (useful for cross-validation).

The theory of convex analysis tells us that necessary and sufficient conditions for a solution to problem (2.5) take the form

$$-\frac{1}{N}\langle\mathbf{x}_j, \mathbf{y} - \mathbf{X}\beta\rangle + \lambda s_j = 0, \; j = 1, \ldots, p. \tag{2.6}$$

Here each s_j is an unknown quantity equal to $\text{sign}(\beta_j)$ if $\beta_j \neq 0$ and some value lying in $[-1, 1]$ otherwise—that is, it is a subgradient for the absolute value function (see Chapter 5 for details). In other words, the solutions $\hat{\beta}$ to problem (2.5) are the same as solutions $(\hat{\beta}, \hat{s})$ to (2.6). This system is a form of the so-called Karush–Kuhn–Tucker (KKT) conditions for problem (2.5). Expressing a problem in subgradient form can be useful for designing

---

[2]This is typically only true for linear regression with squared-error loss; it's not true, for example, for lasso logistic regression.

algorithms for finding its solutions. More details are given in Exercises (2.3) and (2.4).

As an example of the lasso, let us consider the data given in Table 2.1, taken from Thomas (1990). The outcome is the total overall reported crime rate per

Table 2.1 *Crime data: Crime rate and five predictors, for $N = 50$ U.S. cities.*

city	funding	hs	not-hs	college	college4	crime rate
1	40	74	11	31	20	478
2	32	72	11	43	18	494
3	57	70	18	16	16	643
4	31	71	11	25	19	341
5	67	72	9	29	24	773
⋮	⋮	⋮	⋮	⋮		
50	66	67	26	18	16	940

one million residents in 50 U.S cities. There are five predictors: annual police funding in dollars per resident, percent of people 25 years and older with four years of high school, percent of 16- to 19-year olds not in high school and not high school graduates, percent of 18- to 24-year olds in college, and percent of people 25 years and older with at least four years of college. This small example is for illustration only, but helps to demonstrate the nature of the lasso solutions. Typically the lasso is most useful for much larger problems, including "wide" data for which $p \gg N$.

The left panel of Figure 2.1 shows the result of applying the lasso with the bound t varying from zero on the left, all the way to a large value on the right, where it has no effect. The horizontal axis has been scaled so that the maximal bound, corresponding to the least-squares estimates $\tilde{\beta}$, is one. We see that for much of the range of the bound, many of the estimates are exactly zero and hence the corresponding predictor(s) would be excluded from the model. Why does the lasso have this model selection property? It is due to the geometry that underlies the ℓ_1 constraint $\|\beta\|_1 \leq t$. To understand this better, the right panel shows the estimates from *ridge regression*, a technique that predates the lasso. It solves a criterion very similar to (2.3):

$$\underset{\beta_0,\beta}{\text{minimize}} \left\{ \frac{1}{2N} \sum_{i=1}^{N} (y_i - \beta_0 - \sum_{j=1}^{p} x_{ij}\beta_j)^2 \right\}$$

$$\text{subject to } \sum_{j=1}^{p} \beta_j^2 \leq t^2. \tag{2.7}$$

The ridge profiles in the right panel have roughly the same shape as the lasso profiles, but are not equal to zero except at the left end. Figure 2.2 contrasts the two constraints used in the lasso and ridge regression. The residual sum

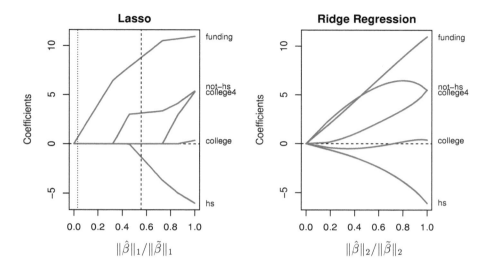

Figure 2.1 *Left: Coefficient path for the lasso, plotted versus the ℓ_1 norm of the coefficient vector, relative to the norm of the unrestricted least-squares estimate $\tilde{\beta}$. Right: Same for ridge regression, plotted against the relative ℓ_2 norm.*

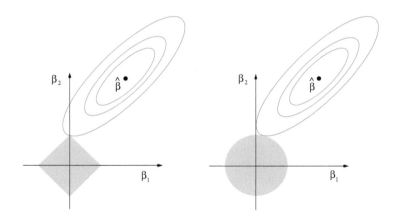

Figure 2.2 *Estimation picture for the lasso (left) and ridge regression (right). The solid blue areas are the constraint regions $|\beta_1|+|\beta_2| \le t$ and $\beta_1^2+\beta_2^2 \le t^2$, respectively, while the red ellipses are the contours of the residual-sum-of-squares function. The point $\widehat{\beta}$ depicts the usual (unconstrained) least-squares estimate.*

Table 2.2 *Results from analysis of the crime data. Left panel shows the least-squares estimates, standard errors, and their ratio (Z-score). Middle and right panels show the corresponding results for the lasso, and the least-squares estimates applied to the subset of predictors chosen by the lasso.*

	LS coef	SE	Z	Lasso	SE	Z	LS	SE	Z
funding	10.98	3.08	3.6	8.84	3.55	2.5	11.29	2.90	3.9
hs	-6.09	6.54	-0.9	-1.41	3.73	-0.4	-4.76	4.53	-1.1
not-hs	5.48	10.05	0.5	3.12	5.05	0.6	3.44	7.83	0.4
college	0.38	4.42	0.1	0.0	-	-	0.0	-	-
college4	5.50	13.75	0.4	0.0	-	-	0.0	-	-

of squares has elliptical contours, centered at the full least-squares estimates. The constraint region for ridge regression is the disk $\beta_1^2 + \beta_2^2 \le t^2$, while that for lasso is the diamond $|\beta_1| + |\beta_2| \le t$. Both methods find the first point where the elliptical contours hit the constraint region. Unlike the disk, the diamond has corners; if the solution occurs at a corner, then it has one parameter β_j equal to zero. When $p > 2$, the diamond becomes a rhomboid, and has many corners, flat edges, and faces; there are many more opportunities for the estimated parameters to be zero (see Figure 4.2 on page 58.)

We use the term *sparse* for a model with few nonzero coefficients. Hence a key property of the ℓ_1-constraint is its ability to yield sparse solutions. This idea can be applied in many different statistical models, and is the central theme of this book.

Table 2.2 shows the results of applying three fitting procedures to the crime data. The lasso bound t was chosen by cross-validation, as described in Section 2.3. The left panel corresponds to the full least-squares fit, while the middle panel shows the lasso fit. On the right, we have applied least-squares estimation to the subset of three predictors with nonzero coefficients in the lasso. The standard errors for the least-squares estimates come from the usual formulas. No such simple formula exists for the lasso, so we have used the bootstrap to obtain the estimate of standard errors in the middle panel (see Exercise 2.6; Chapter 6 discusses some promising new approaches for post-selection inference). Overall it appears that **funding** has a large effect, probably indicating that police resources have been focused on higher crime areas. The other predictors have small to moderate effects.

Note that the lasso sets two of the five coefficients to zero, and tends to shrink the coefficients of the others toward zero relative to the full least-squares estimate. In turn, the least-squares fit on the subset of the three predictors tends to expand the lasso estimates away from zero. The nonzero estimates from the lasso tend to be biased toward zero, so the debiasing in the right panel can often improve the prediction error of the model. This two-stage process is also known as the *relaxed lasso* (Meinshausen 2007).

2.3   Cross-Validation and Inference

The bound t in the lasso criterion (2.3) controls the complexity of the model; larger values of t free up more parameters and allow the model to adapt more closely to the training data. Conversely, smaller values of t restrict the parameters more, leading to sparser, more interpretable models that fit the data less closely. Forgetting about interpretability, we can ask for the value of t that gives the most accurate model for predicting independent test data from the same population. Such accuracy is called the *generalization* ability of the model. A value of t that is too small can prevent the lasso from capturing the main signal in the data, while too large a value can lead to overfitting. In this latter case, the model adapts to the noise as well as the signal that is present in the training data. In both cases, the prediction error on a test set will be inflated. There is usually an intermediate value of t that strikes a good balance between these two extremes, and in the process, produces a model with some coefficients equal to zero.

In order to estimate this best value for t, we can create artificial training and test sets by splitting up the given dataset at random, and estimating performance on the test data, using a procedure known as *cross-validation*. In more detail, we first randomly divide the full dataset into some number of groups $K > 1$. Typical choices of K might be 5 or 10, and sometimes N. We fix one group as the test set, and designate the remaining $K - 1$ groups as the training set. We then apply the lasso to the training data for a range of different t values, and we use each fitted model to predict the responses in the test set, recording the mean-squared prediction errors for each value of t. This process is repeated a total of K times, with each of the K groups getting the chance to play the role of the test data, with the remaining $K - 1$ groups used as training data. In this way, we obtain K different estimates of the prediction error over a range of values of t. These K estimates of prediction error are averaged for each value of t, thereby producing a *cross-validation error curve*.

Figure 2.3 shows the cross-validation error curve for the crime-data example, obtained using $K = 10$ splits. We plot the estimated mean-squared prediction error versus the relative bound $\tilde{t} = \|\widehat{\beta}(t)\|_1 / \|\tilde{\beta}\|_1$, where the estimate $\widehat{\beta}(t)$ correspond to the lasso solution for bound t and $\tilde{\beta}$ is the ordinary least-squares solution. The error bars in Figure 2.3 indicate plus and minus one standard error in the cross-validated estimates of the prediction error. A vertical dashed line is drawn at the position of the minimum ($\tilde{t} = 0.56$) while a dotted line is drawn at the "one-standard-error rule" choice ($\tilde{t} = 0.03$). This is the smallest value of t yielding a CV error no more than one standard error above its minimum value. The number of nonzero coefficients in each model is shown along the top. Hence the model that minimizes the CV error has three predictors, while the one-standard-error-rule model has just one.

We note that the cross-validation process above focused on the bound parameter t. One can just as well carry out cross-validation in the Lagrangian

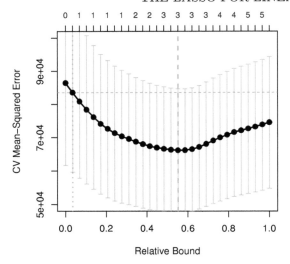

Figure 2.3 *Cross-validated estimate of mean-squared prediction error, as a function of the relative ℓ_1 bound $\tilde{t} = \|\hat{\beta}(t)\|_1/\|\tilde{\beta}\|_1$. Here $\hat{\beta}(t)$ is the lasso estimate corresponding to the ℓ_1 bound t and $\tilde{\beta}$ is the ordinary least-squares solution. Included are the location of the minimum, pointwise standard-error bands, and the "one-standard-error" location. The standard errors are large since the sample size N is only 50.*

form (2.5), focusing on the parameter λ. The two methods will give similar but not identical results, since the mapping between t and λ is data-dependent.

2.4 Computation of the Lasso Solution

The lasso problem is a convex program, specifically a quadratic program (QP) with a convex constraint. As such, there are many sophisticated QP methods for solving the lasso. However there is a particularly simple and effective computational algorithm, that gives insight into how the lasso works. For convenience, we rewrite the criterion in Lagrangian form:

$$\underset{\beta \in \mathbb{R}^p}{\text{minimize}} \left\{ \frac{1}{2N} \sum_{i=1}^{N} (y_i - \sum_{j=1}^{p} x_{ij}\beta_j)^2 + \lambda \sum_{j=1}^{p} |\beta_j| \right\}. \tag{2.8}$$

As before, we will assume that both y_i and the features x_{ij} have been standardized so that $\frac{1}{N}\sum_i y_i = 0$, $\frac{1}{N}\sum_i x_{ij} = 0$, and $\frac{1}{N}\sum_i x_{ij}^2 = 1$. In this case, the intercept term β_0 can be omitted. The Lagrangian form is especially convenient for numerical computation of the solution by a simple procedure known as *coordinate descent*.

Let's first consider a single predictor setting, based on samples $\{(z_i, y_i)\}_{i=1}^N$ (for convenience we have given the name z_i to this single x_{i1}). The problem then is to solve

$$\underset{\beta}{\text{minimize}} \left\{ \frac{1}{2N} \sum_{i=1}^N (y_i - z_i\beta)^2 + \lambda|\beta| \right\}. \tag{2.9}$$

The standard approach to this univariate minimization problem would be to take the gradient (first derivative) with respect to β, and set it to zero. There is a complication, however, because the absolute value function $|\beta|$ does not have a derivative at $\beta = 0$. However we can proceed by direct inspection of the function (2.9), and find that

$$\widehat{\beta} = \begin{cases} \frac{1}{N}\langle \mathbf{z}, \mathbf{y} \rangle - \lambda & \text{if } \frac{1}{N}\langle \mathbf{z}, \mathbf{y} \rangle \ > \ \lambda, \\ 0 & \text{if } \frac{1}{N}|\langle \mathbf{z}, \mathbf{y} \rangle| \ \leq \ \lambda, \\ \frac{1}{N}\langle \mathbf{z}, \mathbf{y} \rangle + \lambda & \text{if } \frac{1}{N}\langle \mathbf{z}, \mathbf{y} \rangle \ < \ -\lambda. \end{cases} \tag{2.10}$$

(Exercise 2.2), which we can write succinctly as

$$\widehat{\beta} = \mathcal{S}_\lambda\left(\tfrac{1}{N}\langle \mathbf{z}, \mathbf{y} \rangle\right). \tag{2.11}$$

Here the *soft-thresholding operator*

$$\mathcal{S}_\lambda(x) = \text{sign}(x)\left(|x| - \lambda\right)_+ \tag{2.12}$$

translates its argument x toward zero by the amount λ, and sets it to zero if $|x| \leq \lambda$.[3] See Figure 2.4 for an illustration. Notice that for standardized data with $\frac{1}{N}\sum_i z_i^2 = 1$, (2.11) is just a soft-thresholded version of the usual least-squares estimate $\widetilde{\beta} = \frac{1}{N}\langle \mathbf{z}, \mathbf{y} \rangle$. One can also derive these results using the notion of subgradients (Exercise 2.3).

Using this intuition from the univariate case, we can now develop a simple coordinatewise scheme for solving the full lasso problem (2.5). More precisely, we repeatedly cycle through the predictors in some fixed (but arbitrary) order (say $j = 1, 2, \ldots, p$), where at the jth step, we update the coefficient β_j by minimizing the objective function in this coordinate while holding fixed all other coefficients $\{\widehat{\beta}_k, k \neq j\}$ at their current values.

Writing the objective in (2.5) as

$$\frac{1}{2N} \sum_{i=1}^N (y_i - \sum_{k\neq j} x_{ik}\beta_k - x_{ij}\beta_j)^2 + \lambda \sum_{k\neq j} |\beta_k| + \lambda|\beta_j|, \tag{2.13}$$

---

[3] t_+ denotes the positive part of $t \in \mathbb{R}$, equal to t if $t > 0$ and 0 otherwise.

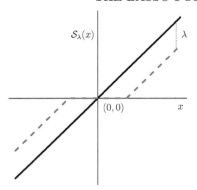

Figure 2.4 *Soft thresholding function* $\mathcal{S}_\lambda(x) = \text{sign}(x)\,(|x| - \lambda)_+$ *is shown in blue (broken lines), along with the 45° line in black.*

we see that solution for each β_j can be expressed succinctly in terms of the *partial residual* $r_i^{(j)} = y_i - \sum_{k \neq j} x_{ik}\widehat{\beta}_k$, which removes from the outcome the current fit from all but the jth predictor. In terms of this partial residual, the jth coefficient is updated as

$$\widehat{\beta}_j = \mathcal{S}_\lambda\left(\tfrac{1}{N}\langle \mathbf{x}_j, \boldsymbol{r}^{(j)}\rangle\right). \tag{2.14}$$

Equivalently, the update can be written as

$$\widehat{\beta}_j \leftarrow \mathcal{S}_\lambda\left(\widehat{\beta}_j + \tfrac{1}{N}\langle \mathbf{x}_j, \boldsymbol{r}\rangle\right), \tag{2.15}$$

where $r_i = y_i - \sum_{j=1}^{p} x_{ij}\widehat{\beta}_j$ are the full residuals (Exercise 2.4). The overall algorithm operates by applying this soft-thresholding update (2.14) repeatedly in a cyclical manner, updating the coordinates of $\widehat{\beta}$ (and hence the residual vectors) along the way.

Why does this algorithm work? The criterion (2.5) is a convex function of β and so has no local minima. The algorithm just described corresponds to the method of *cyclical coordinate descent*, which minimizes this convex objective along each coordinate at a time. Under relatively mild conditions (which apply here), such coordinate-wise minimization schemes applied to a convex function converge to a global optimum. It is important to note that some conditions are required, because there are instances, involving nonseparable penalty functions, in which coordinate descent schemes can become "jammed." Further details are in given in Chapter 5.

Note that the choice $\lambda = 0$ in (2.5) delivers the solution to the ordinary least-squares problem. From the update (2.14), we see that the algorithm does a univariate regression of the partial residual onto each predictor, cycling through the predictors until convergence. When the data matrix $\mathbf{X}$ is of full rank, this point of convergence is the least-squares solution. However, it is not a particularly efficient method for computing it.

In practice, one is often interested in finding the lasso solution not just for a single fixed value of λ, but rather the entire path of solutions over a range of possible λ values (as in Figure 2.1). A reasonable method for doing so is to begin with a value of λ just large enough so that the only optimal solution is the all-zeroes vector. As shown in Exercise 2.1, this value is equal to $\lambda_{max} = \max_j |\frac{1}{N} \langle \mathbf{x}_j, \mathbf{y} \rangle|$. Then we decrease λ by a small amount and run coordinate descent until convergence. Decreasing λ again and using the previous solution as a "warm start," we then run coordinate descent until convergence. In this way we can efficiently compute the solutions over a grid of λ values. We refer to this method as *pathwise coordinate descent*.

Coordinate descent is especially fast for the lasso because the coordinate-wise minimizers are explicitly available (Equation (2.14)), and thus an iterative search along each coordinate is not needed. Secondly, it exploits the sparsity of the problem: for large enough values of λ most coefficients will be zero and will not be moved from zero. In Section 5.4, we discuss computational hedges for guessing the active set, which speed up the algorithm dramatically.

Homotopy methods are another class of techniques for solving the lasso. They produce the entire path of solutions in a sequential fashion, starting at zero. This path is actually piecewise linear, as can be seen in Figure 2.1 (as a function of t or λ). The *least angle regression* (LARS) algorithm is a homotopy method that efficiently constructs the piecewise linear path, and is described in Chapter 5.

2.4.3 Soft-Thresholding and Orthogonal Bases

The soft-thresholding operator plays a central role in the lasso and also in signal denoising. To see this, notice that the coordinate minimization scheme above takes an especially simple form if the predictors are orthogonal, meaning that $\frac{1}{N} \langle \mathbf{x}_j, \mathbf{x}_k \rangle = 0$ for each $j \neq k$. In this case, the update (2.14) simplifies dramatically, since $\frac{1}{N} \langle \mathbf{x}_j, \mathbf{r}^{(j)} \rangle = \frac{1}{N} \langle \mathbf{x}_j, \mathbf{y} \rangle$ so that $\widehat{\beta}_j$ is simply the soft-thresholded version of the univariate least-squares estimate of $\mathbf{y}$ regressed against $\mathbf{x}_j$. Thus, in the special case of an orthogonal design, the lasso has an explicit closed-form solution, and no iterations are required.

Wavelets are a popular form of orthogonal bases, used for smoothing and compression of signals and images. In wavelet smoothing one represents the data in a wavelet basis, and then denoises by soft-thresholding the wavelet coefficients. We discuss this further in Section 2.10 and in Chapter 10.

2.5   Degrees of Freedom

Suppose we have p predictors, and fit a linear regression model using only a subset of k of these predictors. Then if these k predictors were chosen without regard to the response variable, the fitting procedure "spends" k degrees of freedom. This is a loose way of saying that the standard test statistic for testing the hypothesis that all k coefficients are zero has a Chi-squared distribution with k degrees of freedom (with the error variance σ^2 assumed to be known)

However if the k predictors were chosen using knowledge of the response variable, for example to yield the smallest training error among all subsets of size k, then we would expect that the fitting procedure spends more than k degrees of freedom. We call such a fitting procedure *adaptive*, and clearly the lasso is an example of one.

Similarly, a forward-stepwise procedure in which we sequentially add the predictor that most decreases the training error is adaptive, and we would expect that the resulting model uses more than k degrees of freedom after k steps. For these reasons and in general, one cannot simply count as degrees of freedom the number of nonzero coefficients in the fitted model. However, it turns out that for the lasso, one *can* count degrees of freedom by the number of nonzero coefficients, as we now describe.

First we need to define precisely what we mean by the degrees of freedom of an adaptively fitted model. Suppose we have an additive-error model, with

$$y_i = f(x_i) + \epsilon_i, \ i = 1, \ldots, N, \tag{2.16}$$

for some unknown f and with the errors ϵ_i iid $(0, \sigma^2)$. If the N sample predictions are denoted by $\widehat{\mathbf{y}}$, then we define

$$\mathrm{df}(\widehat{\mathbf{y}}) := \frac{1}{\sigma^2} \sum_{i=1}^{N} \mathrm{Cov}\left(\widehat{y}_i, y_i\right). \tag{2.17}$$

The covariance here is taken over the randomness in the response variables $\{y_i\}_{i=1}^{N}$ with the predictors held fixed. Thus, the degrees of freedom corresponds to the total amount of *self-influence* that each response measurement has on its prediction. The more the model fits—that is, adapts—to the data, the larger the degrees of freedom. In the case of a fixed linear model, using k predictors chosen independently of the response variable, it is easy to show that $\mathrm{df}(\widehat{\mathbf{y}}) = k$ (Exercise 2.7). However, under adaptive fitting, it is typically the case that the degrees of freedom is larger than k.

Somewhat miraculously, one can show that for the lasso, with a fixed penalty parameter λ, the number of nonzero coefficients k_λ is an unbiased estimate of the degrees of freedom[4] (Zou, Hastie and Tibshirani 2007, Tibshirani[2] and Taylor 2012). As discussed earlier, a variable-selection method like forward-stepwise regression uses more than k degrees of freedom after k steps. Given the apparent similarity between forward-stepwise regression and the lasso, how can the lasso have this simple degrees of freedom property? The

---

[4] An even stronger statement holds for the LAR path, where the degrees of freedom after k steps is exactly k, under some conditions on $\mathbf{X}$. The LAR path relates closely to the lasso, and is described in Section 5.6.

reason is that the lasso not only selects predictors (which inflates the degrees of freedom), but also shrinks their coefficients toward zero, relative to the usual least-squares estimates. This shrinkage turns out to be just the right amount to bring the degrees of freedom down to k. This result is useful because it gives us a qualitative measure of the amount of fitting that we have done at any point along the lasso path.

In the general setting, a proof of this result is quite difficult. In the special case of an orthogonal design, it is relatively easy to prove, using the fact that the lasso estimates are simply soft-thresholded versions of the univariate regression coefficients for the orthogonal design. We explore the details of this argument in Exercise 2.8. This idea is taken one step further in Section 6.3.1 where we describe the *covariance test* for testing the significance of predictors in the context of the lasso.

2.6 Uniqueness of the Lasso Solutions

We first note that the theory of convex duality can be used to show that when the columns of $\mathbf{X}$ are in general position, then for $\lambda > 0$ the solution to the lasso problem (2.5) is unique. This holds even when $p \geq N$, although then the number of nonzero coefficients in any lasso solution is at most N (Rosset, Zhu and Hastie 2004, Tibshirani$_2$ 2013). Now when the predictor matrix $\mathbf{X}$ is not of full column rank, the least squares fitted values are unique, but the parameter estimates themselves are not. The non-full-rank case can occur when $p \leq N$ due to collinearity, and always occurs when $p > N$. In the latter scenario, there are an infinite number of solutions $\widehat{\beta}$ that yield a perfect fit with zero training error. Now consider the lasso problem in Lagrange form (2.5) for $\lambda > 0$. As shown in Exercise 2.5, the fitted values $\mathbf{X}\widehat{\beta}$ are unique. But it turns out that the solution $\widehat{\beta}$ may not be unique. Consider a simple example with two predictors $\mathbf{x}_1$ and $\mathbf{x}_2$ and response $\mathbf{y}$, and suppose the lasso solution coefficients $\widehat{\beta}$ at λ are $(\widehat{\beta}_1, \widehat{\beta}_2)$. If we now include a third predictor $\mathbf{x}_3 = \mathbf{x}_2$ into the mix, an identical copy of the second, then for any $\alpha \in [0, 1]$, the vector $\tilde{\beta}(\alpha) = (\widehat{\beta}_1, \alpha \cdot \widehat{\beta}_2, (1 - \alpha) \cdot \widehat{\beta}_2)$ produces an identical fit, and has ℓ_1 norm $\|\tilde{\beta}(\alpha)\|_1 = \|\widehat{\beta}\|_1$. Consequently, for this model (in which we might have either $p \leq N$ or $p > N$), there is an infinite family of solutions.

In general, when $\lambda > 0$, one can show that if the columns of the model matrix $\mathbf{X}$ are in *general position*, then the lasso solutions are unique. To be precise, we say the columns $\{\mathbf{x}_j\}_{j=1}^{p}$ are in general position if any affine subspace $\mathbb{L} \subset \mathbb{R}^N$ of dimension $k < N$ contains at most $k + 1$ elements of the set $\{\pm\mathbf{x}_1, \pm\mathbf{x}_2, \ldots \pm \mathbf{x}_p\}$, excluding antipodal pairs of points (that is, points differing only by a sign flip). We note that the data in the example in the previous paragraph are not in general position. If the X data are drawn from a continuous probability distribution, then with probability one the data are in general position and hence the lasso solutions will be unique. As a result, non-uniqueness of the lasso solutions can only occur with discrete-valued data, such as those arising from dummy-value coding of categorical predictors. These results have appeared in various forms in the literature, with a summary given by Tibshirani[2] (2013).

We note that numerical algorithms for computing solutions to the lasso will typically yield valid solutions in the non-unique case. However, the particular solution that they deliver can depend on the specifics of the algorithm. For example with coordinate descent, the choice of starting values can affect the final solution.

2.7 A Glimpse at the Theory

There is a large body of theoretical work on the behavior of the lasso. It is largely focused on the mean-squared-error consistency of the lasso, and recovery of the nonzero support set of the true regression parameters, sometimes called *sparsistency*. For MSE consistency, if β^* and $\hat{\beta}$ are the true and lasso-estimated parameters, it can be shown that as $p, n \to \infty$

$$\|\mathbf{X}(\hat{\beta} - \beta^*)\|_2^2/N \leq C \cdot \|\beta^*\|_1 \sqrt{\log(p)/N} \qquad (2.18)$$

with high probability (Greenshtein and Ritov 2004, Bühlmann and van de Geer 2011, Chapter 6). Thus if $\|\beta^*\|_1 = o(\sqrt{N/\log(p)})$ then the lasso is consistent for prediction. This means that the true parameter vector must be sparse relative to the ratio $N/\log(p)$. The result only assumes that the design $\mathbf{X}$ is fixed and has no other conditions on $\mathbf{X}$. Consistent recovery of the nonzero support set requires more stringent assumptions on the level of cross-correlation between the predictors inside and outside of the support set. Details are given in Chapter 11.

2.8 The Nonnegative Garrote

The *nonnegative garrote* (Breiman 1995)[5] is a two-stage procedure, with a close relationship to the lasso.[6] Given an initial estimate of the regression coefficients $\tilde{\beta} \in \mathbb{R}^p$, we then solve the optimization problem

$$\underset{c \in \mathbb{R}^p}{\text{minimize}} \left\{ \sum_{i=1}^{N} \left(y_i - \sum_{j=1}^{p} c_j x_{ij} \tilde{\beta}_j \right)^2 \right\} \qquad (2.19)$$

$$\text{subject to } c \succeq 0 \text{ and } \|c\|_1 \leq t,$$

where $c \succeq 0$ means the vector has nonnegative coordinates. Finally, we set $\hat{\beta}_j = \hat{c}_j \cdot \tilde{\beta}_j$, $j = 1, \ldots, p$. There is an equivalent Lagrangian form for this procedure, using a penalty $\lambda \|c\|_1$ for some regularization weight $\lambda \geq 0$, plus the nonnegativity constraints.

---

[5]A garrote is a device used for execution by strangulation or by breaking the neck. It is a Spanish word, and is alternately spelled *garrotte* or *garote*. We are using the spelling in the original paper of Breiman (1995).

[6]Breiman's paper was the inspiration for Tibshirani's 1996 lasso paper.

In the original paper (Breiman 1995), the initial $\tilde{\beta}$ was chosen to be the ordinary-least-squares (OLS) estimate. Of course, when $p > N$, these estimates are not unique; since that time, other authors (Yuan and Lin 2007b, Zou 2006) have shown that the nonnegative garrote has attractive properties when we use other initial estimators such as the lasso, ridge regression or the elastic net.

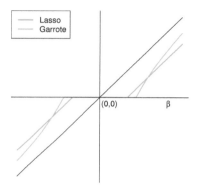

Figure 2.5 *Comparison of the shrinkage behavior of the lasso and the nonnegative garrote for a single variable. Since their λs are on different scales, we used 2 for the lasso and 7 for the garrote to make them somewhat comparable. The garrote shrinks smaller values of β more severely than lasso, and the opposite for larger values.*

The nature of the nonnegative garrote solutions can be seen when the columns of $\mathbf{X}$ are orthogonal. Assuming that t is in the range where the equality constraint $\|c\|_1 = t$ can be satisfied, the solutions have the explicit form

$$\hat{c}_j = \left(1 - \frac{\lambda}{\tilde{\beta}_j^2}\right)_+ , \ j = 1,\ldots,p, \tag{2.20}$$

where λ is chosen so that $\|\hat{c}\|_1 = t$. Hence if the coefficient $\tilde{\beta}_j$ is large, the shrinkage factor will be close to 1 (no shrinkage), but if it is small the estimate will be shrunk toward zero. Figure 2.5 compares the shrinkage behavior of the lasso and nonnegative garrote. The latter exhibits the shrinkage behavior of the nonconvex penalties (next section and Section 4.6). There is also a close relationship between the nonnegative garrote and the *adaptive lasso*, discussed in Section 4.6; see Exercise 4.26.

Following this, Yuan and Lin (2007b) and Zou (2006) have shown that the nonnegative garrote is *path-consistent* under less stringent conditions than the lasso. This holds if the initial estimates are $\sqrt{N}$-consistent, for example those based on least squares (when $p < N$), the lasso, or the elastic net. "Path-consistent" means that the solution path contains the true model somewhere in its path indexed by t or λ. On the other hand, the convergence of the parameter estimates from the nonnegative garrote tends to be slower than that of the initial estimate.

Table 2.3 *Estimators of β_j from (2.21) in the case of an orthonormal model matrix* **X**.

q	Estimator	Formula
0	Best subset	$\tilde{\beta}_j \cdot \mathbb{I}[|\tilde{\beta}_j| > \sqrt{2\lambda}]$
1	Lasso	$\text{sign}(\tilde{\beta}_j)(|\tilde{\beta}_j| - \lambda)_+$
2	Ridge	$\tilde{\beta}_j/(1 + \lambda)$

2.9 ℓ_q Penalties and Bayes Estimates

For a fixed real number $q \geq 0$, consider the criterion

$$\underset{\beta \in \mathbb{R}^p}{\text{minimize}} \left\{ \frac{1}{2N} \sum_{i=1}^{N} (y_i - \sum_{j=1}^{p} x_{ij}\beta_j)^2 + \lambda \sum_{j=1}^{p} |\beta_j|^q \right\}. \qquad (2.21)$$

This is the lasso for $q = 1$ and ridge regression for $q = 2$. For $q = 0$, the term $\sum_{j=1}^{p} |\beta_j|^q$ counts the number of nonzero elements in β, and so solving (2.21) amounts to best-subset selection. Figure 2.6 displays the constraint regions corresponding to these penalties for the case of two predictors ($p = 2$). Both

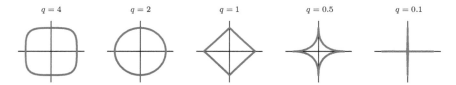

Figure 2.6 *Constraint regions $\sum_{j=1}^{p} |\beta_j|^q \leq 1$ for different values of q. For $q < 1$, the constraint region is nonconvex.*

the lasso and ridge regression versions of (2.21) amount to solving convex programs, and so scale well to large problems. Best subset selection leads to a nonconvex and combinatorial optimization problem, and is typically not feasible with more than say $p = 50$ predictors.

   In the special case of an orthonormal model matrix **X**, all three procedures have explicit solutions. Each method applies a simple coordinate-wise transformation to the least-squares estimate $\tilde{\beta}$, as detailed in Table 2.9. Ridge regression does a proportional shrinkage. The lasso translates each coefficient by a constant factor λ and truncates at zero, otherwise known as soft thresholding. Best-subset selection applies the hard thresholding operator: it leaves the coefficient alone if it is bigger than $\sqrt{2\lambda}$, and otherwise sets it to zero.

   The lasso is special in that the choice $q = 1$ is the smallest value of q (closest to best-subset) that leads to a convex constraint region and hence a convex optimization problem. In this sense, it is the closest convex relaxation of the best-subset selection problem.

There is also a Bayesian view of these estimators. Thinking of $|\beta_j|^q$ as proportional to the negative log-prior density for β_j, the constraint contours represented in Figure 2.6 have the same shape as the equi-contours of the prior distribution of the parameters. Notice that for $q \leq 1$, the prior concentrates more mass in the coordinate directions. The prior corresponding to the $q = 1$ case is an independent double exponential (or Laplace) distribution for each parameter, with joint density $(1/2\tau) \exp(-\|\beta\|_1)/\tau)$ and $\tau = 1/\lambda$. This means that the lasso estimate is the Bayesian MAP (maximum *aposteriori*) estimator using a Laplacian prior, as opposed to the mean of the posterior distribution, which is not sparse. Similarly, if we sample from the posterior distribution corresponding to the Laplace prior, we do not obtain sparse vectors. In order to obtain sparse vectors via posterior sampling, one needs to start with a prior distribution that puts a point mass at zero. Bayesian approaches to the lasso are explored in Section 6.1.

2.10   Some Perspective

The lasso uses an ℓ_1-penalty, and such penalties are now widely used in statistics, machine learning, engineering, finance, and other fields. The lasso was proposed by Tibshirani (1996), and was directly inspired by the nonnegative garrote of Breiman (1995). Soft thresholding was popularized earlier in the context of wavelet filtering by Donoho and Johnstone (1994); this is a popular alternative to Fourier filtering in signal processing, being both "local in time and frequency." Since wavelet bases are orthonormal, wavelet filtering corresponds to the lasso in the orthogonal $\mathbf{X}$ case (Section 2.4.1). Around the same time as the advent of the lasso, Chen, Donoho and Saunders (1998) proposed the closely related *basis pursuit* method, which extends the ideas of wavelet filtering to search for a sparse representation of a signal in over-complete bases using an ℓ_1-penalty. These are unions of orthonormal frames and hence no longer completely mutually orthonormal.

Taking a broader perspective, ℓ_1-regularization has a pretty lengthy history. For example Donoho and Stark (1989) discussed ℓ_1-based recovery in detail, and provided some guarantees for incoherent bases. Even earlier (and mentioned in Donoho and Stark (1989)) there are related works from the 1980s in the geosciences community, for example Oldenburg, Scheuer and Levy (1983) and Santosa and Symes (1986). In the signal processing world, Alliney and Ruzinsky (1994) investigated some algorithmic issues associated with ℓ_1 regularization. And there surely are many other authors who have proposed similar ideas, such as Fuchs (2000). Rish and Grabarnik (2014) provide a modern introduction to sparse methods for machine learning and signal processing.

In the last 10–15 years, it has become clear that the ℓ_1-penalty has a number of good properties, which can be summarized as follows:

Interpretation of the final model: The ℓ_1-penalty provides a natural way to encourage or enforce sparsity and simplicity in the solution.

Statistical efficiency: In the book *The Elements of Statistical Learning* (Hastie, Tibshirani and Friedman 2009), the authors discuss an informal *"bet-on-sparsity principle."* Assume that the underlying true signal is sparse and we use an ℓ_1 penalty to try to recover it. If our assumption is correct, we can do a good job in recovering the true signal. Note that sparsity can hold in the given bases (set of features) or a transformation of the features (e.g., a wavelet bases). But if we are wrong—the underlying truth is not sparse in the chosen bases—then the ℓ_1 penalty will not work well. However in that instance, no method can do well, relative to the Bayes error. There is now a large body of theoretical support for these loose statements: see Chapter 11 for some results.

Computational efficiency: ℓ_1-based penalties are convex and this fact and the assumed sparsity can lead to significant computational advantages. If we have 100 observations and one million features, and we have to estimate one million nonzero parameters, then the computation is very challenging. However, if we apply the lasso, then at most 100 parameters can be nonzero in the solution, and this makes the computation much easier. More details are given in Chapter 5.[7]

In the remainder of this book, we describe many of the exciting developments in this field.

Exercises

Ex. 2.1 Show that the smallest value of λ such that the regression coefficients estimated by the lasso are all equal to zero is given by

$$\lambda_{\max} = \max_j |\frac{1}{N} \langle \mathbf{x}_j, \mathbf{y} \rangle|.$$

Ex. 2.2 Show that the soft-threshold estimator (2.12) yields the solution to the single predictor lasso problem (2.9). (Do not make use of subgradients, and note that $\mathbf{z}$ is standardized).

Ex. 2.3 *Soft thresholding and subgradients.* Since (2.9) is a convex function, it is guaranteed to have a subgradient (see Chapter 5 for more details), and any optimal solution must satisfy the *subgradient* equation

$$-\frac{1}{N} \sum_{i=1}^{N} (y_i - z_i\beta)z_i + \lambda s = 0, \qquad \text{where } s \text{ is a subgradient of } |\beta|. \quad (2.22)$$

For the absolute value function, subgradients take the form $s \in \text{sign}(\beta)$, meaning that $s = \text{sign}(\beta)$ when $\beta \neq 0$ and $s \in [-1, +1]$ when $\beta = 0$. The general

---

[7]Ridge regression also enjoys a similar efficiency in the $p \gg N$ case.

theory of convex optimization, as discussed in Chapter 5, guarantees that any pair $(\widehat{\beta}, \widehat{s})$ that is a solution to the zero subgradient Equation (2.22) with $\widehat{s} \in \text{sign}(\widehat{\beta})$ defines an optimal solution to the original minimization problem (2.9).

Solve Equation (2.22) and hence arrive at solutions (2.10) and (2.11).

Ex. 2.4 Show that the subgradient equations for Problem (2.5) take the form given in (2.6). Hence derive expressions for coordinate descent steps (2.14) and (2.15).

Ex. 2.5 *Uniqueness of fitted values from the lasso.* For some $\lambda \geq 0$, suppose that we have two lasso solutions $\widehat{\beta}, \widehat{\gamma}$ with common optimal value c^*.

(a) Show that it must be the case that $\mathbf{X}\widehat{\beta} = \mathbf{X}\widehat{\gamma}$, meaning that the two solutions must yield the same predicted values. (*Hint:* If not, then use the strict convexity of the function $f(\mathbf{u}) = \|\mathbf{y} - \mathbf{u}\|_2^2$ and convexity of the ℓ_1-norm to establish a contradiction.)

(b) If $\lambda > 0$, show that we must have $\|\widehat{\beta}\|_1 = \|\widehat{\gamma}\|_1$.

(Tibshirani₂ 2013).

Ex. 2.6 Here we use the bootstrap as the basis for inference with the lasso.

(a) For the crime data, apply the bootstrap to estimate the standard errors of the estimated lasso coefficients, as in the middle section of Table 2.2. Use the nonparametric bootstrap, sampling features and outcome values (x_i, y_i) with replacement from the observed data. Keep the bound t fixed at its estimated value from the original lasso fit. Estimate as well the probability that an estimated coefficient is zero.

(b) Repeat part (a), but now re-estimate $\hat{\lambda}$ for each bootstrap replication. Compare the results to those in part (a).

Ex. 2.7 Consider a fixed linear model based on k predictors and fit by least squares. Show that its degrees of freedom (2.17) is equal to k.

Ex. 2.8 *Degrees of freedom for lasso in the orthogonal case.* Suppose that $y_i = \beta_0 + \sum_j x_{ij}\beta_j + \epsilon_i$ where $\epsilon_i \sim N(0, \sigma^2)$, with the x_{ij} fixed (non-random). Assume that the features are centered and also assume they are uncorrelated, so that $\sum_i x_{ij}x_{ik} = 0$ for all j, k. Stein's lemma (Stein 1981) states that for $Y \sim N(\mu, \sigma^2)$ and all absolutely continuous functions g such that $\mathbb{E}|g'(Y)| < \infty$,

$$\mathbb{E}(g(Y)(Y - \mu)) = \sigma^2 \mathbb{E}(g'(Y)). \qquad (2.23)$$

Use this to show that the degrees of freedom (2.17) for the lasso in the orthogonal case is equal to k, the number of nonzero estimated coefficients in the solution.

Ex. 2.9 Derive the solutions (2.20) to the nonnegative garrote criterion (2.19).

Ex. 2.10 *Robust regression view of lasso.* Consider a robust version of the standard linear regression problem, in which we wish to protect ourselves against perturbations of the features. In order to do so, we consider the min-max criterion

$$\underset{\beta}{\text{minimize}} \underset{\mathbf{\Delta} \in \mathcal{U}}{\max} \left\{ \frac{1}{2N} \|\mathbf{y} - (\mathbf{X} + \mathbf{\Delta})\beta\|_2^2 \right\}, \qquad (2.24)$$

where the allowable perturbations $\mathbf{\Delta} := (\boldsymbol{\delta}_1, \dots, \boldsymbol{\delta}_p)$ belong to the subset of $\mathbb{R}^{N \times p}$

$$\mathcal{U} := \left\{ (\boldsymbol{\delta}_1, \boldsymbol{\delta}_2, \dots \boldsymbol{\delta}_p) \mid \|\boldsymbol{\delta}_j\|_2 \leq c_j \text{ for all } j = 1, 2, \dots, p \right\}. \qquad (2.25)$$

Hence each feature value x_{ij} can be perturbed by a maximum amount c_j, with the ℓ_2-norm of the overall perturbation vector for that feature bounded by c_j. The perturbations for different features also act independently of one another. We seek the coefficients that minimize squared error under the "worst" allowable perturbation of the features. We assume that both $\mathbf{y}$ and the columns of $\mathbf{X}$ have been standardized, and have not included an intercept.

Show that the solution to this problem is equivalent to

$$\underset{\beta \in \mathbb{R}^p}{\min} \left\{ \frac{1}{2N} \|\mathbf{y} - \mathbf{X}\beta\|_2^2 + \sum_{j=1}^{p} c_j |\beta_j| \right\}. \qquad (2.26)$$

In the special case $c_j = \lambda$ for all $j = 1, 2, \dots, p$, we thus obtain the lasso, so that it can be viewed as a method for guarding against uncertainty in the measured predictor values, with more uncertainty leading to a greater amount of shrinkage. (See Xu, Caramanis and Mannor (2010) for further details.)

Ex. 2.11 *Robust regression and constrained optimization.* This exercise doesn't involve the lasso itself, but rather a related use of the ℓ_1-norm in regression. We consider the model

$$y_i = \sum_{j=1}^{p} x_{ij}\beta_j + \gamma_i + \epsilon_i$$

with $\epsilon_i \sim N(0, \sigma^2)$ and γ_i, $i = 1, 2, \dots, N$ are unknown constants.

Let $\boldsymbol{\gamma} = (\gamma_1, \gamma_2, \dots, \gamma_N)$ and consider minimization of

$$\underset{\beta \in \mathbb{R}^p, \boldsymbol{\gamma} \in \mathbb{R}^N}{\text{minimize}} \frac{1}{2} \sum_{i=1}^{N} (y_i - \sum_{j=1}^{p} x_{ij}\beta_j - \gamma_i)^2 + \lambda \sum_{1}^{N} |\gamma_i|. \qquad (2.27)$$

The idea is that for each i, γ_i allows y_i to be an outlier; setting $\gamma_i = 0$ means that the observation is not deemed an outlier. The penalty term effectively limits the number of outliers.

(a) Show this problem is jointly convex in β and γ.

(b) Consider Huber's loss function

$$\rho(t; \lambda) = \begin{cases} \lambda|t| - \lambda^2/2 & \text{if } |t| > \lambda \\ t^2/2 & \text{if } |t| \leq \lambda. \end{cases} \tag{2.28}$$

This is a tapered squared-error loss; it is quadratic for $|t| \leq \lambda$ but linear outside of that range, to reduce the effect of outliers on the estimation of β. With the scale parameter σ fixed at one, Huber's robust regression method solves

$$\underset{\beta \in \mathbb{R}^p}{\text{minimize}} \sum_{i=1}^{N} \rho(y_i - \sum_{j=1}^{p} x_{ij}\beta_j; \lambda). \tag{2.29}$$

Show that problems (2.27) and (2.29) have the same solutions $\widehat{\beta}$. (Antoniadis 2007, Gannaz 2007, She and Owen 2011).

Chapter 3

Generalized Linear Models

In Chapter 2, we focused exclusively on linear regression models fit by least squares. Such linear models are suitable when the response variable is quantitative, and ideally when the error distribution is Gaussian. However, other types of response arise in practice. For instance, binary variables can be used to indicate the presence or absence of some attribute (e.g., "cancerous" versus "normal" cells in a biological assay, or "clicked" versus "not clicked" in web browsing analysis); here the binomial distribution is more appropriate. Sometimes the response occurs as counts (e.g., number of arrivals in a queue, or number of photons detected); here the Poisson distribution might be called for. In this chapter, we discuss generalizations of simple linear models and the lasso that are suitable for such applications.

3.1 Introduction

With a binary response coded in the form $Y \in \{0, 1\}$, the linear logistic model is often used: it models the log-likelihood ratio as the linear combination

$$\log \frac{\Pr(Y = 1 \mid X = x)}{\Pr(Y = 0 \mid X = x)} = \beta_0 + \beta^T x, \tag{3.1}$$

where $X = (X_1, X_2, \ldots X_p)$ is a vector of predictors, $\beta_0 \in \mathbb{R}$ is an intercept term, and $\beta \in \mathbb{R}^p$ is a vector of regression coefficients. Inverting this transformation yields an expression for the conditional probability

$$\Pr(Y = 1 \mid X = x) = \frac{e^{\beta_0 + \beta^T x}}{1 + e^{\beta_0 + \beta^T x}}. \tag{3.2}$$

By inspection, without any restriction on the parameters (β_0, β), the model specifies probabilities lying in $(0, 1)$. We typically fit logistic models by maximizing the binomial log-likelihood of the data.

The logit transformation (3.1) of the conditional probabilities is an example of a *link function*. In general, a link function is a transformation of the conditional mean $\mathbb{E}[Y \mid X = x]$—in this case, the conditional probability that $Y = 1$—to a more natural scale on which the parameters can be fit without constraints. As another example, if the response Y represents counts, taking values in $\{0, 1, 2, \ldots\}$, then we need to ensure that the conditional mean is positive. A natural choice is the log-linear model

$$\log \mathbb{E}[Y \mid X = x] = \beta_0 + \beta^T x, \qquad (3.3)$$

with its log link function. Here we fit the parameters by maximizing the Poisson log-likelihood of the data.

The models (3.1) and (3.3) are both special cases of *generalized linear models* (McCullagh and Nelder 1989). These models describe the response variable using a member of the *exponential family*, which includes the Bernoulli, Poisson, and Gaussian as particular cases. A transformed version of the response mean $\mathbb{E}[Y \mid X = x]$ is then approximated by a linear model. In detail, if we use $\mu(x) = \mathbb{E}[Y \mid X = x]$ to denote the conditional mean of Y given $X = x$, then a GLM is based on a model of the form

$$g[\mu(x)] = \underbrace{\beta_0 + \beta^T x}_{\eta(x)}, \qquad (3.4)$$

where $g : \mathbb{R} \to \mathbb{R}$ is a strictly monotonic link function. For example, for a binary response $Y \in \{0, 1\}$, the logistic regression model is based on the choices $\mu(x) = \Pr[Y = 1 \mid X = x]$ and $g(\mu) = \text{logit}(\mu) = \log(\mu/(1 - \mu))$. When the response variable is modeled as a Gaussian, the choices $\mu(x) = \beta_0 + \beta^T x$ and $g(\mu) = \mu$ recover the standard linear model, as discussed in the previous chapter.

Generalized linear models can also be used to model the multicategory responses that occur in many problems, including handwritten digit classification, speech-recognition, document classification, and cancer classification. The multinomial replaces the binomial distribution here, and we use a symmetric log-linear representation:

$$\Pr[Y = k \mid X = x] = \frac{e^{\beta_{0k} + \beta_k^T x}}{\sum_{\ell=1}^{K} e^{\beta_{0\ell} + \beta_\ell^T x}}. \qquad (3.5)$$

Here there are K coefficients for each variable (one per class).

In this chapter, we discuss approaches to fitting generalized linear models that are based on maximizing the likelihood, or equivalently minimizing the negative log-likelihood along with an ℓ_1-penalty

$$\underset{\beta_0, \beta}{\text{minimize}} \left\{ -\frac{1}{N} \mathcal{L}(\beta_0, \beta; \mathbf{y}, \mathbf{X}) + \lambda \|\beta\|_1 \right\}. \qquad (3.6)$$

Here $\mathbf{y}$ is the N-vector of outcomes and $\mathbf{X}$ is the $N \times p$ matrix of predictors, and the specific form of the log-likelihood $\mathcal{L}$ varies according to the GLM. In the special case of Gaussian responses and the standard linear model, we have $\mathcal{L}(\beta_0, \beta; \mathbf{y}, \mathbf{X}) = \frac{1}{2\sigma^2} \|\mathbf{y} - \beta_0 \mathbf{1} - \mathbf{X}\beta\|_2^2 + c$, where c is a constant independent of (β_0, β), so that the optimization problem (3.6) corresponds to the ordinary linear least-squares lasso.

Similar forms of ℓ_1-regularization are also useful for related models. With survival models, the response is the time to failure (death), with possible censoring if subjects are lost to followup. In this context, a popular choice is the Cox proportional hazards model, which takes the form

$$h(t \mid x) = h_0(t)e^{\beta^T x}. \tag{3.7}$$

Here $t \mapsto h(t \mid x)$ is the *hazard function* for an individual with covariates x: the value $h(t \mid x)$ corresponds to the instantaneous probability of failure at time $Y = t$, given survival up to time t. The function h_0 specifies the baseline hazard, corresponding to $x = 0$.

As another example, the support-vector machine (SVM) is a popular classifier in the machine-learning community. Here the goal is to predict a two-class response $y \in \{-1, +1\}$,[1] in the simplest case using a linear classification boundary of the form $f(x) = \beta_0 + \beta^T x$, with the predicted class given by $\text{sign}(f(x))$. Thus, the correctness of a given decision can be determined by checking whether or not the margin $yf(x)$ is positive. The traditional *soft-margin* linear SVM is fit by solving the optimization problem[2]

$$\underset{\beta_0, \beta}{\text{minimize}} \left\{ \frac{1}{N} \sum_{i=1}^{N} \underbrace{[1 - y_i f(x_i)]_+}_{\phi(y_i f(x_i))} + \lambda \|\beta\|_2^2 \right\}. \tag{3.8}$$

The first term, known as *hinge loss*, is designed to penalize the negative margins that represent incorrect classifications. In general, an optimal solution vector $\beta \in \mathbb{R}^p$ to the standard linear SVM (3.8) is not sparse, since the quadratic penalty has no sparsity-enforcing properties. However, replacing the quadratic penalty by the ℓ_1-norm $\|\beta\|_1$ leads to an ℓ_1 linear SVM, which does produce sparse solutions.

In the following sections, we discuss each of these models in more detail. In each case, we provide examples of their applications, discuss some of the issues that arise, as well as computational approaches for fitting the models.

3.2 Logistic Regression

Logistic regression has been popular in biomedical research for half a century, and has recently gained popularity for modeling a wider range of data. In the high-dimensional setting, in which the number of features p is larger than the sample size, it cannot be used without modification. When $p > N$, any linear model is over-parametrized, and regularization is needed to achieve a stable fit. Such high-dimensional models arise in various applications. For example, document classification problems can involve binary features (presence versus

---

[1]For SVMs, it is convenient to code the binary response via the sign function.
[2]This is not the most standard way to introduce the support vector machine. We discuss this topic in more detail in Section 3.6.

absence) over a predefined dictionary of $p = 20,000$ or more words and tokens. Another example is genome-wide association studies (GWAS), where we have genotype measurements at $p = 500,000$ or more "SNPs," and the response is typically the presence/absence of a disease. A SNP (pronounced "snip") is a single-nucleotide polymorphism, and is typically represented as a three-level factor with possible values $\{AA, Aa, aa\}$, where "A" refers to the wild-type, and "a" the mutation.

When the response is binary, it is typically coded as 0/1. Attention then focuses on estimating the conditional probability $\Pr(Y = 1 \mid X = x) = \mathbb{E}[Y \mid X = x]$. Given the logistic model (3.1), the negative log likelihood with ℓ_1-regularization takes the form

$$-\frac{1}{N} \sum_{i=1}^{N} \left\{ y_i \log \Pr(Y = 1 \mid x_i) + (1 - y_i) \log \Pr(Y = 0 \mid x_i) \right\} + \lambda \|\beta\|_1$$

$$= -\frac{1}{N} \sum_{i=1}^{N} \left\{ y_i(\beta_0 + \beta^T x_i) - \log(1 + e^{\beta_0 + \beta^T x_i}) \right\} + \lambda \|\beta\|_1. \quad (3.9)$$

In the machine-learning community, it is more common to code the response Y in terms of sign variables $\{-1, +1\}$ rather than $\{0, 1\}$ values; when using sign variables, the penalized (negative) log-likelihood has the form

$$\frac{1}{N} \sum_{i=1}^{N} \log(1 + e^{-y_i f(x_i \,;\beta_0, \beta)}) + \lambda \|\beta\|_1, \quad (3.10)$$

where $f(x_i \,; \beta_0, \beta) := \beta_0 + \beta^T x_i$. For a given covariate-response pair (x, y), the product $y f(x)$ is referred to as the *margin*: a positive margin means a correct classification, whereas a negative margin means an incorrect classification. From the form of the log-likelihood (3.10), we see that maximizing the likelihood amounts to minimizing a loss function monotone decreasing in the margins. We discuss the interplay of the margin and the penalty in Section 3.6.1.

3.2.1 Example: Document Classification

We illustrate ℓ_1-regularized logistic regression in a domain where it has gained popularity, namely document classification using the 20-Newsgroups corpus (Lang 1995). We use the particular feature set and class definition defined by Koh, Kim and Boyd (2007).[3] There are $N = 11,314$ documents and $p = 777,811$ features, with 52% in the positive class. Only 0.05% of the features are nonzero for any given document.

---

[3] The positive class consists of the 10 groups with names of the form sci.*, comp.* and misc.forsale, and the rest are the negative class. The feature set consists of trigrams, with message headers skipped, no stoplist, and features with less than two documents omitted.

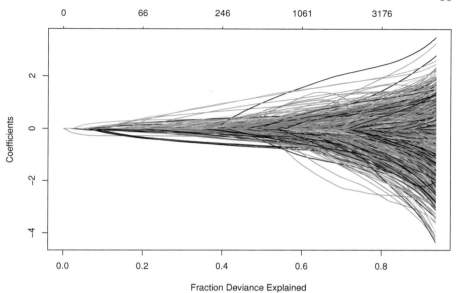

Figure 3.1 *Coefficient paths for an ℓ_1-regularized logistic regression for a document-classification task—the "NewsGroup" data. There are 11K documents roughly divided into two classes, and 0.78M features. Only 0.05% of the features are nonzero. The coefficients are plotted as a function of the fraction of null deviance explained.*

Figure 3.1 shows the coefficient profile, computed using the R package **glmnet**. Although the solutions were computed at 100 values of λ, uniformly spaced on the log scale, we have indexed the solutions by the *fraction of deviance explained*[4] on the training data:

$$\mathrm{D}^2{}_\lambda = \frac{\mathrm{Dev_{null}} - \mathrm{Dev}_\lambda}{\mathrm{Dev_{null}}}. \qquad (3.11)$$

Here the deviance Dev_λ is defined as minus twice the difference in the log-likelihood for a model fit with parameter λ and the "saturated" model (having $\widehat{y} = y_i$). $\mathrm{Dev_{null}}$ is the null deviance computed at the constant (mean) model. Since for these data the classes are separable, the range of λ is chosen so as not to get too close to the saturated fit (where the coefficients would be undefined; see the next section).

The maximum number of nonzero coefficients in any of these models can be shown to be $\min(N, p)$, which is equal $11,314$ in this case. In Figure 3.1, the largest model actually had only $5,277$ coefficients since **glmnet** did not go to the very end of the solution path. Although it might seem more natural to plot against the $\log(\lambda)$ sequence, or perhaps $\|\widehat{\beta}(\lambda)\|_1$, there are problems with both in the $p \gg N$ setting. The former quantity is data and problem dependent,

---

[4]the name D^2 is by analogy with R^2, the fraction of variance explained in regression.

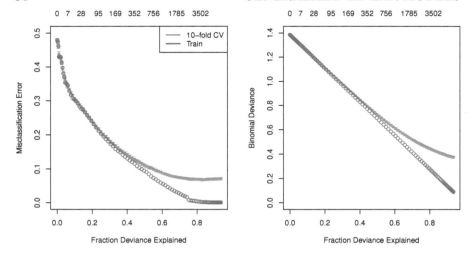

Figure 3.2 *Lasso (ℓ_1)-penalized logistic regression. Tenfold cross-validation curves for the Newsgroup data are shown in red, along with pointwise standard-error bands (not visible). The left plot shows misclassification error; the right plot shows deviance. Also shown in blue is the training error for each of these measures. The number of nonzero coefficients in each model is shown along the top of each plot.*

and gives no indication of the amount of overfitting, whereas for the latter measure, the graph would be dominated by the less interesting right-hand side, in which the coefficients and hence their norm explode.

Figure 3.2 shows the results of tenfold cross-validation for these data, as well as training error. These are also indexed by the fraction of deviance explained on the training data. Figure 3.3 shows the analogous results to those in Figure 3.2, for ridge regression. The cross-validated error rates are about the same as for the lasso. The number of nonzero coefficients in every model is $p = 777,811$ compared to a maximum of $5,277$ in Figure 3.2. However the rank of the ridge regression fitted values is actually $\min(N, p)$ which equals $11,314$ in this case, not much different from that of the lasso fit. Nonetheless, ridge regression might be more costly from a computational viewpoint. We produced the cross-validation results in Figure 3.3 using the `glmnet` package; for ridge the tenfold cross-validation took 8.3 minutes, while for lasso under one minute. A different approach would be to use the kernel trick (Hastie and Tibshirani 2004, for example), but this requires a singular value or similar decomposition of an $11,314 \times 11,314$ matrix.

For this example, using the package `glmnet`, we fit the regularization path in Figure 3.1 at 100 values of λ in 5 secs on a 2.6 GHz Macbook Pro. In examples like this with so many features, dramatic speedups can be achieved by screening the features. For example, the first feature to enter the regularization path achieves $\lambda_{\max} = \max_j |\langle x_j, \mathbf{y} - \bar{\mathbf{p}} \rangle|$, where $\mathbf{y}$ is the vector of binary

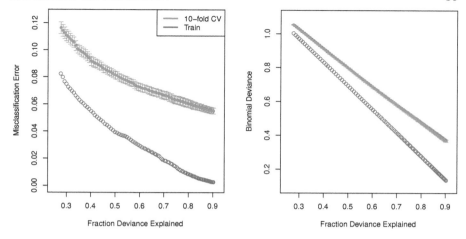

Figure 3.3 *Ridge (ℓ_2)-penalized logistic regression: tenfold cross validation curves for the Newsgroup data are shown in red, along with pointwise standard-error bands. The left plot shows misclassification error; the right plot shows deviance. Also shown in blue is the training error for each of these measures.*

outcomes, and $\bar{\mathbf{p}} = 0.52\,\mathbf{1}$ is a vector of the overall mean. This is the entry value for λ; that is the smallest value for which all coefficients are zero. When computing the solution path from $\lambda_{\max}$ down to a slightly lower value λ_1, we can screen out the vast majority of variables for which this inner-product is substantially lower than λ_1. Once we have computed the solution with the much smaller subset, we can check if any those screened were omitted in error. This can be repeated as we move down the path, using inner-products with the current residuals. This "strong-rule" screening is implemented in the `glmnet` package that we used for the computations in the above example. We discuss strong rules and other computational speedups in more detail in Section 5.10.

3.2.2 Algorithms

Two-class logistic regression is a popular generalization of linear regression, and as a consequence much effort has gone into fitting lasso-penalized logistic models. The objective (3.9) is convex and the likelihood part is differentiable, so in principle finding a solution is a standard task in convex optimization (Koh et al. 2007).

Coordinate descent is both attractive and efficient for this problem, and in the bibliographic notes we give a partial account of the large volume of research on this approach; see also Sections 2.4.2 and 5.4. The `glmnet` package uses a proximal-Newton iterative approach, which repeatedly approximates the negative log-likelihood by a quadratic function (Lee, Sun and Saunders 2014).

In detail, with the current estimate $(\tilde{\beta}_0, \tilde{\beta})$, we form the quadratic function

$$Q(\beta_0, \beta) = \frac{1}{2N} \sum_{i=1}^{N} w_i (z_i - \beta_0 - \beta^T x_i)^2 + C(\tilde{\beta}_0, \tilde{\beta}), \qquad (3.12)$$

where C denotes a constant independent of (β_0, β), and

$$z_i = \tilde{\beta}_0 + \tilde{\beta}^T x_i + \frac{y_i - \tilde{p}(x_i)}{\tilde{p}(x_i)(1 - \tilde{p}(x_i))}, \quad \text{and} \quad w_i = \tilde{p}(x_i)(1 - \tilde{p}(x_i)), \quad (3.13)$$

with $\tilde{p}(x_i)$ being the current estimate for $\Pr(Y = 1 \mid X = x_i)$. Each outer loop then amounts to a weighted lasso regression. By using warm starts on a fine grid of values for λ, typically only a few outer-loop iterations are required, because locally the quadratic approximation is very good. We discuss some of the features of `glmnet` in Sections 3.7 and 5.4.2.

3.3 Multiclass Logistic Regression

Some classification and discrimination problems have $K > 2$ output classes. In machine learning a popular approach is to build all $\binom{K}{2}$ classifiers ("one versus one" or OvO), and then classify to the class that wins the most competitions. Another approach is "one versus all" (OvA) which treats all but one class as the negative examples. Both of these methods can be put on firm theoretical grounds, but also have limitations. OvO can be computationally wasteful, and OvA can suffer from certain masking effects (Hastie et al. 2009, Chapter 4). With multiclass logistic regression, a more natural approach is available. We use the multinomial likelihood and represent the probabilities using the log-linear representation

$$\Pr(Y = k \mid X = x) = \frac{e^{\beta_{0k} + \beta_k^T x}}{\sum_{\ell=1}^{K} e^{\beta_{0\ell} + \beta_\ell^T x}}. \qquad (3.14)$$

This model is over specified, since we can add the linear term $\gamma_0 + \gamma^T x$ to the linear model for each class, and the probabilities are unchanged. For this reason, it is customary to set one of the class models to zero—often the last class—leading to a model with $K - 1$ linear functions to estimate (each a contrast with the last class). The model fit by maximum-likelihood is invariant to the choice of this base class, and the parameter estimates are equivariant (the solution for one base can be obtained from the solution for another).

Here we prefer the redundant but symmetric approach (3.14), because

- we regularize the coefficients, and the regularized solutions are not equivariant under base changes, and

- the regularization automatically eliminates the redundancy (details below).

For observations $\{(x_i, y_i)\}_{i=1}^{N}$, we can write the regularized form of the negative log-likelihood as

$$-\frac{1}{N} \sum_{i=1}^{N} \log \Pr(Y = y_i \mid x_i; \{\beta_{0k}, \beta_k\}_{k=1}^{K}) + \lambda \sum_{k=1}^{K} \|\beta_k\|_1. \tag{3.15}$$

Denote by $\mathbf{R}$ the $N \times K$ *indicator response* matrix with elements $r_{ik} = \mathbb{I}(y_i = k)$. Then we can write the log-likelihood part of the objective (3.15) in the more explicit form

$$\frac{1}{N} \sum_{i=1}^{N} w_i \left[\sum_{k=1}^{K} r_{ik}(\beta_{0k} + \beta_k^T x_i) - \log \left\{ \sum_{k=1}^{K} e^{\beta_{0k} + \beta_k^T x_i} \right\} \right]. \tag{3.16}$$

We have included a weight w_i per observation, where the setting $w_i = 1/N$ is the default. This form allows for *grouped* response data: at each value x_i we have a collection of n_i multicategory responses, with r_{ik} in category k. Alternatively, the rows of $\mathbf{R}$ can be a vector of class proportions, and we can provide $w_i = n_i$ as the observation weights.

As mentioned, the model probabilities and hence the log-likelihood are invariant under a constant shift in the K coefficients for each variable x_j—in other words $\{\beta_{kj} + c_j\}_{k=1}^{K}$ and $\{\beta_{kj}\}_{k=1}^{K}$ produce exactly the same probabilities. It is therefore up to the penalty in the criterion (3.15) to resolve the choice of c_j. Clearly, for any candidate set $\{\tilde{\beta}_{kj}\}_{k=1}^{K}$, the optimal c_j should satisfy

$$c_j = \arg\min_{c \in \mathbb{R}} \left\{ \sum_{k=1}^{K} |\tilde{\beta}_{kj} - c| \right\}. \tag{3.17}$$

Consequently, as shown in Exercise 3.3, for each $j = 1, \ldots, p$, the maximizer of the objective (3.17) is given by the median of $\{\tilde{\beta}_{1j}, \ldots, \tilde{\beta}_{Kj}\}$. Since the intercepts $\{\beta_{0k}\}_{k=1}^{K}$ are not penalized, we do need to resolve their indeterminacy; in the **glmnet** package, they are constrained to sum to zero.

3.3.1 *Example: Handwritten Digits*

As an illustration, we consider the US post-office handwritten digits data (Le Cun, Boser, Denker, Henderson, Howard, Hubbard and Jackel 1990). There are $N = 7291$ training images of the digits $\{0, 1, \ldots, 9\}$, each digitized to a 16×16 gray-scale image. Using the $p = 256$ pixels as features, we fit a 10-class lasso multinomial model. Figure 3.4 shows the training and test misclassification error as a function of the sequence of λ values used. In Figure 3.5 we display the coefficients as images (on average about 25% are nonzero). Some of these can be identified as appropriate contrast functionals for highlighting each digit.

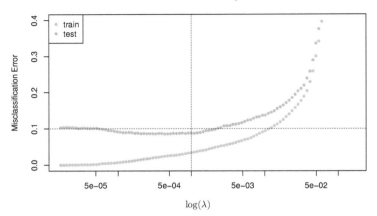

Figure 3.4 *Training and test misclassification errors of a multinomial lasso model fit to the zip code data, plotted as a function of* $\log(\lambda)$. *The minimum test error here is around 0.086, while the minimum training error is 0. We highlight the value* $\lambda = 0.001$, *where we examine the individual class coefficients in Figure 3.5.*

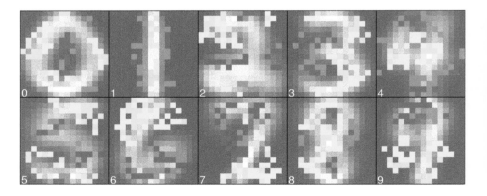

Figure 3.5 *Coefficients of the multinomial lasso, displayed as images for each digit class. The gray background image is the average training example for that class. Superimposed in two colors (yellow for positive, blue for negative) are the nonzero coefficients for each class. We notice that they are nonzero in different places, and create discriminant scores for each class. Not all of these are interpretable.*

3.3.2 Algorithms

Although one could tackle this problem with standard convex-optimization software, we have found coordinate-descent to be particularly effective (Friedman, Hastie, Simon and Tibshirani 2015). In the two-class case, there is an outer Newton loop and an inner weighted least-squares step. The outer loop can be seen as making a quadratic approximation to the log-likelihood, centered at the current estimates $(\tilde{\beta}_{0k}, \tilde{\beta}_k)_{k=1}^{K}$. Here we do the same, except we hold all but one class's parameters fixed when making this approximation. In detail, when updating the parameters $(\beta_{0\ell}, \beta_\ell)$, we form the quadratic function

$$Q_\ell(\beta_{0\ell}, \beta_\ell) = -\frac{1}{2N} \sum_{i=1}^{N} w_{i\ell} \left(z_{i\ell} - \beta_{0\ell} - \beta_\ell^T x_i\right)^2 + C(\{\tilde{\beta}_{0k}, \tilde{\beta}_k\}_{k=1}^{K}), \quad (3.18)$$

where C denotes a constant independent of $(\beta_{0\ell}, \beta_\ell)$, and

$$z_{i\ell} = \tilde{\beta}_{0\ell} + \tilde{\beta}_\ell^T x_i + \frac{r_{i\ell} - \tilde{p}_\ell(x_i)}{\tilde{p}_\ell(x_i)(1 - \tilde{p}_\ell(x_i))}, \quad \text{and} \quad w_{i\ell} = \tilde{p}_\ell(x_i)(1 - \tilde{p}_\ell(x_i))$$

where $\tilde{p}_\ell(x_i)$ is the current estimate for the conditional probability $\Pr(Y = \ell \mid x_i)$. Our approach is similar to the two-class case, except now we have to cycle over the classes as well in the outer loop. For each value of λ, we create an outer loop which cycles over $\ell \in \{1, \ldots, K\}$ and computes the partial quadratic approximation Q_ℓ about the current parameters $(\tilde{\beta}_0, \tilde{\beta})$. Then we use coordinate descent to solve the weighted lasso problem problem

$$\underset{(\beta_{0\ell}, \beta_\ell) \in \mathbb{R}^{p+1}}{\text{minimize}} \{Q(\beta_{0\ell}, \beta_\ell) + \lambda \|\beta_\ell\|_1\}. \quad (3.19)$$

3.3.3 Grouped-Lasso Multinomial

As can be seen in Figure 3.5, the lasso penalty will select different variables for different classes. This can mean that although individual coefficient vectors are sparse, the overall model may not be. In this example, on average there are 25% of the coefficients nonzero per class, while overall 81% of the variables are used.

An alternative approach is to use a grouped-lasso penalty (see Section 4.3) for the set of coefficients $\beta_j = (\beta_{1j}, \beta_{2j}, \ldots, \beta_{Kj})$, and hence replace the criterion (3.15) with the regularized objective

$$-\frac{1}{N} \sum_{i=1}^{N} \log \Pr(Y = y_i \mid X = x_i; \{\beta_j\}_{j=1}^{p}) + \lambda \sum_{j=1}^{p} \|\beta_j\|_2. \quad (3.20)$$

It is important that this criterion involves the sum of the ordinary ℓ_2-norms $\|\cdot\|_2$, as opposed to the squared ℓ_2-norms. In this way, it amounts to imposing a block ℓ_1/ℓ_2 constraint on the overall collection of coefficients: the sum of the ℓ_2-norms over the groups. The effect of this group penalty is to select all the coefficients for a particular variable to be in or out of the model. When included, they are all nonzero in general, and as shown in Exercise 3.6, they will automatically satisfy the constraint $\sum_{k=1}^{K} \beta_{kj} = 0$. Criterion (3.20) is convex, so standard methods can be used to find the optimum. As before, coordinate descent techniques are one reasonable choice, in this case block coordinate descent on each vector β_j, holding all the others fixed; see Exercise 3.7 for the details. The group lasso and variants are discussed in more detail in Chapter 4.3.

3.4   Log-Linear Models and the Poisson GLM

When the response variable Y is nonnegative and represents a count, its mean will be positive and the Poisson likelihood is often used for inference. In this case we typically use the log-linear model (3.3) to enforce the positivity. We assume that for each $X = x$, the response Y follows a Poisson distribution with mean μ satisfying

$$\log \mu(x) = \beta_0 + \beta^T x. \tag{3.21}$$

The ℓ_1-penalized negative log-likelihood is given by

$$-\frac{1}{N} \sum_{i=1}^{N} \left\{ y_i(\beta_0 + \beta^T x_i) - e^{\beta_0 + \beta^T x_i} \right\} + \lambda \|\beta\|_1. \tag{3.22}$$

As with other GLMs, we can fit this model by iteratively reweighted least squares, which amounts to fitting a weighted lasso regression at each outer iteration. Typically, we do not penalize the intercept β_0. It is easy to see that this enforces the constraint that the average fitted value is equal to the mean response—namely, that $\frac{1}{N} \sum_{i=1}^{N} \hat{\mu}_i = \bar{y}$, where $\hat{\mu}_i := e^{\hat{\eta}(x_i)} = e^{\widehat{\beta}_0 + \widehat{\beta}^T x_i}$.

Poisson models are often used to model rates, such as death rates. If the length T_i of the observation window is different for each observation, then the mean count is $\mathbb{E}(y_i \mid X_i = x_i) = T_i \mu(x_i)$ where $\mu(x_i)$ is the rate per unit time interval. In this case, our model takes the form

$$\log(\mathbb{E}(Y \mid X = x, T)) = \log(T) + \beta_0 + \beta^T x. \tag{3.23}$$

The terms $\log(T_i)$ for each observation require no fitting, and are called an *offset*. Offsets play a role in the following example as well.

3.4.1   Example: Distribution Smoothing

The Poisson model is a useful tool for estimating distributions. The following example was brought to our attention by Yoram Singer (Singer and Dubiner 2011). Suppose that we have a sample of N counts $\{y_k\}_{k=1}^{N}$ from an

N-cell multinomial distribution, and let $r_k = y_k / \sum_{\ell=1}^{N} y_\ell$ be the corresponding vector of proportions. For example, in large-scale web applications, these counts might represent the number of people in each county in the USA that visited a particular website during a given week. This vector could be sparse, depending on the specifics, so there is a desire to regularize toward a broader, more stable distribution $\mathbf{u} = \{u_k\}_{k=1}^{N}$ (for example, the same demographic, except measured over a year). Singer and Dubiner (2011) posed the following problem

$$\underset{\mathbf{q} \in \mathbb{R}^N,\, q_k \geq 0}{\text{minimize}} \sum_{k=1}^{N} q_k \log \left(\frac{q_k}{u_k} \right) \quad \text{such that } \|\mathbf{q} - \boldsymbol{r}\|_\infty \leq \delta \text{ and } \sum_{k=1}^{N} q_k = 1.$$

(3.24)

In words, we find the distribution, within a δ tolerance in the ℓ_∞-norm from the observed distribution, that is as close as possible to the nominal distribution $\mathbf{u}$ in terms of Kullback–Leibler (KL) divergence. It can be shown (see

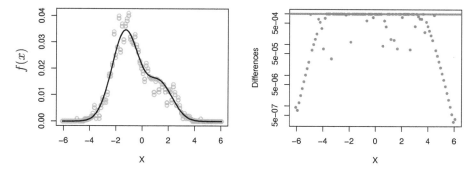

Figure 3.6 *Estimating distributions via the Poisson. In the left panel, the solid black curve is the parent distribution $\mathbf{u}$, here represented as a discretization of a one-dimensional distribution $f(x)$ into 100 cells. The blue points represent the observed distribution, and the orange points represent the distribution recovered by the model. While the observed distribution may have many zero counts, the modeled distribution has the same support as $\mathbf{u}$. The right plot shows the $N = 100$ differences $|\hat{q}_k - r_k|$, which are constrained to be less than $\delta = 0.001$, which is the horizontal orange line.*

Exercise 3.4) that the Lagrange-dual to the optimization problem (3.24) has the form

$$\underset{\beta_0, \boldsymbol{\alpha}}{\text{maximize}} \left\{ \sum_{k=1}^{N} [r_k \log q_k(\beta_0, \alpha_k) - q_k(\beta_0, \alpha_k)] - \delta \|\boldsymbol{\alpha}\|_1 \right\}, \qquad (3.25)$$

where $q_k(\beta_0, \alpha_k) := u_k e^{\beta_0 + \alpha_k}$. This is equivalent to fitting a Poisson GLM with offset $\log(u_k)$, individual parameter α_k per observation, and the extremely sparse design matrix $\mathbf{X} = \mathbf{I}_{N \times N}$. Consequently, it can be fit very efficiently using sparse-matrix methods (see Section 3.7 below). Figure 3.6

shows a simulation example, where the distribution u_k is a discretized continuous distribution (mixture of Gaussians). There are $N = 100$ cells, and a total of $\sum_{k=1}^{N} y_k = 1000$ observations distributed to these cells. As discussed above, the presence of the unpenalized β_0 ensures that $\sum_{k=1}^{N} \hat{q}_k = \sum_{k=1}^{N} r_k = 1$ (see also Exercise 3.5). Although we only show one solution in Figure 3.6, the path gives solutions $\hat{q}_k(\delta)$ that vary smoothly between the background distribution u_k and the observed distribution r_k.

3.5 Cox Proportional Hazards Models

In medical studies, the outcome of interest after treatment is often time to death or time to recurrence of the disease. Patients are followed after their treatment, and some drop out because they move away, or perhaps die from an independent cause. Such outcomes are called *right censored*. Denoting by T the underlying survival time, for each patient we observe the quantity $Y = \min(T, C)$ where C is a *censoring time*. Interest tends to focus on the survivor function $S(t) := \Pr(T > t)$, the probability of surviving beyond a certain time t.

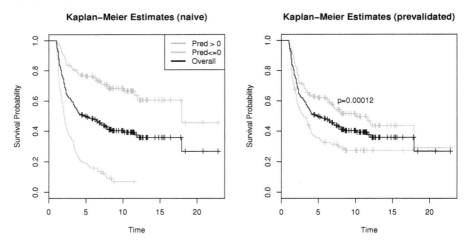

Figure 3.7 *The black curves are the Kaplan–Meier estimates of $S(t)$ for the Lymphoma data. In the left plot, we segment the data based on the predictions from the Cox proportional hazards lasso model, selected by cross-validation. Although the tuning parameter is chosen by cross-validation, the predictions are based on the full training set, and are overly optimistic. The right panel uses* prevalidation *to build a prediction on the entire dataset, with this training-set bias removed. Although the separation is not as strong, it is still significant. The spikes indicate censoring times. The p-value in the right panel comes from the log-rank test.*

The black curves in Figure 3.7 show estimates of $S(t)$ for a population of $N = 240$ Lymphoma patients (Alizadeh et al. 2000). Each of the spikes in the plot indicates a censoring point, meaning a time at which a patient was lost for follow-ups. Although survival curves are useful summaries of such data, when incorporating covariates it is more common to model the *hazard function*, a monotone transformation of S. More specifically, the hazard at time t is given by

$$h(t) = \lim_{\delta \to 0} \frac{\Pr(Y \in (t, t + \delta) \mid Y \geq t)}{\delta} = \frac{f(t)}{S(t)}, \tag{3.26}$$

and corresponds to the instantaneous probability of death at time t, given survival up till t.

We now discuss Cox's proportional hazards model that was used to produce the blue and orange survival curves in Figure 3.7. The proportional hazards model (CPH) is based on the hazard function

$$h(t; x) = h_0(t) e^{\beta^T x}, \tag{3.27}$$

where $h_0(t)$ is a baseline hazard (the hazard for an individual with $x = 0$).

We have data of the form (x_i, y_i, δ_i), where δ_i is binary-valued indicator of whether y_i is a death time or censoring time. For the lymphoma data, there are $p = 7399$ variables, each a measure of gene expression. Of the $N = 240$ samples, a total of 102 samples are right censored. Here we fit an ℓ_1-penalized CPH by solving

$$\underset{\beta}{\text{minimize}} \left\{ -\sum_{\{i \,|\, \delta_i = 1\}} \log \left[\frac{e^{\beta^T x_i}}{\sum_{j \in R_i} e^{\beta^T x_j}} \right] + \lambda \|\beta\|_1 \right\}, \tag{3.28}$$

where for each $i = 1, \ldots, N$, R_i is the *risk set* of individuals who are alive and in the study at time y_i. The first term is the log of the *partial likelihood*, corresponding to the conditional probability in the risk set of the observed death. Note that the baseline hazard does not play a role, an attractive feature of this approach. Here we have assumed that there are no ties, that is, the survival times are all unique. Modification of the partial likelihood is needed in the event of ties.

Figure 3.8 shows the coefficients obtained in fitting the model (3.28) to the Lymphoma data. Since $p \gg N$, the model would "saturate" as $\lambda \downarrow 0$, meaning that some parameters would diverge to $\pm\infty$, and the log partial likelihood would approach 0. We see evidence of this undesirable behavior as λ gets small.

The computations for the Cox model are similar to those for the multinomial model but slightly more complex. Simon, Friedman, Hastie and Tibshirani (2011) give details for an algorithm based on coordinate-descent.

3.5.1   Cross-Validation

All the models in this chapter require a choice of λ, and we typically use K−fold cross-validation where K equal to 5 or 10, as in Figure 3.2. For the

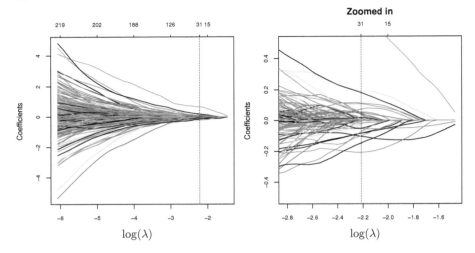

Figure 3.8 *The ℓ_1-regularized coefficient path for the Cox model fit to the Lymphoma data. Since $p \gg N$, the plot has a trumpet shape near the end, corresponding to a saturated model with partial likelihood equal to one. The right-hand plot zooms in on the area of interest, a fairly heavily regularized solution with 31 nonzero coefficients.*

Cox model, we compute the cross-validated deviance, which is minus twice the log partial likelihood. An issue arises in computing the deviance, since if N/K is small, there will not be sufficient observations to compute the risk sets. Here we use a trick due to van Houwelingen et al. (2006). When fold k is left out, we compute the coefficients $\widehat{\beta}^{-k}(\lambda)$, and then compute

$$\widehat{\mathrm{Dev}}_\lambda^k := \mathrm{Dev}[\widehat{\beta}^{-k}(\lambda)] - \mathrm{Dev}^{-k}[\widehat{\beta}^{-k}(\lambda)]. \tag{3.29}$$

The first term on the right uses all N samples in computing the deviance, while the second term omits the fold-k samples. Finally $\mathrm{Dev}_\lambda^{CV} = \sum_{k=1}^{K} \widehat{\mathrm{Dev}}_\lambda^k$ is obtained by subtraction. The point is that each of these terms has sufficient data to compute the deviance, and in the standard cases (that is, any of the other generalized linear models), the estimate would be precisely the deviance on the left-out set.

The deviance in Figure 3.9 was computed in this fashion; we zoom in on the right-hand section. We see that the minimum is achieved at 31 nonzero coefficients. Figure 3.7 shows the effect of the chosen model. We compute $\hat{\eta}(x_i) = x_i^T \hat{\beta}(\lambda_{\min})$ for each observation, and then create two groups by thresholding these scores at zero. The two colored survival curves in the left-hand plot show the difference in survival for the two groups thus formed. They are well separated, which suggests we have derived a powerful signature. However, these scores are biased: we are evaluating their performance on the same data for which they were computed.

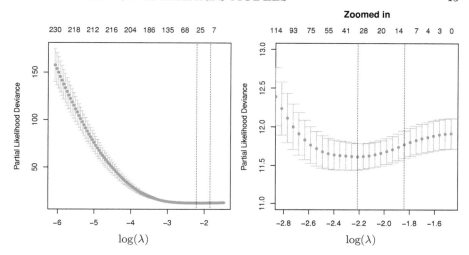

Figure 3.9 *Cross-validated deviance for the lymphoma data, computed by subtrac-*
tions, as described in the text. The right-hand plot zooms in on the area of interest.
The dotted vertical line on the left corresponds to the minimum, and the model we
chose in this case; the one on the right corresponds to the rightmost point on the
curve (simplest model) within one standard error of the minimum. This is a basis
for a more conservative approach to selection. The number of nonzero coefficients is
shown along the top of each plot.

3.5.2 Pre-Validation

In Figure 3.7, we used a variant of cross-validation, known as *pre-validation*
(Tibshirani and Efron 2002), in order to obtain a fair evaluation of the model.
Cross-validation leaves out data in order to obtain a reasonably unbiased
estimate of the error rate of a model. But the error rate is not a very interpretable measure in some settings such as survival modelling. The method of
pre-validation is similar to cross-validation, but instead produces a new set of
"unbiased data" that mimics the performance of the model applied to independent data. The pre-validated data can then be analyzed and displayed. In
computing the score $\hat{\eta}(x_i)^{(k)}$ for the observations in fold k, we use the coefficient vector $\hat{\beta}^{(-k)}$ computed with those observations omitted.[5] Doing this for
all K folds, we obtain the "pre-validated" dataset $\{(\hat{\eta}(x_i)^{(k)}, y_i, \delta_i)\}_{i=1}^{N}$. The
key aspect of this pre-validated data is that each score $\hat{\eta}(x_i)^{(k)}$ is derived independently of its response value (y_i, δ_i). Hence we can essentially treat these
scores as if they were derived from a dataset completely separate from the
"test data" $\{(x_i, y_i, \delta_i)\}_{i=1}^{N}$. In the right-hand panel of Figure 3.7, we have
split the pre-validated scores into two groups and plotted the corresponding

---

[5]Strictly speaking λ should be chosen each time as well, but we did not do that here.

survival curves. Although the curves are not as spread out as in the left-hand plot, they are still significantly different.

3.6 Support Vector Machines

We now turn to a method for binary classification known as the support vector machine (SVM). The idea is shown in Figure 3.10. The decision boundary is the solid line in the middle of the yellow slab. The *margin* is the half-width of the yellow slab. Ideally, all of the blue data points should lie above the slab on the right, and the red points should lie below it on the left. However in the picture, three red points and two blue points lie on the wrong side of their margin. These correspond to the "errors" ξ_i. The SVM decision boundary is chosen to maximize the margin, subject to a fixed budget on the total error $\sum_{i=1}^{N} \xi_i$. The idea is that a decision boundary achieving the largest margin has more space between the classes and will generalize better to test data. This leads to the optimization problem

$$\underset{\beta_0,\, \beta,\, \{\xi_i\}_1^N}{\text{maximize}} \ M \quad \text{subject to } y_i \underbrace{(\beta_0 + \beta^T x_i)}_{f(x_i;\, \beta_0,\, \beta)} \geq M(1 - \xi_i) \ \forall i, \tag{3.30}$$

$$\text{and } \xi_i \geq 0 \ \forall i, \ \sum_{i=1}^{N} \xi_i \leq C, \text{ and } \|\beta\|_2 = 1. \tag{3.31}$$

(See Section 3.6.1 for an explanation of this particular form.)

This problem involves a linear cost function subject to convex constraints, and many efficient algorithms have been designed for its solution. It can be shown to be equivalent to the penalized form (3.8) previously specified on page 31, which we restate here:

$$\underset{\beta_0,\beta}{\text{minimize}} \left\{ \frac{1}{N} \sum_{i=1}^{N} [1 - y_i f(x\,;\beta_0, \beta)]_+ + \lambda\|\beta\|_2^2 \right\}. \tag{3.32}$$

Decreasing λ has a similar effect to decreasing C.[6] The linear SVM can be generalized using a *kernel* to create nonlinear boundaries; it involves replacing the squared ℓ_2-norm in the objective (3.32) by the squared Hilbert norm defined by a symmetric bivariate kernel. Details on this extension can be found elsewhere—for instance, see Hastie et al. (2009), Section 5.8.

Since the criterion (3.32) involves a quadratic penalty, the estimated coefficient vector will not be sparse. However, because the hinge loss function is piecewise linear, it introduces a different kind of sparsity. It can be shown via the dual formulation of the SVM that the solution $\hat{\beta}$ has the form

$$\hat{\beta} = \sum_{i=1}^{N} \hat{\alpha}_i y_i x_i; \tag{3.33}$$

---

[6]Solutions to (3.32) do not have $\|\hat{\beta}\|_2 = 1$, but since a linear classifier is scale invariant, the solution coefficients can be rescaled.

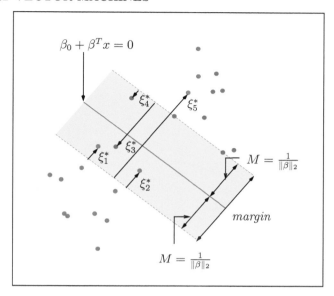

Figure 3.10 *Support vector classifier: The decision boundary is the solid line, while broken lines bound the shaded maximal margin of width $2M = 2/\|\beta\|_2$. The points labelled ξ_j^* are on the wrong side of their margin by an amount $\xi_j^* = M\xi_j$; points on the correct side have $\xi_j^* = 0$. The margin is maximized subject to a total budget $\sum_{i=1}^N \xi_i \leq C$. Hence $\sum_{i=1}^N \xi_j^*$ is the total distance of points on the wrong side of their margin.*

each observation $i \in \{1, \ldots, N\}$ is associated with a nonnegative weight $\hat{\alpha}_i$, and only a subset $\mathcal{V}_\lambda$, referred to as the *support set*, will be associated with nonzero weights.

SVMs are popular in high-dimensional classification problems with $p \gg N$, since the computations are $\mathcal{O}(pN^2)$ for both linear and nonlinear kernels. Additional efficiencies can be realized for linear SVMs, using stochastic subgradient methods (Shalev-Shwartz, Singer and Srebro 2007). They are not, however, sparse in the features. Replacing the ℓ_2 penalty in the objective (3.32) with an ℓ_1 penalty promotes such sparsity, and yields the ℓ_1-*regularized linear SVM*:

$$\underset{\beta_0, \beta}{\text{minimize}} \left\{ \frac{1}{N} \sum_{i=1}^N [1 - y_i f(x_i; \beta_0, \beta)]_+ + \lambda \|\beta\|_1 \right\}. \qquad (3.34)$$

The optimization problem (3.34) is a linear program with many constraints (Zhu, Rosset, Hastie and Tibshirani 2004, Wang, Zhu and Zou 2006), and efficient algorithms can be complex (Exercise 3.9). The solution paths (in fine detail) can have many jumps, and show many discontinuities. For this reason, some authors prefer to replace the usual hinge loss $\phi_{\text{hin}} = (1 - t)_+$

Hinge Loss versus Binomial with L1 Regularization

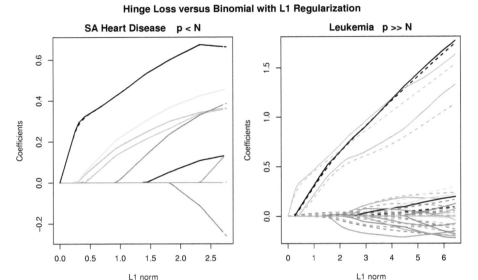

Figure 3.11 *A comparison of the coefficient paths for the ℓ_1-regularized SVM versus logistic regression on two examples. In the left we have the South African heart disease data ($N = 462$ and $p = 9$), and on the right the Leukemia data ($N = 38$ and $p = 6087$). The dashed lines are the SVM coefficients, the solid lines logistic regression. The similarity is striking in the left example, and strong in the right.*

with *squared hinge loss* $\phi_{\mathrm{sqh}}(t) = (1 - t)_+^2$, which is differentiable everywhere (see Exercise 3.8).

The SVM loss function shares many similarities with the binomial loss (Hastie et al. 2009, Section 12.3), and their solutions are not too different. Figure 3.11 compares their ℓ_1 regularization paths on two examples, and supports this claim. In the left-hand plot, they are virtually identical. In the right-hand plot, for more than half of the path, the training data are separated by the solution. As we proceed to the end of the path, the logistic coefficients become less stable than those of the SVM, and can account for the bigger discrepancies.

The support vector machine, on the other hand, is designed for finding maximal-margin solutions for separable data, and its coefficients do not blow up at the least-regularized end of the path. However, in terms of the ℓ_1 penalty, this is at the nonsparse end of the path. In light of this, we do not recommend the ℓ_1 regularized linear SVM as a variable selector, because the corresponding logistic regression problem (3.6) gives very similar solutions when the penalty is active, and the algorithms are more stable.

3.6.1   Logistic Regression with Separable Data

It is a well-known fact that without a penalty on the coefficients, the linear logistic regression model fails when the two classes are linearly separable (Exercise 3.1); the maximum-likelihood estimates for the coefficients are infinite. The problem in this case is that the likelihood is trying to make the probabilities all 1s and 0s, and inspection of (3.2) shows that this cannot be achieved with finite parameters. Once we penalize the criterion as in (3.6) the problem goes away, for as long as $\lambda > 0$, very large coefficients will not be tolerated.

With wide data ($N \ll p$), the classes are almost always separable, unless there are exact ties in covariate space for the two classes. Figure 3.1 shows the logistic-regression coefficient path for a wide-data situation; notice how the coefficients start to fan out near the end of the path. One has to take care at this end of the path, and not allow λ to get too small. In many situations, this end represents the overfit situation, which is not of primary interest. It appears not to be the case in this example, as can be seen in the cross-validation plots in Figure 3.2.

The ends of the path have special meaning in the machine-learning community, since we will see they amount to maximal-margin classifiers. Before giving the details, we review some geometry associated with linear classification. Consider the boundary $\mathcal{B} := \{x \in \mathbb{R}^p \mid f(x) = 0\}$ associated with a linear classifier $f(x) \equiv f(x\,;\beta_0, \beta) = \beta_0 + \beta^T x$. The Euclidean distance from a point x_0 to the boundary is given by

$$\text{dist}_2(x_0, \mathcal{B}) := \inf_{z \in \mathcal{B}} \|z - x_0\|_2 \; = \; \frac{|f(x_0)|}{\|\beta\|_2} \tag{3.35}$$

(Exercise 3.2). Consequently, for a given predictor-response pair (x, y), the quantity $\frac{y\,f(x)}{\|\beta\|_2}$ is the signed Euclidean distance to the boundary: it will be negative if the sign of y disagrees with that of $f(x)$. For separable data, the optimal separating hyperplane $f^*(x) = 0$ solves the optimization problem

$$M_2^* = \max_{\beta_0, \beta} \left\{ \min_{i \in \{1,\dots,N\}} \frac{y_i f(x_i\,;\beta_0, \beta)}{\|\beta\|_2} \right\}. \tag{3.36}$$

In words, it maximizes the Euclidean distance of the closest sample to the boundary.

Rosset et al. (2004) establish an interesting connection between this optimal separating hyperplane and a certain limiting case of ridge-regularized logistic regression. In particular, suppose that we replace the ℓ_1-penalty in the objective (3.10) with a squared ℓ_2-penalty, and solve the problem

$$\underset{\beta_0, \beta}{\text{minimize}} \left\{ \frac{1}{N} \sum_{i=1}^{N} \log(1 + e^{-y_i f(x_i\,;\beta_0, \beta)}) + \lambda\|\beta\|_2^2 \right\}; \tag{3.37}$$

let $(\tilde{\beta}_0(\lambda),\ \tilde{\beta}(\lambda))$ be the optimal solution, specifying a particular linear classifier. We then consider the behavior of this linear classifier as the regularization weight λ vanishes: in particular, it can be shown (Rosset et al. 2004) that

$$\lim_{\lambda \to 0} \left\{ \min_{i \in \{1,...,N\}} \frac{y_i f(x_i; \tilde{\beta}_0(\lambda), \tilde{\beta}(\lambda))}{\|\tilde{\beta}(\lambda)\|_2} \right\} = M_2^*. \tag{3.38}$$

Thus, the end of the ℓ_2-regularized logistic regression path corresponds to the SVM solution. In particular, if $(\breve{\beta}_0, \breve{\beta})$ solves the SVM objective (3.30) with $C = 0$, then

$$\lim_{\lambda \to 0} \frac{\tilde{\beta}(\lambda)}{\|\tilde{\beta}(\lambda)\|_2} = \breve{\beta}. \tag{3.39}$$

How does this translate to the setting of ℓ_1-regularized models? Matters get a little more complicated, since we move into the territory of general projections and dual norms (Mangasarian 1999). The analog of the ℓ_2-distance (3.35) is the quantity

$$\text{dist}_\infty(x_0, \mathcal{B}) := \inf_{z \in \mathcal{B}} \|z - x_0\|_\infty = \frac{|f(x_0)|}{\|\beta\|_1}, \tag{3.40}$$

For a given $\lambda \geq 0$, let $(\widehat{\beta}_0(\lambda), \widehat{\beta}(\lambda))$ denote an optimal solution to the ℓ_1-regularized logistic regression objective (3.10). Then as λ decreases toward zero, we have

$$\lim_{\lambda \to \infty} \left[\min_{i \in \{1,...,N\}} \frac{y_i f(x_i; \widehat{\beta}_0(\lambda), \widehat{\beta}(\lambda))}{\|\widehat{\beta}(\lambda)\|_1} \right] = M_\infty^*, \tag{3.41}$$

so that the worst-case margin of the ℓ_1-regularized logistic regression converges to the ℓ_1-regularized version of the support vector machine, which maximizes the ℓ_∞ margin (3.40).

In summary, then, we can make the following observations:

- At the end of the path, where the solution is most dense, the logistic regression solution coincides with the SVM solution.

- The SVM approach leads to a more stable numerical method for computing the solution in this region.

- In contrast, logistic regression is most useful in the sparser part of the solution path.

3.7 Computational Details and glmnet

Most of the examples in this chapter were fit using the R package glmnet (Friedman et al. 2015). Here we detail some of the options and features in glmnet. Although these are specific to this package, they also would be natural requirements in any other similar software.

Family: The family option allows one to pick the loss-function and the associated model. As of version 1.7, these are `gaussian`, `binomial`, `multinomial` (grouped or not), `poisson`, and `cox`. The `gaussian` family allows for multiple responses (multitask learning), in which case a group lasso is used to select coefficients for each variable, as in the grouped multinomial. Associated with each family is a *deviance* measure, the analog of the residual sum-of-squares for Gaussian errors. Denote by $\hat{\boldsymbol{\mu}}_\lambda$ the N-vector of fitted mean values when the parameter is λ, and $\tilde{\boldsymbol{\mu}}$ the unrestricted or *saturated* fit. Then

$$\text{Dev}_\lambda \doteq 2[\ell(\mathbf{y}, \tilde{\boldsymbol{\mu}}) - \ell(\mathbf{y}, \hat{\boldsymbol{\mu}}_\lambda)]. \tag{3.42}$$

Here $\ell(\mathbf{y}, \boldsymbol{\mu})$ is the log-likelihood of the model $\boldsymbol{\mu}$, a sum of N terms. The *null deviance* is $\text{Dev}_{\text{null}} = \text{Dev}_\infty$; typically this means $\hat{\boldsymbol{\mu}}_\infty = \bar{y}\mathbf{1}$, or in the case of the `cox` family $\hat{\boldsymbol{\mu}}_\infty = \mathbf{0}$. Glmnet reports D^2, the fraction of deviance explained, as defined in (3.11) on page 33.

Penalties: For all models, the `glmnet` algorithm admits a range of elastic-net penalties ranging from ℓ_2 to ℓ_1. The general form of the penalized optimization problem is

$$\underset{\beta_0, \beta}{\text{minimize}} \left\{ -\frac{1}{N}\ell(\mathbf{y}; \beta_0, \beta) + \lambda \sum_{j=1}^p \gamma_j \left\{ (1-\alpha)\beta_j^2 + \alpha|\beta_j| \right\} \right\}. \tag{3.43}$$

This family of penalties is specified by three sets of real-valued parameters:

- The parameter λ determines the overall complexity of the model. By default, the `glmnet` algorithm generates a sequence of 100 values for λ that cover the whole path (on the log scale), with care taken at the lower end for saturated fits.

- The elastic-net parameter $\alpha \in [0, 1]$ provides a mix between ridge regression and the lasso. Although one can select α via cross-validation, we typically try a course grid of around three to five values of α.

- For each $j = 1, 2, \ldots, p$, the quantity $\gamma_j \geq 0$ is a penalty modifier. When $\gamma_j = 0$, the jth variable is always included; when $\gamma_j = \text{inf}$ it is always excluded. Typically $\gamma_j = 1$ (the default), and all variables are treated as equals.

Coefficient bounds: With coordinate descent, it is very easy to allow for upper and lower bounds on each coefficient in the model. For example, we might ask for a nonnegative lasso. In this case, if a coefficient exceeds an upper or lower bound during the coordinate-descent cycle, it is simply set to the bound.

Offset: All the models allow for an *offset* term. This is a real valued number o_i for each observation, that gets added to the linear predictor, and is not associated with any parameter:

$$\eta(x_i) = o_i + \beta_0 + \beta^T x_i. \tag{3.44}$$

The offset has many uses. Sometimes we have a previously-fit model $h(z)$ (where z might include or coincide with x), and we wish to see if augmenting it with a linear model offers improvement. We would supply $o_i = h(z_i)$ for each observation.

For Poisson models the offset allows us to model rates rather than mean counts, if the observation period differs for each observation. Suppose we observe a count Y over period t, then $\mathbb{E}[Y \mid T = t, X = x] = t\mu(x)$, where $\mu(x)$ is the rate per unit time. Using the log link, we would supply $o_i = \log(t_i)$ for each observation. See Section 3.4.1 for an example.

Matrix input and weights: Binomial and multinomial responses are typically supplied as a 2 or K-level factor. As an alternative **glmnet** allows the response to be supplied in matrix form. This allows for *grouped* data, where at each x_i we see a multinomial sample. In this case the rows of the $N \times K$ response matrix represent counts in each category. Alternatively the rows can be proportions summing to one. For the latter case, supplying an observation weight equal to the total count for each observation is equivalent to the first form. Trivially an indicator response matrix is equivalent to supplying the data as a factor, in *ungrouped* form.

Sparse model matrices **X**: Often when $p \gg N$ is very large, there are many zeros in the input matrix **X**. For example, in document models, each feature vector $x_i \in \mathbb{R}^p$ might count the number of times each word in a very large dictionary occurs in a document. Such vectors and matrices can be stored efficiently by only storing the nonzero values, and then row and column indices of where they occur. Coordinate descent is ideally suited to capitalize on such sparsity, since it handles the variables one-at-a-time, and the principal operation is an inner-product. For example, in Section 3.4.1, the model-matrix **X** = **I** is the extremely sparse $N \times N$ identity matrix. Even with $N = 10^6$, the program can compute the relaxation path at 100 values of δ in only 27 seconds.

Bibliographic Notes

Generalized linear models were proposed as a comprehensive class of models by Nelder and Wedderburn (1972); see the book by McCullagh and Nelder (1989) for a thorough account. Application of the lasso to logistic regression was proposed in Tibshirani (1996); coordinate descent methods for logistic, multinomial, and Poisson regression were developed in Friedman, Hastie, Hoefling and Tibshirani (2007), Friedman, Hastie and Tibshirani (2010b), Wu and Lange (2008), and Wu, Chen, Hastie, Sobel and Lange (2009). Pre-validation was proposed by Tibshirani and Efron (2002). Boser, Guyon and Vapnik (1992) described the support vector machine, with a thorough treatment in Vapnik (1996).

Exercises

Ex. 3.1 Consider a linear logistic regression model with separable data, meaning that the data can be correctly separated into two classes by a hyperplane. Show that the likelihood estimates are unbounded, and that the log-likelihood objective reaches its maximal value of zero. Are the fitted probabilities well-defined?

Ex. 3.2 For a response variable $y \in \{-1, +1\}$ and a linear classification function $f(x) = \beta_0 + \beta^T x$, suppose that we classify according to $\text{sign}(f(x))$. Show that the signed Euclidean distance of the point x with label y to the decision boundary is given by

$$\frac{1}{\|\beta\|_2} y f(x). \tag{3.45}$$

Ex. 3.3 Here we show that for the multinomial model, the penalty used automatically imposes a normalization on the parameter estimates. We solve this problem for a general elastic-net penalty (Section 4.2). For some parameter $\alpha \in [0, 1]$ consider the problem

$$c_j(\alpha) = \underset{t \in \mathbb{R}}{\arg\min} \left\{ \sum_{\ell=1}^K \left[\frac{1}{2}(1 - \alpha)(\beta_{j\ell} - t)^2 + \alpha |\beta_{j\ell} - t| \right] \right\}. \tag{3.46}$$

Let $\bar{\beta}_j = \frac{1}{K} \sum_{\ell=1}^K \beta_{j\ell}$ be the sample mean, and let $\tilde{\beta}_j$ be a sample median. (For simplicity, assume that $\bar{\beta}_j \leq \tilde{\beta}_j$). Show that

$$\bar{\beta}_j \leq c_j(\alpha) \leq \tilde{\beta}_j \quad \text{for all } \alpha \in [0, 1] \tag{3.47}$$

with the lower inequality achieved if $\alpha = 0$, and the upper inequality achieved if $\alpha = 1$.

Ex. 3.4 Derive the Lagrange dual (3.25) of the *maximum-entropy* problem (3.24). Note that positivity is automatically enforced, since the log function in the objective (3.24) serves as a barrier. (*Hint:* It may help to introduce additional variables $w_i = p_i - r_i$, and now minimize the criterion (3.24) with respect to both $\{p_i, w_i\}_{i=1}^N$, subject to the additional constraints that $w_i = p_i - r_i$.)

Ex. 3.5 Recall the dual (3.25) of the maximum entropy problem, and the associated example motivating it. Suppose that for each cell, we also measure the value x_k corresponding to the mid-cell ordinate on the continuous domain x. Consider the model

$$q_k = u_k e^{\beta_0 + \sum_{m=1}^M \beta_m x_k^m + \alpha_k}, \tag{3.48}$$

and suppose that we fit it using the penalized log-likelihood (3.25) without penalizing any of the coefficients. Show that for the estimated distribution $\hat{\mathbf{q}} = \{\hat{q}_k\}_{k=1}^N$, the moments of X up to order M match those of the empirical distribution $\mathbf{r} = \{r_k\}_{k=1}^N$.

Ex. 3.6 Consider the group-lasso-regularized version of multinomial regression (3.20). Suppose that for a particular value of λ, the coefficient $\widehat{\beta}_{kj}$ is *not equal* to 0. Show that $\widehat{\beta}_{\ell j} \neq 0$ for all $\ell \in (1, \ldots, K)$, and moreover that $\sum_{\ell=1}^{K} \widehat{\beta}_{\ell j} = 0$.

Ex. 3.7 This problem also applies to the group-lasso-regularized form of multinomial regression (3.20). Suppose that for a particular value of λ, and the fitted probabilities are $\hat{\pi}_i = (\hat{\pi}_{i1}, \ldots, \hat{\pi}_{iK})^T$. Similarly let $r_i = (r_{i1}, \ldots, r_{iK})^T$ be the observed proportions. Suppose we consider including an additional variable (vector) Z with observed values z_i, and wish to update the fit. Let $g = \sum_{i=1}^{N} z_i(r_i - \hat{\pi}_i)$. Show that if $\|g\|_2 < \lambda$, then the coefficients of Z are zero, and the model remains unchanged.

Ex. 3.8 The *squared hinge loss function* $\phi_{\mathrm{sqh}}(t) := (1 - t)_+^2$ can be used as a margin-based loss function $\phi(y f(x))$ for binary classification problems.

(a) Show that ϕ_{sqh} is differentiable everywhere.

(b) Suppose $Y \in \{-1, +1\}$ with $\Pr(Y = +1) = \pi \in (0, 1)$. Find the function $f : \mathbb{R}^p \to \mathbb{R}$ that minimizes (for each $x \in \mathbb{R}^p$) the criterion

$$\underset{f}{\text{minimize}}\, \mathbb{E}_Y \left[\phi_{\mathrm{sqh}}(Y f(x)) \right] \tag{3.49}$$

(c) Repeat part (b) using the usual hinge loss $\phi_{\mathrm{hin}}(t) = (1 - t)_+$.

Ex. 3.9 Given binary responses $y_i \in \{-1, +1\}$, consider the ℓ_1-regularized SVM problem

$$(\widehat{\beta}_0,\, \widehat{\beta}) = \underset{\beta_0, \beta}{\arg\min} \left\{ \sum_{i=1}^{N} \{1 - y_i f(x_i\,;\beta_0, \beta)\}_+ + \lambda \sum_{j=1}^{p} |\beta_j| \right\}, \tag{3.50}$$

where $f(x\,;\beta_0, \beta) := \beta_0 + \beta^T x$. In this exercise, we compare solutions of this problem to those of weighted ℓ_2-regularized SVM problem: given nonnegative weights $\{w_j\}_{j=1}^{p}$, we solve

$$(\tilde{\beta}_0,\, \tilde{\beta}) = \underset{\beta_0, \beta}{\arg\min} \left\{ \sum_{i=1}^{N} \{1 - y_i f(x_i\,;\beta_0, \beta)\}_+ + \frac{\lambda}{2} \sum_{j=1}^{p} w_j \beta_j^2 \right\}. \tag{3.51}$$

(a) Show that if we solve the problem (3.51) with $w_j = 1/|\widehat{\beta}_j|$, then $(\tilde{\beta}_0, \tilde{\beta}) = (\widehat{\beta}_0, \widehat{\beta})$.

(b) For a given weight sequence $\{w_j\}_{j=1}^{p}$ with $w_j \in (0, \infty)$ for all $j = 1, \ldots, p$, show how to solve the criterion (3.51) using a regular unweighted SVM solver. What do you do if $w_j = \infty$ for some subset of indices?

(c) In light of the preceding parts, suggest an iterative algorithm for the problem (3.50) using a regular SVM solver.

Chapter 4

Generalizations of the Lasso Penalty

4.1 Introduction

In the previous chapter, we considered some generalizations of the lasso obtained by varying the loss function. In this chapter, we turn to some useful variations of the basic lasso ℓ_1-penalty itself, which expand the scope of the basic model. They all inherit the two essential features of the standard lasso, namely the shrinkage and selection of variables, or groups of variables.

Such generalized penalties arise in a wide variety of settings. For instance, in microarray studies, we often find groups of correlated features, such as genes that operate in the same biological pathway. Empirically, the lasso sometimes does not perform well with highly correlated variables. By combining a squared ℓ_2-penalty with the ℓ_1-penalty, we obtain the *elastic net*, another penalized method that deals better with such correlated groups, and tends to select the correlated features (or not) together. In other applications, features may be structurally grouped. Examples include the dummy variables that are used to code a multilevel categorical predictor, or sets of coefficients in a multiple regression problem. In such settings, it is natural to select or omit all the coefficients within a group together. The *group lasso* and the *overlap group lasso* achieve these effects by using sums of (un-squared) ℓ_2 penalties. Another kind of structural grouping arises from an underlying index set such as time; our parameters might each have an associated time stamp. We might then ask for time-neighboring coefficients to be the same or similar. The *fused lasso* is a method naturally tailored to such situations.

Finally, a variety of nonparametric smoothing methods operate implicitly with large groups of variables. For example, each term in an additive smoothing-spline model has an associated cubic-spline basis. The grouped lasso extends naturally to these situations as well; the COSSO and the SPAM families are examples of such nonparametric models. In summary, all these variants deal with different kinds of groupings of the features in natural ways, and it is the goal of this chapter to explore them in some more detail.

4.2   The Elastic Net

The lasso does not handle highly correlated variables very well; the coefficient paths tend to be erratic and can sometimes show wild behavior. Consider a simple but extreme example, where the coefficient for a variable X_j with a particular value for λ is $\hat{\beta}_j > 0$. If we augment our data with an *identical* copy $X_{j'} = X_j$, then they can share this coefficient in infinitely many ways— any $\tilde{\beta}_j + \tilde{\beta}_{j'} = \hat{\beta}_j$ with both pieces positive—and the loss and ℓ_1 penalty are indifferent. So the coefficients for this pair are not defined. A quadratic penalty, on the other hand, will divide $\hat{\beta}_j$ exactly equally between these two twins (see Exercise 4.1). In practice, we are unlikely to have an identical

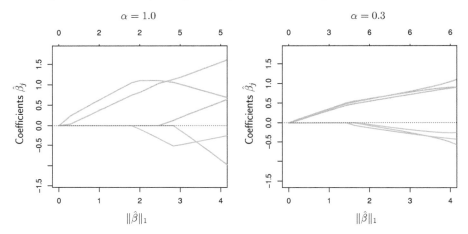

Figure 4.1 *Six variables, highly correlated in groups of three. The lasso estimates ($\alpha = 1$), as shown in the left panel, exhibit somewhat erratic behavior as the regularization parameter λ is varied. In the right panel, the elastic net with ($\alpha = 0.3$) includes all the variables, and the correlated groups are pulled together.*

pair of variables, but often we do have groups of very correlated variables. In microarray studies, groups of genes in the same biological pathway tend to be expressed (or not) together, and hence measures of their expression tend to be strongly correlated. The left panel of Figure 4.1 shows the lasso coefficient path for such a situation. There are two sets of three variables, with pairwise correlations around 0.97 in each group. With a sample size of $N = 100$, the data were simulated as follows:

$$Z_1,\ Z_2 \sim N(0,1)\ \text{independent,}$$
$$Y = 3 \cdot Z_1 - 1.5Z_2 + 2\varepsilon,\ \text{with}\ \varepsilon \sim N(0,1),$$
$$X_j = Z_1 + \xi_j/5,\ \text{with}\ \xi_j \sim N(0,1)\ \text{for}\ j = 1,2,3,\ \text{and} \qquad (4.1)$$
$$X_j = Z_2 + \xi_j/5,\ \text{with}\ \xi_j \sim N(0,1)\ \text{for}\ j = 4,5,6.$$

As shown in the left panel of Figure 4.1, the lasso coefficients do not reflect the relative importance of the individual variables.

The *elastic net* makes a compromise between the ridge and the lasso penalties (Zou and Hastie 2005); it solves the convex program

$$\underset{(\beta_0, \beta) \in \mathbb{R} \times \mathbb{R}^p}{\text{minimize}} \quad \left\{ \frac{1}{2} \sum_{i=1}^N (y_i - \beta_0 - x_i^T \beta)^2 + \lambda \left[\frac{1}{2}(1-\alpha)\|\beta\|_2^2 + \alpha \|\beta\|_1 \right] \right\}, \quad (4.2)$$

where $\alpha \in [0, 1]$ is a parameter that can be varied. By construction, the penalty applied to an individual coefficient (disregarding the regularization weight $\lambda > 0$) is given by

$$\frac{1}{2}(1-\alpha)\beta_j^2 + \alpha|\beta_j|. \quad (4.3)$$

When $\alpha = 1$, it reduces to the ℓ_1-norm or lasso penalty, and with $\alpha = 0$, it reduces to the squared ℓ_2-norm, corresponding to the ridge penalty.[1]

Returning to Figure 4.1, the right-hand panel shows the elastic-net coefficient path with $\alpha = 0.3$. We see that in contrast to the lasso paths in the left panel, the coefficients are selected approximately together in their groups, and also approximately share their values equally. Of course, this example is idealized, and in practice the group structure will not be so cleanly evident. But by adding some component of the ridge penalty to the ℓ_1-penalty, the elastic net automatically controls for strong within-group correlations. Moreover, for any $\alpha < 1$ and $\lambda > 0$, the elastic-net problem (4.2) is *strictly convex*: a unique solution exists irrespective of the correlations or duplications in the X_j.

Figure 4.2 compares the constraint region for the elastic net (left image) to that of the lasso (right image) when there are three variables. We see that the elastic-net ball shares attributes of the ℓ_2 ball and the ℓ_1 ball: the sharp corners and edges encourage selection, and the curved contours encourage sharing of coefficients. See Exercise 4.2 for further exploration of these properties.

The elastic net has an additional tuning parameter α that has to be determined. In practice, it can be viewed as a higher-level parameter, and can be set on subjective grounds. Alternatively, one can include a (coarse) grid of values of α in a cross-validation scheme.

The elastic-net problem (4.2) is convex in the pair $(\beta_0, \beta) \in \mathbb{R} \times \mathbb{R}^p$, and a variety of different algorithms can be used to solve it. Coordinate descent is particularly effective, and the updates are a simple extension of those for the lasso in Chapter 2. We have included an unpenalized intercept in the model, which can be dispensed with at the onset; we simply center the covariates x_{ij}, and then the optimal intercept is $\widehat{\beta}_0 = \bar{y} = \frac{1}{N}\sum_{j=1}^N y_j$. Having solved for the optimal $\widehat{\beta}_0$, it remains to compute the optimal vector $\widehat{\beta} = (\widehat{\beta}_1, \dots, \widehat{\beta}_p)$. It can be verified (Exercise 4.3) that the coordinate descent update for the jth

---

[1] The $\frac{1}{2}$ in the quadratic part of the elastic-net penalty (4.3) leads to a more intuitive soft-thresholding operator in the optimization.

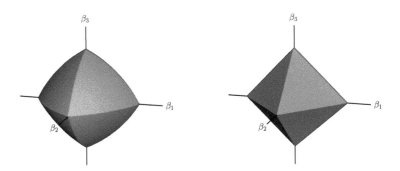

Figure 4.2 *The elastic-net ball with $\alpha = 0.7$ (left panel) in $\mathbb{R}^3$, compared to the ℓ_1 ball (right panel). The curved contours encourage strongly correlated variables to share coefficients (see Exercise 4.2 for details).*

coefficient takes the form

$$\widehat{\beta}_j = \frac{\mathcal{S}_{\lambda\alpha}\left(\sum_{i=1}^N r_{ij} x_{ij}\right)}{\sum_{i=1}^N x_{ij}^2 + \lambda(1 - \alpha)}, \tag{4.4}$$

where $\mathcal{S}_\mu(z) := \text{sign}(z)\,(z - \mu)_+$ is the soft-thresholding operator, and $r_{ij} := y_i - \widehat{\beta}_0 - \sum_{k \neq j} x_{ik}\widehat{\beta}_k$ is the partial residual. We cycle over the updates (4.4) until convergence. Friedman et al. (2015) give more details, and provide an efficient implementation of the elastic net penalty for a variety of loss functions.

4.3 The Group Lasso

There are many regression problems in which the covariates have a natural group structure, and it is desirable to have all coefficients within a group become nonzero (or zero) simultaneously. The various forms of group lasso penalty are designed for such situations. A leading example is when we have qualitative factors among our predictors. We typically code their levels using a set of dummy variables or contrasts, and would want to include or exclude this group of variables together. We first define the group lasso and then develop this and other motivating examples.

Consider a linear regression model involving J groups of covariates, where for $j = 1, \ldots, J$, the vector $Z_j \in \mathbb{R}^{p_j}$ represents the covariates in group j. Our goal is to predict a real-valued response $Y \in \mathbb{R}$ based on the collection of covariates $(Z_1, \ldots, Z_J)$. A linear model for the regression function $\mathbb{E}(Y \mid Z)$

takes the form $\theta_0 + \sum_{j=1}^{J} Z_j^T \theta_j$, where $\theta_j \in \mathbb{R}^{p_j}$ represents a group of p_j regression coefficients.[2]

Given a collection of N samples $\{(y_i, z_{i1}, z_{i,2}, \ldots, z_{i,J})\}_{i=1}^{N}$, the group lasso solves the convex problem

$$\underset{\theta_0 \in \mathbb{R}, \theta_j \in \mathbb{R}^{p_j}}{\text{minimize}} \left\{ \frac{1}{2} \sum_{i=1}^{N} \left(y_i - \theta_0 - \sum_{j=1}^{J} z_{ij}^T \theta_j \right)^2 + \lambda \sum_{j=1}^{J} \|\theta_j\|_2 \right\}, \qquad (4.5)$$

where $\|\theta_j\|_2$ is the Euclidean norm of the vector θ_j.

This is a group generalization of the lasso, with the properties:

- depending on $\lambda \geq 0$, either the entire vector $\hat{\theta}_j$ will be zero, or all its elements will be nonzero;[3]
- when $p_j = 1$, then we have $\|\theta_j\|_2 = |\theta_j|$, so if all the groups are singletons, the optimization problem (4.5) reduces to the ordinary lasso.

Figure 4.3 compares the constraint region for the group lasso (left image) to that of the lasso (right image) when there are three variables. We see that the group lasso ball shares attributes of both the ℓ_2 and ℓ_1 balls.

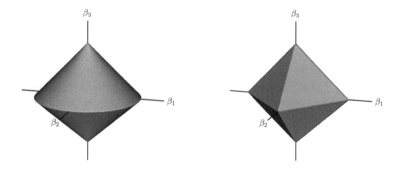

Figure 4.3 *The group lasso ball (left panel) in* $\mathbb{R}^3$, *compared to the* ℓ_1 *ball (right panel). In this case, there are two groups with coefficients* $\theta_1 = (\beta_1, \beta_2) \in \mathbb{R}^2$ *and* $\theta_2 = \beta_3 \in \mathbb{R}^1$.

In the formulation (4.5), all groups are equally penalized, a choice which leads larger groups to be more likely to be selected. In their original proposal, Yuan and Lin (2006) recommended weighting the penalties for each group according to their size, by a factor $\sqrt{p_j}$. In their case, the group matrices $\mathbf{Z}_j$ were orthonormal; for general matrices one can argue for a factor

---

[2]To avoid confusion, we use Z_j and θ_j to represent groups of variables and their coefficients, rather than the X_j and β_j we have used for scalars.

[3]Nonzero for generic problems, although special structure could result in some coefficients in a group being zero, just as they can for linear or ridge regression.

$\|\mathbf{Z}_j\|_F$ (Exercise 4.5). These choices are somewhat subjective, and are easily accommodated; for simplicity, we omit this modification in our presentation.

We now turn to some examples to illustrate applications of the group lasso (4.5).

Example 4.1. Regression with multilevel factors. When a predictor variable in a linear regression is a multilevel factor, we typically include a separate coefficient for each level of the factor. Take the simple case of one continuous predictor X and a three-level factor G with levels g_1, g_2, and g_3. Our linear model for the mean is

$$\mathbb{E}(Y \mid X, G) = X\beta + \sum_{k=1}^{3} \theta_k \, \mathbb{I}_k[G], \qquad (4.6)$$

where $\mathbb{I}_k[G]$ is a 0-1 valued indicator function for the event $\{G = g_k\}$. The model (4.6) corresponds to a linear regression in X with different intercepts θ_k depending on the level of G.

By introducing a vector $Z = (Z_1, Z_2, Z_3)$ of three *dummy variables* with $Z_k = \mathbb{I}_k[G]$, we can write this model as a standard linear regression

$$\mathbb{E}(Y \mid X, G) = \mathbb{E}(Y \mid X, Z) = X\beta + Z^T\theta, \qquad (4.7)$$

where $\theta = (\theta_1, \theta_2, \theta_3)$. In this case Z is a group variable that represents the single factor G. If the variable G—as coded by the vector Z—has no predictive power, then the full vector $\theta = (\theta_1, \theta_2, \theta_3)$ should be zero. On the other hand, when G is useful for prediction, then at least generically, we expect that all coefficients of θ are likely to be nonzero. More generally we can have a number of such single and group variables, and so have models of the form

$$\mathbb{E}(Y \mid X, G_1, \ldots, G_J) = \beta_0 + X^T\beta + \sum_{j=1}^{J} Z_j^T\theta_j. \qquad (4.8)$$

When selecting variables for such a model we would typically want to include or exclude groups at a time, rather than individual coefficients, and the group lasso is designed to enforce such behavior.

With unpenalized linear regression with factors, one has to worry about aliasing; in the example here, the dummy variables in a set add to one, which is aliased with the intercept term. One would then use contrasts to code factors that enforce, for example, that the coefficients in a group sum to zero. With the group lasso this is not a concern, because of the ℓ_2 penalties. We use the symmetric full representation as above, because the penalty term ensures that the coefficients in a group sum to zero (see Exercise 4.4).   ◇

Variables can be grouped for other reasons. For example, in gene-expression arrays, we might have a set of highly correlated genes from the same biological pathway. Selecting the group amounts to selecting a pathway. Figure 4.4 shows the coefficient path for a group-lasso fit to some genomic data for splice-site detection (Meier, van de Geer and Bühlmann 2008, Section 5). The data arise from human DNA, and each observation consists of seven bases with values $\{A, G, C, T\}^7$. Some of the observations are at exon-intron boundaries (splice sites), and others not, coded in a binary response; see Burge and Karlin (1977) for further details about these data. The regression problem is to predict the binary response Y using the seven four-level factors G_j as predictors, and we use a training sample of 5610 observations in each class.

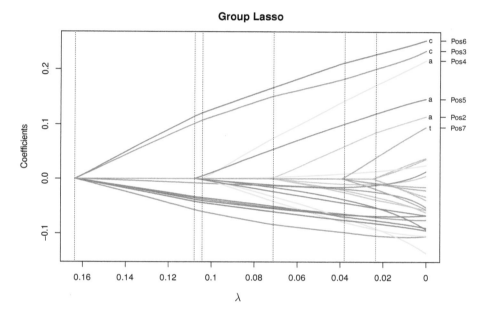

Figure 4.4 *Coefficient profiles from the group lasso, fit to splice-site detection data. The coefficients come in groups of four, corresponding to the nucleotides A, G, C, T. The vertical lines indicate when a group enters. On the right-hand side we label some of the variables; for example, "Pos6" and the level "c". The coefficients in a group have the same color, and they always average zero.*

Example 4.2. Multivariate regression. Sometimes we are interested in predicting a multivariate response $Y \in \mathbb{R}^K$ on the basis of a vector $X \in \mathbb{R}^p$ of predictors (also known as *multitask learning*). Given N observations $\{(y_i, x_i)\}_{i=1}^N$, we let $\mathbf{Y} \in \mathbb{R}^{N \times K}$ and $\mathbf{X} \in \mathbb{R}^{N \times p}$ be matrices with y_i and x_i, respectively, as their i^{th} row. If we assume a linear model for the full collection of data, then it can be written in the form

$$\mathbf{Y} = \mathbf{X}\mathbf{\Theta} + \mathbf{E} \tag{4.9}$$

where $\mathbf{\Theta} \in \mathbb{R}^{p \times K}$ is a matrix of coefficients, and $\mathbf{E} \in \mathbb{R}^{N \times K}$ a matrix of errors.

One way to understand the model (4.9) is as a coupled collection of K standard regression problems in $\mathbb{R}^p$, each sharing the same covariates, in which the kth column $\theta_k \in \mathbb{R}^p$ of $\boldsymbol{\Theta}$ is the coefficient vector for the kth problem. Thus, in principle, we could fit a separate regression coefficient vector θ_k for each of the K different problems, using the lasso in the case of a sparse linear model. In many applications, the different components of the response vector $Y \in \mathbb{R}^K$ are strongly related, so that one would expect that the underlying regression vectors would also be related. For instance, in collaborative filtering applications, the different components of Y might represent a given user's preference scores for different categories of objects, such as books, movies, music, and so on, all of which are closely related. For this reason, it is natural—and often leads to better prediction performance—to solve the K regression problems jointly, imposing some type of group structure on the coefficients. In another example, each response might be the daily return of an equity in a particular market sector; hence we have multiple equities, and all being predicted by the same market signals.

As one example, in the setting of sparsity, we might posit that there is an unknown subset $S \subset \{1, 2, \ldots, p\}$ of the covariates that are relevant for prediction, and this same subset is *preserved* across all K components of the response variable. In this case, it would be natural to consider a group lasso penalty, in which the p groups are defined by the rows $\{\theta'_j \in \mathbb{R}^K, \ j = 1, \ldots, p\}$ of the full coefficient matrix $\boldsymbol{\Theta} \in \mathbb{R}^{p \times K}$. Using this penalty, we then solve the regularized least-squares problem

$$\underset{\boldsymbol{\Theta} \in \mathbb{R}^{p \times K}}{\text{minimize}} \left\{ \frac{1}{2} \|\mathbf{Y} - \mathbf{X}\boldsymbol{\Theta}\|_{\text{F}}^2 + \lambda \left(\sum_{j=1}^{p} \|\theta'_j\|_2 \right) \right\}, \qquad (4.10)$$

where $\| \cdot \|_{\text{F}}$ denotes the Frobenius norm.[4] This problem is a special case of the general group lasso (4.5), in which $J = p$, and $p_j = K$ for all groups j. $\diamond$

4.3.1 Computation for the Group Lasso

Turning to computational issues associated with the group lasso, let us rewrite the relevant optimization problem (4.5) in a more compact matrix-vector notation:

$$\underset{(\theta_1, \ldots, \theta_J)}{\text{minimize}} \left\{ \frac{1}{2} \|\mathbf{y} - \sum_{j=1}^{J} \mathbf{Z}_j \theta_j\|_2^2 + \lambda \sum_{j=1}^{J} \|\theta_j\|_2 \right\}. \qquad (4.11)$$

For simplicity we ignore the intercept θ_0, since in practice we can center all the variables and the response, and it goes away. For this problem, the zero subgradient equations (see Section 5.2.2 take the form

$$-\mathbf{Z}_j^T (\mathbf{y} - \sum_{\ell=1}^{J} \mathbf{Z}_\ell \widehat{\theta}_\ell) + \lambda \widehat{s}_j = 0, \quad \text{for } j = 1, \cdots J, \qquad (4.12)$$

---

[4]The Frobenius norm of a matrix is simply the ℓ_2-norm applied to its entries.

where $\widehat{s}_j \in \mathbb{R}^{p_j}$ is an element of the subdifferential of the norm $\|\cdot\|_2$ evaluated at $\widehat{\theta}_j$. As verified in Exercise 5.5 on page 135, whenever $\widehat{\theta}_j \neq 0$, then we necessarily have $\widehat{s}_j = \widehat{\theta}_j/\|\widehat{\theta}_j\|_2$, whereas when $\widehat{\theta}_j = 0$, then $\widehat{s}_j$ is any vector with $\|\widehat{s}_j\|_2 \leq 1$. One method for solving the zero subgradient equations is by holding fixed all block vectors $\{\widehat{\theta}_k, \; k \neq j\}$, and then solving for $\widehat{\theta}_j$. Doing so amounts to performing block coordinate descent on the group lasso objective function. Since the problem is convex, and the penalty is block separable, it is guaranteed to converge to an optimal solution (Tseng 1993). With all $\{\widehat{\theta}_k, \; k \neq j\}$ fixed, we write

$$-\mathbf{Z}_j^T(\mathbf{r}_j - \mathbf{Z}_j\widehat{\theta}_j) + \lambda\widehat{s}_j = 0, \tag{4.13}$$

where $\mathbf{r}_j = \mathbf{y} - \sum_{k \neq j} \mathbf{Z}_k\widehat{\theta}_k$ is the jth *partial residual*. From the conditions satisfied by the subgradient $\widehat{s}_j$, we must have $\widehat{\theta}_j = 0$ if $\|\mathbf{Z}_j^T\mathbf{r}_j\|_2 < \lambda$, and otherwise the minimizer $\widehat{\theta}_j$ must satisfy

$$\widehat{\theta}_j = \left(\mathbf{Z}_j^T\mathbf{Z}_j + \frac{\lambda}{\|\widehat{\theta}_j\|_2}\mathbf{I}\right)^{-1}\mathbf{Z}_j^T\mathbf{r}_j. \tag{4.14}$$

This update is similar to the solution of a ridge regression problem, except that the underlying penalty parameter depends on $\|\widehat{\theta}_j\|_2$. Unfortunately, Equation (4.14) does not have a closed-form solution for $\widehat{\theta}_j$ unless $\mathbf{Z}_j$ is orthonormal. In this special case, we have the simple update

$$\widehat{\theta}_j = \left(1 - \frac{\lambda}{\|\mathbf{Z}_j^T\mathbf{r}_j\|_2}\right)_+ \mathbf{Z}_j^T\mathbf{r}_j, \tag{4.15}$$

where $(t)_+ := \max\{0, t\}$ is the positive part function. See Exercise 4.6 for further details.

Although the original authors (Yuan and Lin 2006) and many others since have made the orthonormality assumption, it has implications that are not always reasonable (Simon and Tibshirani 2012). Exercise 4.8 explores the impact of this assumption on the dummy coding used here for factors. In the general (nonorthonormal case) one has to solve (4.14) using iterative methods, and it reduces to a very simple one-dimensional search (Exercise 4.7).

An alternative approach is to apply the composite gradient methods of Section 5.3.3 to this problem. Doing so leads to an algorithm that is also iterative within each block; at each iteration the block-optimization problem is approximated by an easier problem, for which an update such as (4.15) is possible. In detail, the algorithm would iterate until convergence the updates

$$\omega \leftarrow \widehat{\theta}_j + \nu \cdot \mathbf{Z}_j^T(\mathbf{r}_j - \mathbf{Z}_j\widehat{\theta}_j), \text{ and} \tag{4.16a}$$

$$\widehat{\theta}_j \leftarrow \left(1 - \frac{\nu\lambda}{\|\omega\|_2}\right)_+ \omega, \tag{4.16b}$$

where ν is a step-size parameter. See Exercise 4.9 for details of this derivation.

4.3.2 Sparse Group Lasso

When a group is included in a group-lasso fit, all the coefficients in that group are nonzero. This is a consequence of the ℓ_2 norm. Sometimes we would like sparsity both with respect to which groups are selected, and which coefficients are nonzero within a group. For example, although a biological pathway may be implicated in the progression of a particular type of cancer, not all genes in the pathway need be active. The *sparse group lasso* is designed to achieve such within-group sparsity.

In order to achieve within-group sparsity, we augment the basic group lasso (4.11) with an additional ℓ_1-penalty, leading to the convex program

$$\underset{\{\theta_j \in \mathbb{R}^{p_j}\}_{j=1}^{J}}{\text{minimize}} \left\{ \frac{1}{2}\|\mathbf{y} - \sum_{j=1}^{J} \mathbf{Z}_j \theta_j\|_2^2 + \lambda \sum_{j=1}^{J} \left[(1-\alpha)\|\theta_j\|_2 + \alpha\|\theta_j\|_1\right]\right\}, \quad (4.17)$$

with $\alpha \in [0,1]$. Much like the elastic net of Section 4.2, the parameter α creates a bridge between the group lasso ($\alpha = 0$) and the lasso ($\alpha = 1$). Figure 4.5 contrasts the group lasso constraint region with that of the sparse group lasso for the case of three variables. Note that in the two horizontal axes, the constraint region resembles that of the elastic net.

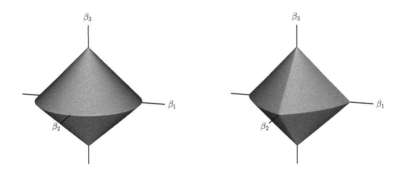

Figure 4.5 *The group lasso ball (left panel) in $\mathbb{R}^3$, compared to the sparse group-lasso ball with $\alpha = 0.5$ (right panel). Depicted are two groups with coefficients $\theta_1 = (\beta_1, \beta_2) \in \mathbb{R}^2$ and $\theta_2 = \beta_3 \in \mathbb{R}^1$.*

Since the optimization problem (4.17) is convex, its optima are specified by zero subgradient equations, similar to (4.13) for the group lasso. More precisely, any optimal solution must satisfy the condition

$$-\mathbf{Z}_j^T (\mathbf{y} - \sum_{\ell=1}^{J} \mathbf{Z}_\ell \widehat{\theta}_\ell) + \lambda(1-\alpha) \cdot \widehat{s}_j + \lambda\alpha\widehat{t}_j = 0, \text{ for } j = 1, \cdots, J, \quad (4.18)$$

where $\widehat{s}_j \in \mathbb{R}^{p_j}$ belongs to the subdifferential of the Euclidean norm at $\widehat{\theta}_j$, and $\widehat{t}_j \in \mathbb{R}^{p_j}$ belongs to the subdifferential of the ℓ_1-norm at $\widehat{\theta}_j$; in particular, we have each $\widehat{t}_{jk} \in \text{sign}(\theta_{jk})$ as with the usual lasso.

We once again solve these equations via block-wise coordinate descent, although the solution is a bit more complex than before. As in Equation (4.13), with r_j the partial residual in the jth coordinate, it can be seen that $\widehat{\theta}_j = 0$ if and only if the equation

$$\mathbf{Z}_j^T r_j = \lambda(1 - \alpha)\widehat{s}_j + \lambda\alpha\,\widehat{t}_j \tag{4.19}$$

has a solution with $\|\widehat{s}_j\|_2 \leq 1$ and $\widehat{t}_{jk} \in [-1, 1]$ for $k = 1, \ldots, p_j$. Fortunately, this condition is easily checked, and we find that (Exercise 4.12)

$$\widehat{\theta}_j = 0 \quad \text{if and only if} \quad \|\mathcal{S}_{\lambda\alpha}(\mathbf{Z}_j^T r_j)\|_2 \leq \lambda(1 - \alpha), \tag{4.20}$$

where $\mathcal{S}_{\lambda\alpha}(\,\cdot\,)$ is the soft-thresholding operator applied here component-wise to its vector argument $\mathbf{Z}_j^T r_j$. Notice the similarity with the conditions for the group lasso (4.13), except here we use the soft-thresholded gradient $\mathcal{S}_{\lambda\alpha}(\mathbf{Z}_j^T r_j)$. Likewise, if $\mathbf{Z}_j^T \mathbf{Z}_j = \mathbf{I}$, then as shown in Exercise 4.13, we have

$$\widehat{\theta}_j = \left(1 - \frac{\lambda(1 - \alpha)}{\|\mathcal{S}_{\lambda\alpha}(\mathbf{Z}_j^T r_j)\|_2}\right)_+ \mathcal{S}_{\lambda\alpha}(\mathbf{Z}_j^T r_j). \tag{4.21}$$

In the general case when the $\mathbf{Z}_j$ are not orthonormal and we have checked that $\widehat{\theta}_j \neq 0$, finding $\widehat{\theta}_j$ amounts to solving the subproblem

$$\underset{\theta_j \in \mathbb{R}^{p_j}}{\text{minimize}} \left\{ \frac{1}{2}\|r_j - \mathbf{Z}_j\theta_j\|_2^2 + \lambda(1 - \alpha)\|\theta_j\|_2 + \lambda\alpha\|\theta_j\|_1 \right\}. \tag{4.22}$$

Here we can again use generalized gradient descent (Section (5.3.3)) to produce a simple iterative algorithm to solve each block, as in Equation (4.16a). The algorithm would iterate until convergence the sequence

$$\omega \leftarrow \widehat{\theta}_j + \nu \cdot \mathbf{Z}_j^T (r_j - \mathbf{Z}_j\widehat{\theta}_j), \text{ and} \tag{4.23a}$$

$$\theta_j \leftarrow \left(1 - \frac{\nu\lambda(1 - \alpha)}{\|\mathcal{S}_{\lambda\alpha}(\omega)\|_2}\right)_+ \mathcal{S}_{\lambda\alpha}(\omega), \tag{4.23b}$$

where ν is the step size. See Exercise 4.10 for the details.

4.3.3   The Overlap Group Lasso

Sometimes variables can belong to more than one group: for example, genes can belong to more than one biological pathway. The *overlap group lasso* is a modification that allows variables to contribute to more than one group.

To gain some intuition, consider the case of $p = 5$ variables partitioned into two groups, say of the form

$$Z_1 = (X_1, X_2, X_3), \text{ and } Z_2 = (X_3, X_4, X_5). \tag{4.24}$$

Here X_3 belongs to both groups. The overlap group lasso simply replicates a variable in whatever group it appears, and then fits the ordinary group lasso as before. In this particular example, the variable X_3 is replicated, and we fit coefficient vectors $\theta_1 = (\theta_{11}, \theta_{12}, \theta_{13})$ and $\theta_2 = (\theta_{21}, \theta_{22}, \theta_{23})$ using the group lasso (4.5), using a group penalty $\|\theta_1\|_2 + \|\theta_2\|_2$. In terms of the original variables, the coefficient $\widehat{\beta}_3$ of X_3 is given by the sum $\widehat{\beta}_3 = \widehat{\theta}_{13} + \widehat{\theta}_{21}$. As a consequence, the coefficient $\widehat{\beta}_3$ can be nonzero if either (or both) of the coefficients $\widehat{\theta}_{13}$ or $\widehat{\theta}_{21}$ are nonzero. Hence, all else being equal, the variable X_3 has a better chance of being included in the model than the other variables, by virtue of belonging to two groups.

Rather than replicate variables, it is tempting to simply replicate the coefficients in the group-lasso penalty. For instance, for the given grouping above, with $X = (X_1, \ldots, X_5)$, and $\beta = (\beta_1, \ldots, \beta_5)$, suppose that we define

$$\theta_1 = (\beta_1, \beta_2, \beta_3), \quad \text{and} \quad \theta_2 = (\beta_3, \beta_4, \beta_5), \tag{4.25}$$

and then apply the group-lasso penalty $\|\theta_1\|_2 + \|\theta_2\|_2$ as before. However, this approach has a major drawback. Whenever $\widehat{\theta}_1 = 0$ in any optimal solution, then we must necessarily have $\widehat{\beta}_3 = 0$ in *both groups*. Consequently, in this particular example, the only possible sets of nonzero coefficients are $\{1, 2\}$, $\{4, 5\}$, and $\{1, 2, 3, 4, 5\}$; the original groups $\{1, 2, 3\}$ and $\{3, 4, 5\}$ are not considered as possibilities, since if either group appears, then both groups must appear.[5] As a second practical point, the penalty in this approach is not separable, and hence coordinate-descent algorithms may fail to converge to an optimal solution (see Section 5.4 for more details).

Jacob, Obozinski and Vert (2009) recognized this problem, and hence proposed the replicated variable approach (4.24) or *overlap group lasso*. For our motivating example, the possible sets of nonzero coefficients for the overlap group lasso are $\{1, 2, 3\}$, $\{3, 4, 5\}$, and $\{1, 2, 3, 4, 5\}$. In general, the sets of possible nonzero coefficients always correspond to groups, or the unions of groups. They also defined an implicit penalty on the original variables that yields the replicated variable approach as its solution, which we now describe.

Denote by $\nu_j \in \mathbb{R}^p$ a vector which is zero everywhere except in those positions corresponding to the members of group j, and let $\mathcal{V}_j \subseteq \mathbb{R}^p$ be the subspace of such possible vectors. In terms of the original variables, $X = (X_1, \cdots, X_p)$, the coefficient vector is given by the sum $\beta = \sum_{j=1}^{J} \nu_j$, and hence the overlap group lasso solves the problem

$$\underset{\nu_j \in \mathcal{V}_j,\, j=1,\ldots,J}{\text{minimize}} \left\{ \frac{1}{2} \|\mathbf{y} - \mathbf{X}(\sum_{j=1}^{J} \nu_j)\|_2^2 + \lambda \sum_{j=1}^{J} \|\nu_j\|_2 \right\}. \tag{4.26}$$

This optimization problem can be re-cast in the terms of the original β vari-

---

[5]More generally, the replicated-variable approach always yields solutions in which the sets of zero coefficients are unions of groups, so that the sets of nonzeros must be the intersections of complements of groups.

ables by defining a suitable penalty function. With

$$\Omega_{\mathcal{V}}(\beta) := \inf_{\substack{\nu_j \in \mathcal{V}_j \\ \beta = \sum_{j=1}^{J} \nu_j}} \sum_{j=1}^{J} \|\nu_j\|_2, \tag{4.27}$$

it can then be shown (Jacob et al. 2009) that solving problem (4.26) is equivalent to solving

$$\underset{\beta \in \mathbb{R}^p}{\text{minimize}} \left\{ \frac{1}{2} \|\mathbf{y} - \mathbf{X}\beta\|_2^2 + \lambda \, \Omega_{\mathcal{V}}(\beta) \right\}. \tag{4.28}$$

This equivalence is intuitively obvious, and underscores the mechanism underlying this penalty; the contributions to the coefficient for a variable are distributed among the groups to which it belongs in a norm-efficient manner.

Figure 4.6 contrasts the group lasso constraint region with that of the overlap group lasso when there are three variables. There are two rings corresponding to the two groups, with X_2 in both groups.

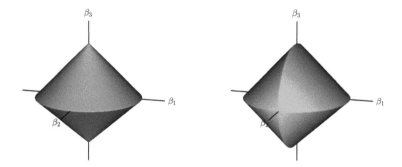

Figure 4.6 *The group-lasso ball (left panel) in $\mathbb{R}^3$, compared to the overlap-group-lasso ball (right panel). Depicted are two groups in both. In the left panel the groups are $\{X_1, X_2\}$ and X_3; in the right panel the groups are $\{X_1, X_2\}$ and $\{X_2, X_3\}$. There are two rings corresponding to the two groups in the right panel. When β_2 is close to zero, the penalty on the other two variables is much like the lasso. When β_2 is far from zero, the penalty on the other two variables "softens" and resembles the ℓ_2 penalty.*

Example 4.3. Interactions and hierarchy. The overlap-group lasso can also be used to enforce *hierarchy* when selecting interactions in linear models. What this means is that interactions are allowed in the model only in the presence of both of their main effects. Suppose Z_1 represents the p_1 dummy variables for the p_1 levels of factor G_1; likewise Z_2 the p_2 dummy variables for G_2. A linear model with Z_1 and Z_2 is a *main-effects* model. Now let $Z_{1:2} = Z_1 \star Z_2$, a $p_1 \times p_2$ vector of dummy variables (the vector of all pairwise products). Lim and Hastie (2014) consider the following formulation for a pair of such categorical variables[6]

$$\underset{\mu,\alpha,\tilde{\alpha}}{\text{minimize}} \left\{ \frac{1}{2} \left\| \mathbf{y} - \mu\mathbf{1} - \mathbf{Z}_1\alpha_1 - \mathbf{Z}_2\alpha_2 - [\mathbf{Z}_1 \ \mathbf{Z}_2 \ \mathbf{Z}_{1:2}] \begin{bmatrix} \tilde{\alpha}_1 \\ \tilde{\alpha}_2 \\ \alpha_{1:2} \end{bmatrix} \right\|_2^2 \right.$$
$$\left. + \lambda \left(\|\alpha_1\|_2 + \|\alpha_2\|_2 + \sqrt{p_2\|\tilde{\alpha}_1\|_2^2 + p_1\|\tilde{\alpha}_2\|_2^2 + \|\alpha_{1:2}\|_2^2} \right) \right\} \quad (4.29)$$

subject to the constraints

$$\sum_{i=1}^{p_1} \alpha_1^i = 0, \quad \sum_{j=1}^{p_2} \alpha_2^j = 0, \quad \sum_{i=1}^{p_1} \tilde{\alpha}_1^i = 0, \quad \sum_{j=1}^{p_2} \tilde{\alpha}_2^j = 0, \quad (4.30)$$

$$\sum_{i=1}^{p_1} \alpha_{1:2}^{ij} = 0 \text{ for fixed } j, \quad \sum_{j=1}^{p_2} \alpha_{1:2}^{ij} = 0 \text{ for fixed } i. \quad (4.31)$$

The summation constraints are standard in hierarchical ANOVA formulations. Notice that the main effect matrices $\mathbf{Z}_1$ and $\mathbf{Z}_2$ each have two different coefficient vectors α_j and $\tilde{\alpha}_j$, creating an overlap in the penalties, and their ultimate coefficient is the sum $\theta_j = \alpha_j + \tilde{\alpha}_j$. The $\sqrt{p_2\|\tilde{\alpha}_1\|_2^2 + p_1\|\tilde{\alpha}_2\|_2^2 + \|\alpha_{1:2}\|_2^2}$ term results in estimates that satisfy strong hierarchy, because either $\widehat{\tilde{\alpha}}_1 = \widehat{\tilde{\alpha}}_2 = \widehat{\alpha}_{1:2} = 0$ or *all* are nonzero, i.e., interactions are always present with both main effects. They show that the solution to the above constrained problem (4.29)–(4.31) is equivalent to the solution to the simpler unconstrained problem

$$\underset{\mu,\beta}{\text{minimize}} \left\{ \frac{1}{2} \|\mathbf{y} - \mu\mathbf{1} - \mathbf{Z}_1\beta_1 - \mathbf{Z}_2\beta_2 - \mathbf{Z}_{1:2}\beta_{1:2}\|_2^2 \right.$$
$$\left. + \lambda \left(\|\beta_1\|_2 + \|\beta_2\|_2 + \|\beta_{1:2}\|_2 \right) \right\} \quad (4.32)$$

(Exercise 4.14). In other words, a linear model in $Z_{1:2}$ is the full interaction model (i.e., interactions with main effects implicitly included). A group lasso in Z_1, Z_2, and $Z_{1:2}$ will hence result in a hierarchical model; whenever $Z_{1:2}$ is in the model, the pair of main effects is implicitly included. In this case the variables do not strictly overlap, but their subspaces do. A different approach to the estimation of hierarchical interactions is the *hierNet* proposal of Bien, Taylor and Tibshirani (2013).   ◇

---

[6]This extends naturally to more than two pairs, as well as other loss functions, e.g., logistic regression, as well as interactions between factors and quantitative variables.

4.4 Sparse Additive Models and the Group Lasso

Suppose we have a zero-mean response variable $Y \in \mathbb{R}$, and a vector of predictors $X \in \mathbb{R}^J$, and that we are interested in estimating the regression function $f(x) = \mathbb{E}(Y \mid X = x)$. It is well-known that nonparametric regression suffers from the curse of dimensionality, so that approximations are essential. Additive models are one such approximation, and effectively reduce the estimation problem to that of many one-dimensional problems. When J is very large, this may not be sufficient; the class of sparse additive models limits these approximations further, by encouraging many of the components to be zero. Methods for estimating sparse additive models are closely related to the group lasso.

4.4.1 Additive Models and Backfitting

We begin by introducing some background on the class of additive models, which are based on approximating the regression function by sums of the form

$$f(x) = f(x_1, \ldots, x_J) \approx \sum_{j=1}^{J} f_j(x_j), \tag{4.33}$$

$$f_j \in \mathcal{F}_j, \ j = 1, \ldots, J,$$

where the $\mathcal{F}_j$ are a fixed set of univariate function classes. Typically, each $\mathcal{F}_j$ is assumed to be a subset of $L^2(\mathbb{P}_j)$ where $\mathbb{P}_j$ is the distribution of covariate X_j, and equipped with the usual squared $L^2(\mathbb{P}_j)$ norm $\|f_j\|_2^2 := \mathbb{E}[f_j^2(X_j)]$.

In the population setting, the best additive approximation to the regression function $\mathbb{E}(Y|X = x)$, as measured in the $L^2(\mathbb{P})$ sense, solves the problem

$$\underset{f_j \in \mathcal{F}_j, \ j=1,\ldots,J}{\text{minimize}} \ \mathbb{E}\Big[\big(Y - \sum_{j=1}^{J} f_j(X_j)\big)^2\Big]. \tag{4.34}$$

The optimal solution $(\tilde{f}_1, \ldots, \tilde{f}_J)$ is characterized by the *backfitting equations*, namely

$$\tilde{f}_j(x_j) = \mathbb{E}\Big[Y - \sum_{k \neq j} \tilde{f}_k(X_k) \mid X_j = x_j\Big], \ \text{for } j = 1, \ldots, J. \tag{4.35}$$

More compactly, this update can be written in the form $\tilde{f}_j = \mathcal{P}_j(R_j)$, where $\mathcal{P}_j$ is the conditional-expectation operator in the jth coordinate, and the quantity $R_j := Y - \sum_{k \neq j} \tilde{f}_k(X_k)$ is the jth partial residual.

Given data $\{(x_i, y_i)\}_{i=1}^N$, a natural approach is to replace the population operator $\mathcal{P}_j$ with empirical versions, such as scatterplot smoothers $\mathcal{S}_j$, and then solve a data-based version version of the updates (4.35) by coordinate descent or *backfitting* (Hastie and Tibshirani 1990). Hence we repeatedly cycle over the coordinates $j = 1, \ldots, J$, and update each function estimate $\hat{f}_j$ using the smooth of the partial residuals

$$\widehat{f}_j \leftarrow \mathcal{S}_j(\mathbf{y} - \sum_{k \neq j} \widehat{\mathbf{f}}_k), \ j = 1, \dots, J, \tag{4.36}$$

until the fitted functions $\widehat{f}_j$ stabilize. In (4.36) $\widehat{\mathbf{f}}_k$ is the fitted function $\widehat{f}_k$ evaluated at the N sample values $(x_{1k}, \dots, x_{Nk})$. The operator $\mathcal{S}_j$ represents an algorithm that takes a response vector $\mathbf{r}$, *smooths* it against the vector $\mathbf{x}_j$, and returns the function $\widehat{f}_j$. Although $\mathcal{S}_j$ will have its own tuning parameters and *bells and whistles*, for the moment we regard it as a black-box that estimates a conditional expectation using data.

4.4.2 Sparse Additive Models and Backfitting

An extension of the basic additive model is the notion of a *sparse additive model*, in which we assume that there is a subset $S \subset \{1, 2, \dots, J\}$ such that the regression function $f(x) = \mathbb{E}(Y \mid X = x)$ satisfies an approximation of the form $f(x) \approx \sum_{j \in S} f_j(x_j)$. Ravikumar, Liu, Lafferty and Wasserman (2009) proposed a natural extension of the backfitting equations, motivated by a sparse analog of the population level problem (4.34). For a given sparsity level $k \in \{1, \dots, J\}$, the best k-sparse approximation to the regression function is given by

$$\underset{\substack{|S|=k \\ f_j \in \mathcal{F}_j, j=1,\dots,J}}{\text{minimize}} \ \mathbb{E}\Big(Y - \sum_{j \in S} f_j(X_j)\Big)^2. \tag{4.37}$$

Unfortunately, this criterion is nonconvex and computationally intractable, due to combinatorial number—namely $\binom{J}{k}$—of possible subsets of size k. Suppose that instead we measure the sparsity of an additive approximation $f = \sum_{j=1}^{J} f_j$ via the sum $\sum_{j=1}^{J} \|f_j\|_2$, where we recall that $\|f_j\|_2 = \sqrt{\mathbb{E}[f_j^2(X_j)]}$ is the $L^2(\mathbb{P}_j)$ norm applied to component j. For a given regularization parameter $\lambda \geq 0$, this relaxed notion defines an alternative type of best sparse approximation, namely one that minimizes the penalized criterion

$$\underset{f_j \in \mathcal{F}_j, \ j=1,\dots,J}{\text{minimize}} \ \left\{ \mathbb{E}\Big(Y - \sum_{j=1}^{J} f_j(X_j)\Big)^2 + \lambda \sum_{j=1}^{J} \|f_j\|_2 \right\}. \tag{4.38}$$

Since this objective is a convex functional of $(f_1, \dots, f_J)$, Lagrangian duality ensures that it has an equivalent representation involving an explicit constraint on the norm $\sum_{j=1}^{J} \|f_j\|_2$. See Exercise 4.15.

Ravikumar et al. (2009) show that any optimal solution $(\tilde{f}_1, \dots, \tilde{f}_J)$ to the penalized problem (4.38) is characterized by the *sparse backfitting equations*

$$\tilde{f}_j = \left(1 - \frac{\lambda}{\|\mathcal{P}_j(R_j)\|_2}\right)_+ \mathcal{P}_j(R_j), \tag{4.39}$$

where the residual R_j and the conditional expectation operator $\mathcal{P}_j$ were defined in the text after the ordinary backfitting equations (4.35).

In parallel with our earlier development, given data $\{(x_i, y_i)\}_1^N$, these population-level updates suggest the natural data-driven analog, in which we replace the population operator $\mathcal{P}_j$ with the scatterplot smoother $\mathcal{S}_j$, and then perform the updates

$$\widetilde{f}_j = \mathcal{S}_j\left(\mathbf{y} - \sum_{k \neq j}\widehat{\mathbf{f}}_k\right), \quad \text{and} \quad \widehat{f}_j = \left(1 - \frac{\lambda}{\|\widetilde{\mathbf{f}}\|_2}\right)_+ \widetilde{f}_j, \qquad (4.40)$$

for $j = 1, \ldots, J$, again iterating until convergence. Figure 4.7 illustrates the performance of the SPAM updates (4.40) on some air-pollution data. We use smoothing-splines, with a fixed degree of freedom $df = 5$ for each coordinate (Hastie and Tibshirani 1990).

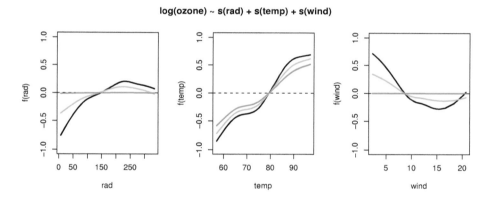

log(ozone) ~ s(rad) + s(temp) + s(wind)

Figure 4.7 *A sequence of three SPAM models fit to some air-pollution data. The response is the log of ozone concentration, and there are three predictors: radiation, temperature, and wind speed. Smoothing splines were used in the additive model fits, each with df = 5 The three curves in each plot correspond to $\lambda = 0$ (black curves), $\lambda = 2$ (orange curves), and $\lambda = 4$ (red curves). We see that while the shrinkage leaves the functions of* temp *relatively untouched, it has a more dramatic effect on* **rad** *and* **wind***.*

We can make a more direct connection with the grouped lasso if the smoothing method for variable X_j is a projection on to a set of basis functions. Consider

$$f_j(\cdot) = \sum_{\ell=1}^{p_j} \psi_{j\ell}(\cdot)\beta_{j\ell}, \qquad (4.41)$$

where the $\{\psi_{j\ell}\}_1^{p_j}$ are a family of basis functions in X_j, such as cubic splines with a collection of knots along the range of X_j. Let $\mathbf{\Psi}_j$ be the $N \times p_j$ matrix of evaluations of the $\psi_{j\ell}$, and assume that $\mathbf{\Psi}_j^T \mathbf{\Psi}_j = \mathbf{I}_{p_j}$. Then for any coefficient vector $\theta_j = (\beta_{j1}, \ldots, \beta_{jp_j})^T$ and corresponding fitted vector $\mathbf{f}_j = \boldsymbol{\Psi}_j \theta_j$, we have $\|\mathbf{f}_j\|_2 = \|\theta_j\|_2$. In this case it is easy to show that the updates (4.40) are equivalent to those for a group lasso with predictor matrix $\boldsymbol{\Psi} := \begin{bmatrix} \boldsymbol{\Psi}_1 & \boldsymbol{\Psi}_2 & \cdots & \boldsymbol{\Psi}_J \end{bmatrix}$ and a corresponding block vector of coefficients $\theta := \begin{bmatrix} \theta_1 & \theta_2 & \cdots \theta_J \end{bmatrix}$ (see Exercise 4.16 for more details).

4.4.3 Approaches Using Optimization and the Group Lasso

Although the population-level sparse backfitting equations (4.39) do solve an optimization problem, in general, the empirical versions (4.40) do not, but rather are motivated by analogy to the population version. We now discuss the *Component Selection and Smoothing Operator* or COSSO for short, which does solve a data-defined optimization problem. The COSSO method (Lin and Zhang 2003) is a predecessor to the SPAM method, and operates in the world of reproducing kernel Hilbert spaces, with a special case being the smoothing spline model.

We begin by recalling the traditional form of an additive smoothing-spline model, obtained from the optimization of a penalized objective function:

$$\underset{f_j \in \mathcal{H}_j,\, j=1,\ldots,J}{\text{minimize}} \left\{ \frac{1}{N} \sum_{i=1}^{N} (y_i - \sum_{j=1}^{J} f_j(x_{ij}))^2 + \lambda \sum_{j=1}^{J} \frac{1}{\gamma_j} \|f_j\|_{\mathcal{H}_j}^2 \right\}. \quad (4.42)$$

Here $\|f_j\|_{\mathcal{H}_j}$ is an appropriate Hilbert-space norm for the jth coordinate. Typically, the Hilbert space $\mathcal{H}_j$ is chosen to enforce some type of smoothness, in which context the parameter $\lambda \geq 0$ corresponds to overall smoothness, and the parameters $\gamma_j \geq 0$ are coordinate specific modifiers. For example, a roughness norm for a cubic smoothing spline on $[0, 1]$ is

$$\|g\|_{\mathcal{H}}^2 := \left(\int_0^1 g(t)dt \right)^2 + \left(\int_0^1 g'(t)dt \right)^2 + \int_0^1 g''(t)^2 dt. \quad (4.43)$$

When this particular Hilbert norm is used in the objective function (4.42), each component $\widehat{f}_j$ of the optimal solution is a cubic spline with knots at the unique sample values of X_j. The solution can be computed by the backfitting updates (4.36), where each $\mathcal{S}_j$ is a type of cubic spline smoother with penalty λ/γ_j.

Instead of the classical formulation (4.42), the COSSO method is based on the objective function

$$\underset{f_j \in \mathcal{H}_j,\, j=1,\ldots,J}{\text{minimize}} \left\{ \frac{1}{N} \sum_{i=1}^{N} (y_i - \sum_{j=1}^{J} f_j(x_{ij}))^2 + \tau \sum_{j=1}^{J} \|f_j\|_{\mathcal{H}_j} \right\}. \quad (4.44)$$

As before, the penalties are norms rather than squared norms, and as such result in coordinate selection for sufficiently large τ. Note that, unlike the usual penalty for a cubic smoothing spline, the norm in (4.43) includes a linear component; this ensures that the entire function is zero when the term is selected out of the model, rather than just its nonlinear component. Despite the similarity with the additive spline problem (4.38), the structure of the penalty $\|f_j\|_{\mathcal{H}_j}$ means that the solution is not quite as simple as the sparse backfitting equations (4.40).

Equipped with the norm (4.43), the space $\mathcal{H}_j$ of cubic splines is a particular instance of a reproducing-kernel Hilbert space (RKHS) on the unit interval $[0, 1]$. Any such space is characterized by a symmetric positive definite kernel function $\mathcal{R}_j : [0, 1] \times [0, 1] \to \mathbb{R}$ with the so-called reproducing property. In particular, we are guaranteed for each $x \in [0, 1]$, the function $\mathcal{R}_j(\cdot, x)$ is a member of $\mathcal{H}_j$, and moreover that $\langle \mathcal{R}(\cdot, x), f \rangle_{\mathcal{H}_j} = f(x)$ for all $f \in \mathcal{H}_j$. Here $\langle \cdot, \cdot \rangle_{\mathcal{H}_j}$ denotes the inner product on the Hilbert space $\mathcal{H}_j$.

Using the reproducing property, it can be shown (Exercise 4.17) that the jth coordinate function $\widehat{f}_j$ in any optimal COSSO solution can be written in the form $\widehat{f}_j(\cdot) = \sum_{i=1}^{N} \widehat{\theta}_{ij} \mathcal{R}_j(\cdot, x_{ij})$, for a suitably chosen weight vector $\widehat{\theta}_j \in \mathbb{R}^N$. Moreover, it can be shown that $\widehat{f}_j$ has Hilbert norm $\|\widehat{f}_j\|_{\mathcal{H}_j}^2 = \widehat{\theta}_j^T \mathbf{R}_j \widehat{\theta}_j$, where $\mathbf{R}_j \in \mathbb{R}^{N \times N}$ is a *Gram matrix* defined by the kernel—in particular, with entries $(\mathbf{R}_j)_{ii'} = \mathcal{R}_j(x_{ij}, x_{i'j})$. Consequently, the COSSO problem (4.44) can be rewritten as a more general version of the group lasso: in particular, it is equivalent to the optimization problem

$$\operatorname*{minimize}_{\theta_j \in \mathbb{R}^N,\, j=1,\ldots,J} \left\{ \frac{1}{N} \|\mathbf{y} - \sum_{j=1}^{J} \mathbf{R}_j \theta_j\|_2^2 + \tau \sum_{j=1}^{J} \sqrt{\theta_j^T \mathbf{R}_j \theta_j} \right\}, \qquad (4.45)$$

as verified in Exercise 4.17.

We are now back in a parametric setting, and the solution is a more general version of the group lasso (4.14). It can be shown that any optimal solution $(\widehat{\theta}_1, \ldots, \widehat{\theta}_J)$ is specified by the fixed point equations

$$\widehat{\theta}_j = \begin{cases} \mathbf{0} & \text{if } \sqrt{r_j^T \mathbf{R}_j r_j} < \tau \\ \left(\mathbf{R}_j + \dfrac{\tau}{\sqrt{\theta_j^T \mathbf{R}_j \widehat{\theta}_j}} \mathbf{I} \right)^{-1} r_j & \text{otherwise,} \end{cases} \qquad (4.46)$$

where $r_j := \mathbf{y} - \sum_{k \neq j} \mathbf{R}_k \widehat{\theta}_k$ corresponds to the jth partial residual. Although θ_j appears in both sides of the Equation (4.46), it can be solved with a one-time SVD of $\mathbf{R}_j$ and a simple one-dimensional search; see Exercise 4.7 for the details.

Lin and Zhang (2003) propose an alternative approach, based on introducing a vector $\gamma \in \mathbb{R}^J$ of auxiliary variables, and then considering the joint optimization problem

$$\underset{\substack{\gamma \geq 0 \\ f_j \in \mathcal{H}_j,\, j=1,\ldots J}}{\text{minimize}} \left\{ \frac{1}{N} \sum_{i=1}^{N}(y_i - \sum_{j=1}^{J} f_j(x_{ij}))^2 + \sum_{j=1}^{J} \frac{1}{\gamma_j} \|f_j\|_{\mathcal{H}_j}^2 + \lambda \sum_{j=1}^{J} \gamma_j \right\}.$$

(4.47)

As shown in Exercise 4.18, if we set $\lambda = \tau^4/4$ in the lifted problem (4.47), then the $\widehat{\theta}$ component of any optimal solution coincides with an optimal solution of the original COSSO (4.44).

The reformulation (4.47) is useful, because it naturally leads to a convenient algorithm that alternates between two steps:

- For γ_j fixed, the problem is a version of our earlier objective (4.42), and results in an additive-spline fit.

- With the fitted additive spline fixed, updating the vector of coefficients $\gamma = (\gamma_1, \ldots, \gamma_J)$ amounts to a nonnegative lasso problem. More precisely, for each $j = 1, \ldots, J$, define the vector $\mathbf{g}_j = \mathbf{R}_j \boldsymbol{\theta}_j / \gamma_j \in \mathbb{R}^N$, where $\mathbf{f}_j = \mathbf{R}_j \boldsymbol{\theta}_j$ is the fitted vector for the jth function using the current value of γ_j. Then we update the vector $\gamma = (\gamma_1, \ldots, \gamma_J)$ by solving

$$\min_{\gamma \geq 0} \left\{ \frac{1}{N} \|\mathbf{y} - \mathbf{G}\gamma\|_2^2 + \lambda\|\gamma\|_1 \right\},$$

(4.48)

where $\mathbf{G}$ is the $N \times J$ matrix with columns $\{\mathbf{g}_j, j = 1, \ldots, J\}$. These updates are a slightly different form than that given in Lin and Zhang (2003); full details are mapped out in Exercise 4.19.

When applied with the cubic smoothing-spline norm (4.43), the COSSO is aimed at setting component functions f_j to zero. There are many extensions to this basic idea. For instance, given a univariate function g, we might instead represent each univariate function in the form $g(t) = \alpha_0 + \alpha_1 t + h(t)$, and focus the penalty on departures from linearity using the norm

$$\|h\|_{\mathcal{H}}^2 := \int_0^1 h''(t)^2 dt,$$

(4.49)

In this setting, a variant of COSSO can select between nonlinear and linear forms for each component function.

We discuss penalties for additive models further in Section 4.4.4, in particular the benefits of using more than one penalty in this context.

4.4.4 Multiple Penalization for Sparse Additive Models

As we have seen thus far, there are multiple ways of enforcing sparsity for a nonparametric problem. Some methods, such as the SPAM back-fitting procedure, are based on a combination of the ℓ_1-norm with the empirical L^2-norm—namely, the quantity

$$\|f\|_{N,1} := \sum_{j=1}^{J} \|f_j\|_N, \tag{4.50}$$

where $\|f_j\|_N^2 := \frac{1}{N}\sum_{i=1}^{N} f_j^2(x_{ij})$ is the squared empirical L^2-norm for the univariate function f_j.[7] Other methods, such as the COSSO method, enforce sparsity using a combination of the ℓ_1-norm with the Hilbert norm

$$\|f\|_{\mathcal{H},1} := \sum_{j=1}^{J} \|f_j\|_{\mathcal{H}_j}. \tag{4.51}$$

Which of these two different regularizers is to be preferred for enforcing sparsity in the nonparametric setting?

Instead of focusing on only one regularizer, one might consider the more general family of estimators

$$\min_{\substack{f_j \in \mathcal{H}_j \\ j=1,\dots,J}} \left\{ \frac{1}{N}\sum_{i=1}^{N}(y_i - \sum_{j=1}^{J} f_j(x_{ij}))^2 + \lambda_{\mathcal{H}} \sum_{j=1}^{J} \|f_j\|_{\mathcal{H}_j} + \lambda_N \sum_{j=1}^{J} \|f_j\|_N \right\}, \tag{4.52}$$

parametrized by the pair of nonnegative regularization weights $(\lambda_{\mathcal{H}}, \lambda_N)$. If we set $\lambda_N = 0$, then the optimization problem (4.52) reduces to the COSSO estimator, whereas for $\lambda_{\mathcal{H}} = 0$, we obtain a method closely related to the SPAM estimator. For any nonnegative $(\lambda_{\mathcal{H}}, \lambda_N)$, the optimization problem (4.52) is convex. When the underlying univariate Hilbert spaces $\mathcal{H}_j$ are described by a reproducing kernel, then the problem (4.52) can be re-formulated as a second-order cone program, and is closely related to the group lasso. Whenever the Hilbert space $\mathcal{H}_j$ is defined by a reproducing kernel $\mathcal{R}_j$, then the jth coordinate function $\widehat{f}_j$ in any optimal solution again takes the form $\widehat{f}_j(\cdot) = \sum_{i=1}^{N} \widehat{\theta}_{ij} \mathcal{R}_j(\cdot, x_{ij})$ for a vector of weights $\widehat{\theta}_j \in \mathbb{R}^N$. This fact allows us to reduce the solution of the infinite-dimensional problem (4.52) to the simpler problem

$$\min_{\substack{\boldsymbol{\theta}_j \in \mathbb{R}^N \\ j=1,\dots,J}} \left\{ \frac{1}{N}\|y - \sum_{j=1}^{J} \mathbf{R}_j \boldsymbol{\theta}_j\|_2^2 + \lambda_{\mathcal{H}} \sum_{j=1}^{J} \sqrt{\boldsymbol{\theta}_j^T \mathbf{R}_j \boldsymbol{\theta}_j} + \lambda_N \sum_{j=1}^{J} \sqrt{\boldsymbol{\theta}_j^T \mathbf{R}_j^2 \boldsymbol{\theta}_j} \right\}. \tag{4.53}$$

As before, for each coordinate $j \in \{1, \dots, J\}$, the matrix $\mathbf{R}_j \in \mathbb{R}^{N \times N}$ is the kernel Gram matrix, with entries $[\mathbf{R}_j]_{ii'} = \mathcal{R}_j(x_{ij}, x_{i'j})$. See Exercise 4.20 for more details on this reduction.

---

[7]$\|f_j\|_{N,1}$ is the same as the $\|\mathbf{f}_j\|_2$ used in Section 4.4.2; here we are using a more generalizable notation.

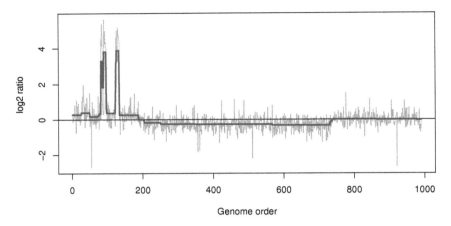

Figure 4.8 *Fused lasso applied to CGH data. Each spike represents the copy number of a gene in a tumor sample, relative to that of a control (on the log base-2 scale). The piecewise-constant green curve is the fused lasso estimate.*

The optimization problem (4.53) is an instance of a second-order cone program, and can be solved efficiently by a variant of the methods previously described. But why it is useful to impose two forms of regularization? As shown by Raskutti, Wainwright and Yu (2012), the combination of these two regularizers yields an estimator that is minimax-optimal, in that its convergence rate—as a function of sample size, problem dimension, and sparsity—is the fastest possible.

4.5   The Fused Lasso

Consider the gray spikes in Figure 4.8, the results of a *comparative genomic hybridization (CGH)* experiment. Each of these represents the (log base 2) relative copy number of a gene in a cancer sample relative to a control sample; these copy numbers are plotted against the chromosome order of the gene. These data are very noisy, so that some kind of smoothing is essential. Biological considerations dictate that it is typically segments of a chromosome—rather than individual genes—that are replicated, Consequently, we might expect that the underlying vector of true copy numbers to be piecewise-constant over contiguous regions of a chromosome. The *fused lasso signal approximator* exploits such structure within a signal, and is the solution of the following optimization problem

$$\underset{\boldsymbol{\theta}\in\mathbb{R}^N}{\text{minimize}}\left\{\frac{1}{2}\sum_{i=1}^{N}(y_i-\theta_i)^2+\lambda_1\sum_{i=1}^{N}|\theta_i|+\lambda_2\sum_{i=2}^{N}|\theta_i-\theta_{i-1}|\right\}. \qquad (4.54)$$

The first penalty is the familiar ℓ_1-norm, and serves to shrink the θ_i toward zero. Since the observation index i orders the data (in this case along the chromosome), the second penalty encourages neighboring coefficients θ_i to be similar, and will cause some to be identical (also known as *total-variation denoising*). Notice that (4.54) does not include a constant term θ_0; the coefficient θ_i represents the response y_i directly, and for these kinds of problems zero is a natural origin. (See Exercise 4.21 for further exploration of this intercept issue.) The green curve in Figure 4.8 is fit to these data using the fused lasso.

There are more general forms of the fused lasso; we mention two here.

- We can generalize the notion of neighbors from a linear ordering to more general neighborhoods, for examples adjacent pixels in an image. This leads to a penalty of the form

$$\lambda_2 \sum_{i \sim i'} |\theta_i - \theta_{i'}|, \tag{4.55}$$

where we sum over all neighboring pairs $i \sim i'$.

- In (4.54) every observation is associated with a coefficient. More generally we can solve

$$\underset{(\beta_0, \beta) \in \mathbb{R} \times \mathbb{R}^p}{\text{minimize}} \left\{ \frac{1}{2} \sum_{i=1}^{N} (y_i - \beta_0 - \sum_{j=1}^{p} x_{ij} \beta_j)^2 \right.$$

$$\left. + \lambda_1 \sum_{j=1}^{p} |\beta_j| + \lambda_2 \sum_{j=2}^{p} |\beta_j - \beta_{j-1}| \right\}. \tag{4.56}$$

Here the covariates x_{ij} and their coefficients β_j are indexed along some sequence j for which neighborhood clumping makes sense; (4.54) is clearly a special case.

4.5.1 Fitting the Fused Lasso

Problem (4.54) and its relatives are all convex optimization problems, and so all have well-defined solutions. As in other problems of this kind, here we seek efficient *path algorithms* for finding solutions for a range of values for the tuning parameters. Although coordinate descent is one of our favorite algorithms for lasso-like problems, it need not work for the fused lasso (4.54), because the difference penalty is not a separable function of the coordinates. Consequently, coordinate descent can become "stuck" at a nonoptimal point as illustrated in Figure 5.8 on page 111. This separability condition is discussed in more detail in Section 5.4.

We begin by considering the structure of the optimal solution $\widehat{\theta}(\lambda_1, \lambda_2)$ of the fused lasso problem (4.54) as a function of the two regularization parameters λ_1 and λ_2. The following result due to Friedman et al. (2007) provides some useful insight into the behavior of this optimum:

Lemma 4.1. For any $\lambda'_1 > \lambda_1$, we have

$$\widehat{\theta}_i(\lambda'_1, \lambda_2) = \mathcal{S}_{\lambda'_1 - \lambda_1}\big(\widehat{\theta}_i(\lambda_1, \lambda_2)\big) \text{ for each } i = 1, \ldots, N, \qquad (4.57)$$

where $\mathcal{S}$ is the soft-thresholding operator $\mathcal{S}_\lambda(z) := \text{sign}(z)(|z| - \lambda)_+$.
One important special case of Lemma 4.1 is the equality

$$\widehat{\theta}_i(\lambda_1, \lambda_2) = \mathcal{S}_{\lambda_1}\big(\widehat{\theta}_i(0, \lambda_2)\big) \text{ for each } i = 1, \ldots, N. \qquad (4.58)$$

Consequently, if we solve the fused lasso with $\lambda_1 = 0$, all other solutions can be obtained immediately by soft thresholding. This useful reduction also applies to the more general versions of the fused lasso (4.55). On the basis of Lemma 4.1, it suffices to focus our attention on solving the problem[8]

$$\underset{\boldsymbol{\theta} \in \mathbb{R}^N}{\text{minimize}} \left\{ \frac{1}{2} \sum_{i=1}^{N} (y_i - \theta_i)^2 + \lambda \sum_{i=2}^{N} |\theta_i - \theta_{i-1}| \right\}. \qquad (4.59)$$

We consider several approaches to solving (4.59).

4.5.1.1   *Reparametrization*

One simple approach is to reparametrize problem (4.59) so that the penalty is additive. In detail, suppose that we consider a linear transformation of the form $\boldsymbol{\gamma} = \mathbf{M}\boldsymbol{\theta}$ for an invertible matrix $\mathbf{M} \in \mathbb{R}^{N \times N}$ such that

$$\gamma_1 = \theta_1, \text{ and } \gamma_i = \theta_i - \theta_{i-1} \text{ for } i = 2, \ldots, N. \qquad (4.60)$$

In these transformed coordinates, the problem (4.59) is equivalent to the ordinary lasso problem

$$\underset{\boldsymbol{\gamma} \in \mathbb{R}^N}{\text{minimize}} \left\{ \frac{1}{2} \|\mathbf{y} - \mathbf{X}\boldsymbol{\gamma}\|^2 + \lambda \|\boldsymbol{\gamma}\|_1 \right\}, \text{ with } \mathbf{X} = \mathbf{M}^{-1}. \qquad (4.61)$$

In principle, the reparametrize problem (4.61) can be solved using any efficient algorithm for the lasso, including coordinate descent, projected gradient descent or the LARS procedure. However, $\mathbf{X}$ is a lower-triangular matrix with all nonzero entries equal to 1, and hence has large correlations among the "variables." Neither coordinate-descent nor LARS performs well under these circumstances (see Exercise 4.22). So despite the fact that reparametrization appears to solve the problem, it is not recommended, and more efficient algorithms exist, as we now discuss.

---

[8]Here we have adopted the notation λ (as opposed to λ_2) for the regularization parameter, since we now have only one penalty.

The one-dimensional fused lasso (4.59) has an interesting property, namely that as the regularization parameter λ increases, pieces of the optimal solution can only be joined together, not split apart. More precisely, letting $\widehat{\boldsymbol{\theta}}(\lambda)$ denote the optimal solution to the convex program (4.59) as a function of λ, we have:

Lemma 4.2. Monotone fusion. Suppose that for some value of λ and some index $i \in \{1, \ldots, N-1\}$, the optimal solution satisfies $\widehat{\theta}_i(\lambda) = \widehat{\theta}_{i+1}(\lambda)$. Then for all $\lambda' > \lambda$, we also have $\widehat{\theta}_i(\lambda') = \widehat{\theta}_{i+1}(\lambda')$.

Friedman et al. (2007) observed that this fact greatly simplifies the construction of the fused lasso solution path. One starts with $\lambda = 0$, for which there are no fused groups, and then computes the smallest value of λ that causes a fused group to form. The parameter estimates for this group are then fused together (i.e., constrained to be equal) for the remainder of the path. Along the way, a simple formula is available for the estimate within each fused group, so that the resulting procedure is quite fast, requiring $\mathcal{O}(N)$ operations. However, we note that the monotone-fusion property in Lemma 4.2 is special to the one-dimensional fused lasso (4.59). For example, it does not hold for the general fused lasso (4.56) with a model matrix $\mathbf{X}$, nor for the two-dimensional fused lasso (4.55). See Friedman et al. (2007) and Hoefling (2010) for more details on this approach.

Tibshirani₂ and Taylor (2011) take a different approach, and develop path algorithms for the *convex duals* of fused lasso problems. Here we illustrate their approach on the problem (4.59), but note that their methodology applies to the general problem (4.56) as well.

We begin by observing that problem (4.59) can be written in an equivalent *lifted* form

$$\underset{(\boldsymbol{\theta}, \mathbf{z}) \in \mathbb{R}^N \times \mathbb{R}^{N-1}}{\text{minimize}} \left\{ \frac{1}{2} \|\mathbf{y} - \boldsymbol{\theta}\|_2^2 + \lambda \|\mathbf{z}\|_2 \right\} \text{ subject to } \mathbf{D}\boldsymbol{\theta} = \mathbf{z}, \qquad (4.62)$$

where we have introduced a vector $\mathbf{z} \in \mathbb{R}^{N-1}$ of auxiliary variables, and $\mathbf{D}$ is a $(N-1) \times N$ matrix of first differences. Now consider the Lagrangian associated with the lifted problem, namely

$$L(\boldsymbol{\theta}, \mathbf{z}; \mathbf{u}) := \frac{1}{2} \|\mathbf{y} - \boldsymbol{\theta}\|_2^2 + \lambda \|\mathbf{z}\| + \mathbf{u}^T (\mathbf{D}\boldsymbol{\theta} - \mathbf{z}), \qquad (4.63)$$

where $\mathbf{u} \in \mathbb{R}^{N-1}$ is a vector of Lagrange multipliers. A straightforward computation shows that the Lagrangian dual function $\mathcal{Q}$ takes form

$$\mathcal{Q}(\mathbf{u}) := \underset{(\boldsymbol{\theta}, \mathbf{z}) \in \mathbb{R}^N \times \mathbb{R}^{N-1}}{\inf} L(\boldsymbol{\theta}, \mathbf{z}; \mathbf{u}) = \begin{cases} -\frac{1}{2} \|\mathbf{y} - \mathbf{D}^T \mathbf{u}\|^2 & \text{if } \|\mathbf{u}\|_\infty \leq \lambda, \\ -\infty & \text{otherwise.} \end{cases}$$

$$(4.64)$$

The Lagrangian dual problem is to maximize $\mathcal{Q}(\mathbf{u})$, and given an optimal solution $\widehat{\mathbf{u}} = \widehat{\mathbf{u}}(\lambda)$, we can recover an optimal solution $\widehat{\boldsymbol{\theta}} = \widehat{\boldsymbol{\theta}}(\lambda)$ to the original problem by setting $\widehat{\boldsymbol{\theta}} = \mathbf{y} - \mathbf{D}^T\widehat{\mathbf{u}}$. See Exercise 4.23 for the details of these duality calculations.

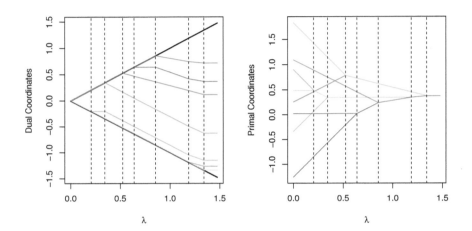

Figure 4.9 *The dual path algorithm in action on a small example. The left panel shows the progress of $\widehat{\mathbf{u}}(\lambda)$, while the right panel shows $\widehat{\boldsymbol{\theta}}(\lambda)$. We see that in the dual coordinates, as a parameter hits the boundary, an* unfusing *occurs in the primal coordinates.*

When the regularization parameter λ is sufficiently large, the dual maximization, or equivalently the problem of minimizing $-\mathcal{Q}(\mathbf{u})$, reduces to an unrestricted linear regression problem, with optimal solution

$$\mathbf{u}^* := (\mathbf{D}\mathbf{D}^T)^{-1}\mathbf{D}\mathbf{y}. \qquad (4.65)$$

The restrictions kick in when λ decreases to the critical level $\|\mathbf{u}^*\|_\infty$. Tibshirani[2] and Taylor (2011) show that as we decrease λ, once elements $\widehat{u}_j(\lambda)$ of the optimal solution hit the bound λ, then they are guaranteed to never leave the bound. This property leads to a very straightforward path algorithm, similar in spirit to LARS in Section 5.6; see Figure 4.9 for an illustration of the dual path algorithm in action. Exercise 4.23 explores some of the details.

4.5.1.4   Dynamic Programming for the Fused Lasso

Dynamic programming is a computational method for solving difficult problems by breaking them down into simpler subproblems. In the case of the one-dimensional fused lasso, the linear ordering of the variables means that fixing any variable breaks down the problem into two separate subproblems to the left and right of the fixed variable. In the "forward pass," we move from left to right, fixing one variable and solving for the variable to its left, as a function of this fixed variable. When we reach the right end, a backward pass then gives the complete solution.

Johnson (2013) proposed this dynamic programming approach to the fused lasso. In more detail, we begin by separating off terms in (4.59) that depend on θ_1, and rewrite the objective function (4.59) in the form

$$f(\boldsymbol{\theta}) = \underbrace{\frac{1}{2}(y_1 - \theta_1)^2 + \lambda|\theta_2 - \theta_1|}_{g(\theta_1, \theta_2)} + \left\{ \frac{1}{2}\sum_{i=2}^{N}(y_i - \theta_i)^2 + \lambda\sum_{i=3}^{N}|\theta_i - \theta_{i-1}| \right\}.$$

(4.66)

This decomposition shows the subproblem to be solved in the first step of the forward pass: we compute $\widehat{\theta}_1(\theta_2) := \arg\min_{\theta_1 \in \mathbb{R}} g(\theta_1, \theta_2)$, We have thus eliminated the first variable, and can now focus on the reduced objective function $f_2 : \mathbb{R}^{N-1} \to \mathbb{R}$ given by

$$f_2(\theta_2, \ldots \theta_N) = f(\widehat{\theta}_1(\theta_2), \theta_2, \ldots \theta_N).$$

(4.67)

We can then iterate the procedure, maximizing over θ_2 to obtain $\widehat{\theta}_2(\theta_3)$, and so on until we obtain $\widehat{\theta}_N$. Then we back-substitute to obtain $\widehat{\theta}_{N-1} = \widehat{\theta}_{N-1}(\widehat{\theta}_N)$, and so on for the sequences $\widehat{\theta}_{N-2}, \ldots, \widehat{\theta}_2, \widehat{\theta}_1$.

If each parameter θ_i can take only one of K distinct values, then each of the minimizers $\widehat{\theta}_j(\theta_{j+1})$ can be easily computed and stored as a $K \times K$ matrix. In the continuous case, the functions to be minimized are piecewise linear and quadratic, and care must be taken to compute and store the relevant information in an efficient manner; see Johnson (2013) for the details. The resulting algorithm is the fastest that we are aware of, requiring just $\mathcal{O}(N)$ operations, and considerably faster than the path algorithm described above. Interestingly, if we change the ℓ_1 difference penalty to an ℓ_0, this approach can still be applied, despite the fact that the problem is no longer convex. Exercise 4.24 asks the reader to implement the discrete case.

4.5.2 Trend Filtering

The first-order absolute difference penalty in the fused lasso can be generalized to use a higher-order difference, leading to the problem

$$\underset{\boldsymbol{\theta} \in \mathbb{R}^N}{\text{minimize}} \left\{ \frac{1}{2}\sum_{i=1}^{N}(y_i - \theta_i)^2 + \lambda \cdot \|\mathbf{D}^{(k+1)}\boldsymbol{\theta}\|_1 \right\}.$$

(4.68)

This is known as *trend filtering*. Here $\mathbf{D}^{(k+1)}$ is a matrix of dimension $(N - k - 1) \times N$ that computes discrete differences of order $k + 1$. The fused

lasso uses first-order differences ($k = 0$), while higher-order differences encourage higher-order smoothness. In general, trend filtering of order k results
in solutions that are piecewise polynomials of degree k. Linear trend filtering ($k = 1$) is especially attractive, leading to piecewise-linear solutions. The

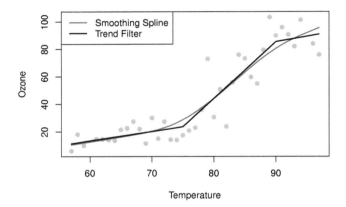

Figure 4.10 *A piecewise-linear function fit to some air-pollution data using trend
filtering. For comparison, a smoothing spline with the same degrees of freedom is
included.*

knots in the solution need not be specified but fall out of the convex optimization procedure. Kim, Koh, Boyd and Gorinevsky (2009) propose an efficient
interior point algorithm for this problem. Tibshirani₂ (2014) proves that the
trend filtering estimate adapts to the local level of smoothness much better
than smoothing splines, and displays a surprising similarity to locally-adaptive
regression splines. Further, he shows that the estimate converges to the true
underlying function at the minimax rate for functions whose k^{th} derivative is of
bounded variation (a property not shared by linear estimators such as smoothing splines). Furthermore, Tibshirani₂ and Taylor (2011) show that a solution
with m knots has estimated degrees of freedom given by df $= m + k + 1$.[9]

Figure 4.10 shows a piecewise-linear function fit by trend filtering to some
air-pollution data. As a comparison, we include the fit of a smoothing spline,
with the same effective df $= 4$. While the fits are similar, it appears that trend
filtering has found natural changepoints in the data.

In (4.68) it is assumed that the observations occur at evenly spaced positions. The penalty can be modified (Tibshirani₂ 2014) to accommodate arbitrary (ordered) positions x_i as follows:

$$\underset{\boldsymbol{\theta}\in\mathbb{R}^N}{\text{minimize}} \left\{ \frac{1}{2}\sum_{i=1}^{N}(y_i - \theta_i)^2 + \lambda \cdot \sum_{i=1}^{N-2}\left| \frac{\theta_{i+2} - \theta_{i+1}}{x_{i+2} - x_{i+1}} - \frac{\theta_{i+1} - \theta_i}{x_{i+1} - x_i} \right| \right\} \quad (4.69)$$

---

[9]This is an unbiased estimate of the degrees of freedom; see Section 2.5.

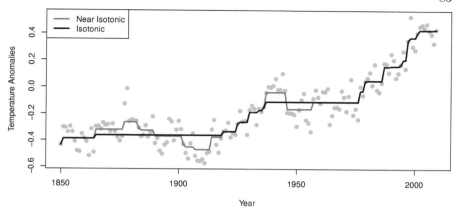

Figure 4.11 *Near isotonic fit to global-warming data, showing annual temperature anomalies. The value of λ was chosen by cross-validation, and the fit appears to support the evidence of nonmonotone behavior seen in the data.*

It compares the empirical slopes for adjacent pairs, and encourages them to be the same. This is the penalty that was used in Figure 4.10, since the **Temperature** values are not uniformly spaced.

4.5.3 Nearly Isotonic Regression

Tibshirani[2], Hoefling and Tibshirani (2011) suggest a simple modification of the one-dimensional fused lasso that encourages the solution to be monotone. It is based on a relaxation of isotonic regression. In the classical form of isotonic regression, we estimate $\theta \in \mathbb{R}^N$ by solving the constrained minimization problem

$$\underset{\theta \in \mathbb{R}^N}{\text{minimize}} \left\{ \sum_{i=1}^{N} (y_i - \theta_i)^2 \right\} \text{ subject to } \theta_1 \leq \theta_2 \leq \ldots \leq \theta_N. \qquad (4.70)$$

The resulting solution gives the best monotone (nondecreasing) fit to the data. Monotone nonincreasing solutions can be obtaining by first flipping the signs of the data. There is a unique solution to problem (4.70), and it can be obtained using the *pool adjacent violators algorithm* (Barlow, Bartholomew, Bremner and Brunk 1972), or PAVA for short.

Nearly isotonic regression is a natural relaxation, in which we introduce a regularization parameter $\lambda \geq 0$, and instead solve the penalized problem

$$\underset{\theta \in \mathbb{R}^N}{\text{minimize}} \left\{ \frac{1}{2} \sum_{i=1}^{N} (y_i - \theta_i)^2 + \lambda \sum_{i=1}^{N-1} (\theta_i - \theta_{i+1})_+ \right\}. \qquad (4.71)$$

The penalty term penalizes adjacent pairs that violate the monotonicity property, that is, having $\theta_i > \theta_{i+1}$. When $\lambda = 0$, the solution interpolates the data, and letting $\lambda \to \infty$, we recover the solution to the classical isotonic regression problem (4.70). Intermediate value of λ yield nonmonotone solutions that trade off monotonicity with goodness-of-fit; this trade-off allows one to assess the validity of the monotonicity assumption for the given data sequence. Figure 4.11 illustrates the method on data on annual temperature anomalies from 1856 to 1999, relative to the 1961–1990 mean. The solution to the nearly isotonic problem (4.71) can be obtained from a simple modification of the path algorithm discussed previously, a procedure that is analogous to the PAVA algorithm for problem (4.70); see Tibshirani$_2$ et al. (2011) for details.

4.6   Nonconvex Penalties

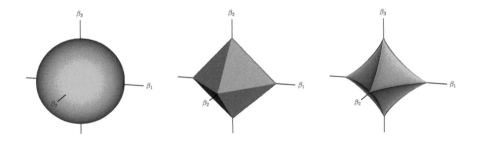

Figure 4.12 *The ℓ_q unit balls in $\mathbb{R}^3$ for $q = 2$ (left), $q = 1$ (middle), and $q = 0.8$ (right). For $q < 1$ the constraint regions are nonconvex. Smaller q will correspond to fewer nonzero coefficients, and less shrinkage. The nonconvexity leads to combinatorially hard optimization problems.*

By moving from an ℓ_2 penalty to ℓ_1, we have seen that for the same *effective df* the lasso selects a subset of variables to have nonzero coefficients, and shrinks their coefficients less. When p is large and the number of relevant variables is small, this may not be enough; in order to reduce the set of chosen variables sufficiently, lasso may end up over-shrinking the retained variables. For this reason there has been interest in nonconvex penalties.

The natural choice might be the ℓ_q penalty, for $0 \le q \le 1$, with the limiting ℓ_0 corresponding to best-subset selection. Figure 4.12 compares the ℓ_q unit balls for $q \in \{2, 1, 0.8\}$. The spiky nonconvex nature of the ball on the right implies edges and coordinate axes will be favored in selection under such constraints. Unfortunately, along with nonconvexity comes combinatorial computational complexity; even the simplest case of ℓ_0 can be solved exactly only for $p \approx 40$ or less. For this and related statistical reasons alternative nonconvex penalties have been proposed. These include the SCAD (Fan and Li 2001, *smoothly clipped absolute deviation*) and MC+ (Zhang 2010, *minimax concave*) penalties. Figure 4.13 shows four members of the MC+ penalty family in $\mathbb{R}^1$, indexed by the nonconvexity parameter $\gamma \in (1, \infty)$; this family

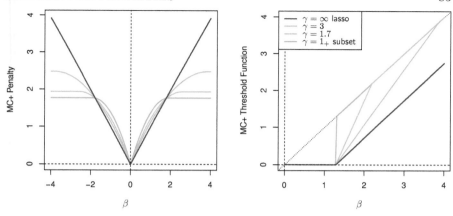

Figure 4.13 *Left: The MC+ family of nonconvex sparsity penalties, indexed by a sparsity parameter $\gamma \in (1, \infty)$. Right: piecewise-linear and continuous threshold functions associated with MC+ (only the north-east quadrant is shown), making this penalty family suitable for coordinate descent algorithms.*

bridges the gap between lasso ($\gamma = \infty$) and best-subset ($\gamma = 1_+$). The penalty functions are piecewise quadratic (see Exercise 4.25), and importantly the corresponding threshold functions are piecewise linear and continuous. In detail, for squared-error loss we pose the (nonconvex) optimization problem

$$\underset{\beta \in \mathbb{R}^p}{\text{minimize}} \left\{ \frac{1}{2} \|\mathbf{y} - \mathbf{X}\beta\|_2^2 + \sum_{j=1}^p P_{\lambda,\gamma}(\beta_j) \right\}, \tag{4.72}$$

with the MC+ penalty on each coordinate defined by

$$P_{\lambda,\gamma}(\theta) := \int_0^{|\theta|} \left(1 - \frac{x}{\lambda\gamma} \right)_+ dx. \tag{4.73}$$

With coordinate descent in mind, we consider solving a one-dimensional version of (4.72) (in standardized form)

$$\underset{\beta \in \mathbb{R}^1}{\text{minimize}} \left\{ \frac{1}{2}(\beta - \tilde{\beta})^2 + \lambda \int_0^{|\beta|} \left(1 - \frac{x}{\lambda\gamma} \right)_+ dx \right\}. \tag{4.74}$$

The solution is unique[10] for $\gamma > 1$ and is given by

$$\mathcal{S}_{\lambda,\gamma}(\tilde{\beta}) = \begin{cases} 0 & \text{if } |\tilde{\beta}| \leq \lambda \\ \text{sign}(\tilde{\beta}) \left(\frac{|\tilde{\beta}| - \lambda}{1 - \frac{1}{\gamma}} \right) & \text{if } \lambda < |\tilde{\beta}| \leq \lambda\gamma \\ \tilde{\beta} & \text{if } |\tilde{\beta}| > \lambda\gamma \end{cases} \tag{4.75}$$

---

[10]Despite the nonconvexity, there is a unique solution in $\mathbb{R}^1$; this is not necessarily the case for the p-dimensional problem (4.72).

The right panel in Figure 4.13 shows examples of (4.75). Large values of $\tilde{\beta}$ are left alone, small values are set to zero, and intermediate values are shrunk. As γ gets smaller, the intermediate zone gets narrower, until eventually it becomes the hard-thresholding function of best subset (orange curve in figure). By contrast, the threshold functions for the ℓ_q family ($q < 1$) are discontinuous in $\tilde{\beta}$.

Mazumder, Friedman and Hastie (2011) exploit the continuity of $\mathcal{S}_{\lambda,\gamma}$ (in both λ and γ) in a coordinate-descent scheme for fitting solution paths for the entire MC+ family. Starting with the lasso solution, their R package sparsenet (Mazumder, Hastie and Friedman 2012) moves down a sequence in γ toward sparser models, and for each fits a regularization path in λ. Although it cannot claim to solve the nonconvex problem (4.72), this approach is both very fast and appears to find good solutions.

Zou (2006) proposed the *adaptive lasso* as a means for fitting models sparser than lasso. Using a pilot estimate $\tilde{\beta}$, the adaptive lasso solves

$$\underset{\beta \in \mathbb{R}^p}{\text{minimize}} \left\{ \frac{1}{2}\|\mathbf{y} - \mathbf{X}\beta\|_2^2 + \lambda \sum_{j=1}^p w_j |\beta_j| \right\}, \qquad (4.76)$$

where $w_j = 1/|\tilde{\beta}_j|^\nu$. The adaptive lasso penalty can be seen as an approximation to the ℓ_q penalties with $q = 1 - \nu$. One advantage of the adaptive lasso is that given the pilot estimates, the criterion (4.76) is convex in β. Furthermore, if the pilot estimates are $\sqrt{N}$ consistent, Zou (2006) showed that the method recovers the true model under more general conditions than does the lasso. If $p < N$ one can use the least-squares solutions as the pilot estimates. When $p \geq N$, the least-squares estimates are not defined, but the univariate regression coefficients can be used for the pilot estimates and result in good recovery properties under certain conditions (Huang, Ma and Zhang 2008). Exercise 4.26 explores the close connections between the adaptive lasso and the nonnegative garrote of Section 2.8.

We end this section by mentioning a practical alternative to nonconvex optimization for sparse model-path building. Forward-stepwise methods (Hastie et al. 2009, Chapter 3) are very efficient, and are hard to beat in terms of finding good, sparse subsets of variables. Forward stepwise is a greedy algorithm— at each step fixing the identity of the terms already in the model, and finding the best variable to include among those remaining. The theoretical properties of forward-stepwise model paths are less well understood, partly because of the algorithmic definition of the procedure, as opposed to being a solution to an optimization problem.

Bibliographic Notes

The elastic net was proposed by Zou and Hastie (2005), who also distinguished between the naive version, similar to the one presented here, and a debiased version that attempts to undo the biasing effect of the ridge shrinkage. Friedman et al. (2015) build a system of coordinate-descent algorithms for fitting elastic-net penalized generalized linear models, implemented in the R package glmnet. Yuan and Lin (2006) introduced the group lasso, and their paper has stimulated much research. Meier et al. (2008) extended the group lasso to logistic regression problems, whereas Zhao, Rocha and Yu (2009) describe a more general family of structured penalties, including the group lasso as a special case. A line of theoretical work has sought to understand when the group lasso estimator has lower statistical error than the ordinary lasso. Huang and Zhang (2010) and Lounici, Pontil, Tsybakov and van de Geer (2009) establish error bounds for the group lasso, which show how it outperforms the ordinary lasso in certain settings. Negahban, Ravikumar, Wainwright and Yu (2012) provide a general framework for analysis of M-estimators, including the group lasso as a particular case as well as more general structured penalties. Obozinski, Wainwright and Jordan (2011) characterize multivariate regression problems for which the group lasso does (or does not) yield better variable selection performance than the ordinary lasso.

The overlap group lasso was introduced by Jacob et al. (2009), and the sparse group lasso by Puig, Wiesel and Hero (2009) and Simon, Friedman, Hastie and Tibshirani (2013). Various algorithms have been developed for solving the group and overlap group lassos, as well as a variety of structured generalizations; see Bach, Jenatton, Mairal and Obozinski (2012) for a good review.

Additive models were proposed by Stone (1985) as a means of side-stepping the curse of dimensionality in nonparametric regression; see Hastie and Tibshirani (1990) for further background on (generalized) additive models. The COSSO model was developed by Lin and Zhang (2003) in the context of reproducing kernel Hilbert spaces, and ANOVA spline decompositions. The books by Wahba (1990) and Gu (2002) provide further background on splines and RKHSs. Ravikumar et al. (2009) followed up with the SPAM model, which is somewhat simpler and more general, and established certain forms of high-dimensional consistency for their estimator. Meier, van de Geer and Bühlmann (2009) studied a related family of estimators, based on explicit penalization with the empirical L^2-norm, corresponding to the doubly penalized estimator with $\lambda_{\mathcal{H}} = 0$. Koltchinski and Yuan (2008, 2010) analyzed the COSSO estimator, as well as the doubly penalized estimator (4.52). Raskutti et al. (2009, 2012) derived minimax bounds for sparse additive models, and also show that the doubly penalized estimator (4.52) can achieve these bounds for various RKHS families, including splines as a special case.

The fused lasso was introduced by Tibshirani, Saunders, Rosset, Zhu and Knight (2005). Various algorithms have been proposed for versions of the fused lasso, including the methods of Hoefling (2010), Johnson (2013), and Tibshirani2 and Taylor (2011).

The MC+ threshold function was first described in Gao and Bruce (1997) in the context of wavelet shrinkage. There has been a lot of activity in nonconvex penalties for sparse modeling. Zou and Li (2008) develop local linear approximation algorithms for tackling the nonconvex optimization problems. These and other approaches are discussed in Mazumder et al. (2011).

Exercises

Ex. 4.1 Suppose we have two identical variables $X_1 = X_2$, and a response Y, and we perform a ridge regression (see (2.7) in Section 2.2) with penalty $\lambda > 0$. Characterize the coefficient estimates $\widehat{\beta}_j(\lambda)$.

Ex. 4.2 Consider a slightly noisy version of the identical twins example in the beginning of Section 4.2, where the two variables are strongly positively correlated. Draw a schematic of the contours of the loss function and the penalty function, and demonstrate why the elastic net encourages coefficient sharing more than does the lasso.

Ex. 4.3 Consider the elastic-net problem (4.2).

(a) Show how to simplify the calculation of $\widehat{\beta}_0$ by centering each of the predictors, leading to $\widehat{\beta}_0 = \bar{y}$ (for all values of λ). How does one convert back to the estimate of $\widehat{\beta}_0$ for uncentered predictors?

(b) Verify the soft-thesholding expression (4.4) for the update of $\widehat{\beta}_j$ by coordinate descent.

Ex. 4.4 Consider the solution to the group lasso problem (4.5) when some of the variables are factors. Show that when there is an intercept in the model, the optimal coefficients for each factor sum to zero.

Ex. 4.5 This exercise investigates the penalty modifier for the group lasso. Consider the *entry* criterion $\|\mathbf{Z}_j^T \mathbf{r}_j\|_2 < \lambda$ for the group lasso (Section 4.3.1). Suppose $\mathbf{r}_j$ is i.i.d noise with mean $\mathbf{0}$ and covariance $\sigma^2 \mathbf{I}$—a null situation. Show that

$$\mathbb{E}\|\mathbf{Z}_j^T \mathbf{r}_j\|_2^2 = \sigma^2 \|\mathbf{Z}\|_F^2. \qquad (4.77)$$

Hence argue that to make comparisons fair among the penalty terms in the group lasso, one should replace $\lambda \sum_{j=1}^J \|\theta_j\|_2$ in Equation (4.5) with

$$\lambda \sum_{j=1}^J \tau_j \|\theta_j\|_2, \qquad (4.78)$$

where $\tau_j = \|\mathbf{Z}_j\|_F$. Show that when $\mathbf{Z}_j$ is orthonormal, this results in $\tau_j = \sqrt{p_j}$.

Ex. 4.6 Show that under the orthonormality condition $\mathbf{Z}_j^T \mathbf{Z}_j = \mathbf{I}$, the update (4.15) solves the fixed point Equation (4.13).

Ex. 4.7 Consider the block-wise solution vector (4.14) for the group lasso. If $\|\hat{\theta}_j\|$ is known, we can write the solution in closed form. Let $\mathbf{Z}_j = \mathbf{U}\mathbf{D}\mathbf{V}^T$ be the singular value decomposition of $\mathbf{Z}_j$. Let $r^* = \mathbf{U}^T \mathbf{r}_j \in \mathbb{R}^{p_j}$. Show that $\phi = \|\hat{\theta}_j\|$ solves

$$\sum_{\ell=1}^{p_j} \frac{r_\ell^{*2} d_\ell^2}{(d_\ell^2 \phi + \lambda)^2} = 1, \tag{4.79}$$

where d_ℓ is the ℓth diagonal element of $\mathbf{D}$. Show how to use a golden search strategy to solve for ϕ. Write an R function to implement this algorithm, along with the golden search.

Ex. 4.8 Discuss the impact of the normalization $\mathbf{Z}_j^T \mathbf{Z}_j = \mathbf{I}$ in the context of a matrix of dummy variables representing a factor with p_j levels. Does the use of contrasts rather than dummy variables alleviate the situation?

Ex. 4.9 Using the approach outlined in Section 5.3.3, derive the generalized gradient update (4.16a) for the group lasso. Write a R function to implement this algorithm (for a single group). Include an option to implement Nesterov acceleration.

Ex. 4.10 Using the approach outlined in Section 5.3.3, derive the generalized gradient update (4.23) for the sparse group lasso.

Ex. 4.11 Run a series of examples of increasing dimension to compare the performance of your algorithms in Exercises 4.7 and 4.9. Make sure they are producing the same solutions. Compare their computational speed—for instance, the command `system.time()` can be used in R.

Ex. 4.12 Consider the condition (4.19) for $\hat{\theta}_j$ to be zero for the sparse group lasso. Define

$$\begin{aligned} J(t) &= \frac{1}{\lambda(1-\alpha)}\|\mathbf{Z}_j^T \mathbf{r}_j - \lambda\alpha \cdot t\|_2 \tag{4.80} \\ &= \|s\|_2. \end{aligned}$$

Now solve

$$\min_{t:t_k \in [-1,1]} J(t), \tag{4.81}$$

and show that this leads to the condition $\hat{\theta}_j = 0$ if and only if $\|\hat{g}_j\|_2 \le \lambda(1-\alpha)$ with $\hat{g}_j = \mathcal{S}_{\lambda\alpha}(\mathbf{Z}_j^T \mathbf{r}_j)$.

Ex. 4.13 Show that if $\mathbf{Z}_j^T \mathbf{Z}_j = \mathbf{I}$, then (4.21) solves (4.12).

Ex. 4.14 Consider the hierarchical interaction formulation in Example 4.3, and
the optimization problem (4.29)–(4.31).

(a) Give an argument why the multipliers p_1 and p_2 make sense in the third
penalty.

(b) Suppose we augment the third matrix in (4.29) with a vector of ones
$[1 \; \mathbf{Z}_1 \; \mathbf{Z}_2 \; \mathbf{Z}_{1:2}]$, and augment the parameter vector with $\tilde{\mu}$. We now replace
the third group penalty term with

$$\sqrt{p_1 p_2 \tilde{\mu}^2 + p_2 \|\tilde{\alpha}_1\|_2^2 + p_1 \|\tilde{\alpha}_2\|_2^2 + \|\alpha_{1:2}\|_2^2}.$$

Show that for any $\lambda > 0$, $\widehat{\tilde{\mu}} = 0$.

(c) Show that the solution to (4.29)–(4.31) is equivalent to the solution to
(4.32) for any $\lambda > 0$. Show how to map the solution to the latter to the
solution to the former.

Ex. 4.15 Consider a criterion for sparse additive models:

$$
\begin{aligned}
\underset{\beta \in \mathbb{R}^J, \, \{f_j \in \mathcal{F}_j\}_1^J}{\text{minimize}} \quad & \mathbb{E}(Y - \sum_{j=1}^{J} \beta_j f_j(X_j))^2 \\
\text{subject to} \quad & \|f_j\|_2 = 1 \; \forall j \qquad\qquad (4.82) \\
& \sum_{j=1}^{J} |\beta_j| \le t.
\end{aligned}
$$

Although evocative, this criterion is not convex, but rather biconvex in β and
$\{f_j\}_1^J$. Show that one can absorb the β_j into f_j and that solving (4.82) is
equivalent to solving the convex (4.38):

$$\underset{f_j \in \mathcal{F}_j, \, j=1,\ldots,J}{\text{minimize}} \left\{ \mathbb{E}\left[\left(Y - \sum_{j=1}^{J} f_j(X_j) \right)^2 \right] + \lambda \sum_{j=1}^{J} \|f_j\|_2 \right\}.$$

(Ravikumar et al. 2009)

Ex. 4.16 The SPAM backfitting equations (4.40) are in terms of function
updates, where $\hat{f}_j$ is a fitted function returned by a smoothing operator $\mathcal{S}_j$,
and the N-vector form $\mathbf{f}_j$ is f_j evaluated at the N samples values of X_j.
Suppose that the smoothing operator $\mathcal{S}_j$ fits a linear expansion of the form

$$f_j(\cdot) = \sum_{\ell=1}^{p_j} \beta_{j\ell} \, \psi_{j\ell}(\cdot), \qquad\qquad (4.83)$$

where $\theta_j = \begin{bmatrix} \beta_{j1} & \beta_{j2} & \cdots & \beta_{jp_j} \end{bmatrix}$ is the coefficient vector.

(a) Suppose that the basis matrices are orthonormal: $\mathbf{\Psi}_j^T \mathbf{\Psi}_j = \mathbf{I}_{p_j}$. Show that the SPAM backfitting equations are equivalent to the ordinary group-lasso estimating equations in terms of the parameters θ_j.

(b) What changes if $\mathbf{\Psi}_j$ is not orthonormal?

Ex. 4.17 In this exercise, we show that any optimal solution to the COSSO problem (4.44) is a member of $\mathcal{H}_0$, the linear span of the kernel functions $\{\mathcal{R}(\cdot, x_i), i = 1, \ldots, N\}$. We use the fact that any function $f \in \mathcal{H}$ has a decomposition of the form $g + h$, where $g \in \mathcal{H}_0$ and h is orthogonal to $\mathcal{H}_0$, meaning that $\langle h, f_0 \rangle_{\mathcal{H}}$ for all $f_0 \in \mathcal{H}_0$.

(a) For a function of the form $f = g + h$ as above, show that the term $\frac{1}{N} \sum_{i=1}^{N} (y - f(x_i))^2$ depends only on g. (*Hint:* The kernel reproducing property could be useful here.)

(b) Show that the penalty term is only increased by including a component $h \neq 0$. Conclude that any optimal solution $\widehat{f}$ must belong to $\mathcal{H}_0$.

Ex. 4.18 Verify that the solutions for f_j in (4.47) with $\lambda = \tau^4/4$ coincide with the solutions in (4.44).

Ex. 4.19 Consider the additive model criterion (4.42), and assume associated with each function f_j is a reproducing kernel $\mathcal{R}_j$, leading to a data criterion

$$\underset{\substack{\boldsymbol{\theta}_j \in \mathbb{R}^N \\ j=1,\ldots J}}{\text{minimize}} \left\{ \left\| \mathbf{y} - \sum_{j=1}^{J} \mathbf{R}_j \boldsymbol{\theta}_j \right\|^2 + \lambda \sum_{j=1}^{J} \frac{1}{\gamma_j} \boldsymbol{\theta}_j^T \mathbf{R}_j \boldsymbol{\theta}_j \right\} \tag{4.84}$$

(The $1/N$ has been absorbed into λ).

(a) Define $\tilde{\mathbf{R}}_j = \gamma_j \mathbf{R}_j$ and $\tilde{\boldsymbol{\theta}}_j = \boldsymbol{\theta}_j / \gamma_j$. In this new parametrization, show that the estimating equations for $\tilde{\boldsymbol{\theta}}_j$ are

$$-\tilde{\mathbf{R}}_j (\mathbf{y} - \mathbf{f}_+) + \lambda \tilde{\mathbf{R}}_j \tilde{\boldsymbol{\theta}}_j = 0, \ j = 1, \ldots, J, \tag{4.85}$$

where $\mathbf{f}_+ = \sum_{j=1}^{J} \mathbf{f}_j$, and $\mathbf{f}_j = \tilde{\mathbf{R}}_j \tilde{\boldsymbol{\theta}}_j$.

(b) Show that these can be rewritten as

$$\tilde{\boldsymbol{\theta}}_j = (\tilde{\mathbf{R}}_j + \lambda \mathbf{I})^{-1} \mathbf{r}_j, \quad \text{and} \tag{4.86a}$$

$$\tilde{\mathbf{f}}_j = \tilde{\mathbf{R}}_j (\tilde{\mathbf{R}}_j + \lambda \mathbf{I})^{-1} \mathbf{r}_j, \tag{4.86b}$$

where $\mathbf{r}_j = \mathbf{y} - \mathbf{f}_+ + \mathbf{f}_j$.

(c) Define $\tilde{\mathbf{R}}_+ = \sum_{j=1}^{J} \tilde{\mathbf{R}}_j = \sum_{j=1}^{J} \gamma_j \mathbf{R}_j$. Show that

$$\mathbf{f}_+ = \tilde{\mathbf{R}}_+ (\tilde{\mathbf{R}}_+ + \lambda \mathbf{I})^{-1} \mathbf{y} = \tilde{\mathbf{R}}_+ \mathbf{c} \tag{4.87a}$$

$$\mathbf{c} = (\tilde{\mathbf{R}}_+ + \lambda \mathbf{I})^{-1} \mathbf{y}. \tag{4.87b}$$

Compare with the previous item.

(d) Show that $\tilde{\boldsymbol{\theta}}_j = \mathbf{c} \, \forall \, j$. So even though there are J N-dimensional parameters $\tilde{\boldsymbol{\theta}}_j$ in this representation, their estimates are all the same.

This shows that given γ_j, $\mathbf{f}_j = \gamma_j \mathbf{R}_j \mathbf{c} = \gamma_j \mathbf{g}_j$, and justifies the second step (4.46) in the alternating algorithm for fitting the COSSO model (see Section 4.4).

Ex. 4.20 Show that any optimal solution to the doubly regularized estimator (4.52) takes the form $\widehat{f}_j(\cdot) = \sum_{i=1}^N \widehat{\theta}_{ij} \mathcal{R}(\cdot, x_{ij})$, where the optimal weights $(\widehat{\theta}_j, j = 1, \ldots, J)$ are obtained by solving the convex program (4.53).

Ex. 4.21 Consider the fused lasso problem (4.56). Characterize $\widehat{\beta}_0$. Show that if we center the predictors and the response by subtracting their sample means, we can omit the term β_0 and the estimates $\widehat{\beta}_j$ are unaffected. Now consider a version of the fused-lasso signal approximator (4.54) with a constant term θ_0 included:

$$\underset{\theta_0, \boldsymbol{\theta}}{\text{minimize}} \sum_{i=1}^N (y_i - \theta_0 - \theta_i)^2 + \lambda_1 \sum_{i=1}^N |\theta_i| + \lambda_2 \sum_{i=2}^N |\theta_i - \theta_{i-1}|. \qquad (4.88)$$

Characterize $\widehat{\theta}_0$, and show that median$(\widehat{\theta}_i) = 0$.

Ex. 4.22 Consider the matrix $\mathbf{M}$ corresponding to the linear transformation (4.60).

(a) Show that its inverse $\mathbf{M}^{-1}$ is lower triangular with all ones on and below the diagonal.

(b) Explore the pairwise correlations between the columns of such a matrix for the CGH data of Figure 4.8.

(c) Using `glmnet` with `maxdf=200`, and `type="naive"`, fit model (4.61), and show that the fitted values correspond to the parameters of interest. Compare the performance of `lars` for the same task. Using a soft-thresholding post-processor, try to match Figure 4.8.

Ex. 4.23 Derive the dual optimization problem (4.64) in Section 4.5.1.3. Suppose the kth element of $\hat{\mathbf{u}}(\lambda)$ has reached the bound at $\lambda = \lambda_k$, and let the set B hold their indices, and $\mathbf{s}$ a vector of their signs. Show that the solution to (4.64) at λ_k also solves

$$\underset{\mathbf{u}_{-B}}{\text{minimize}} \frac{1}{2} \|\mathbf{y} - \lambda_k \mathbf{D}_B^T \mathbf{s} - \mathbf{D}_{-B}^T \mathbf{u}_{-B}\|^2, \qquad (4.89)$$

with solution $\hat{\mathbf{u}}_B(\lambda) = \lambda \mathbf{s}$ and $\hat{\mathbf{u}}_{-B}(\lambda) = (\mathbf{D}_{-B} \mathbf{D}_{-B}^T)^{-1} \mathbf{D}_{-B}(\mathbf{y} - \lambda \mathbf{D}_B^T \mathbf{s})$ and $\lambda = \lambda_k$. By definition each of the elements of $\hat{\mathbf{u}}_{-B}(\lambda)$ has absolute value less than λ_k. Show that the solution is piecewise-linear in $\lambda < \lambda_k$, and remains the same until the next element of $\hat{\mathbf{u}}_{-B}(\lambda)$ hits the boundary. Show that one can determine exactly for which element and value of λ this will be.

Ex. 4.24 Here we use dynamic programming to fit the fused lasso.

(a) Implement the dynamic programming approach to the fused lasso, in the simple case where each β_i can take one of K distinct values.

(b) Do the same as in (a), replacing the ℓ_1 difference penalty with an ℓ_0 difference penalty. Compare the two procedures on the CGH data.

Ex. 4.25 Derive the threshold function (4.75) for the uni-dimensional MC+ criterion (4.74) in Section 4.6.

Ex. 4.26 Show that with $\nu = 1$ in (4.76), the adaptive-lasso solutions are similar to those of the nonnegative garrote (2.19). In particular, if we constrain the adaptive lasso solutions to have the same sign as the pilot estimates, then they are the same as the solutions to the garrote with a suitably chosen regularization parameter.

Chapter 5

Optimization Methods

5.1 Introduction

In this chapter, we present an overview of some basic optimization concepts and algorithms for convex problems, with an emphasis on aspects of particular relevance to regularized estimators such as the lasso. At the algorithmic level, we focus primarily on *first-order methods*, since they are especially useful for large-scale optimization problems. We begin with an overview of some basic optimality theory for convex programs, and then move on to consider various types of iterative algorithms. Although we limit our focus mainly to convex problems, we do touch upon algorithms for biconvex problems later in the chapter.

5.2 Convex Optimality Conditions

An important class of optimization problems involves convex cost functions and convex constraints. A set $\mathcal{C} \subseteq \mathbb{R}^p$ is *convex* if for all $\beta, \beta' \in \mathcal{C}$ and all scalars $s \in [0, 1]$, all vectors of the form $\beta(s) = s\beta + (1 - s)\beta'$ also belong to $\mathcal{C}$. A function $f : \mathbb{R}^p \to \mathbb{R}$ is convex means that for any two vectors β, β' in the domain of f and any scalar $s \in (0, 1)$, we have

$$f(\beta(s)) = f(s\beta + (1 - s)\beta') \leq sf(\beta) + (1 - s)f(\beta'). \qquad (5.1)$$

In geometric terms, this inequality implies that the chord joining the $f(\beta)$ and $f(\beta')$ lies above the graph of f, as illustrated in Figure 5.1(a). This inequality guarantees that a convex function cannot have any local minima that are not also globally minimal, as illustrated Figure 5.1(b).

5.2.1 Optimality for Differentiable Problems

Consider the constrained optimization problem

$$\underset{\beta \in \mathbb{R}^p}{\text{minimize}} f(\beta) \quad \text{such that } \beta \in \mathcal{C}, \qquad (5.2)$$

where $f : \mathbb{R}^p \to \mathbb{R}$ is a convex objective function to be minimized, and $\mathcal{C} \subset \mathbb{R}^p$ is a convex constraint set. When the cost function f is differentiable, then a

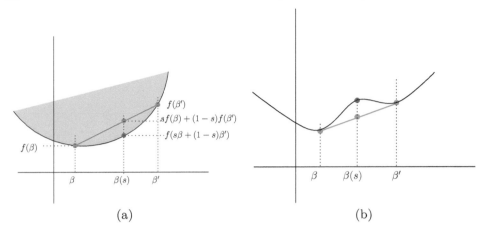

(a)   (b)

Figure 5.1 *(a) For a convex function, the line $sf(\beta) + (1 - s)f(\beta')$ always lies above the function value $f(s\beta + (1-s)\beta')$. (b) A nonconvex function that violates the inequality (5.1). Without convexity, there may be local minima that are not globally minima, as shown by the point β'.*

necessary and sufficient condition for a vector $\beta^* \in \mathcal{C}$ to be a global optimum is that

$$\langle \nabla f(\beta^*), \beta - \beta^* \rangle \geq 0 \tag{5.3}$$

for all $\beta \in \mathcal{C}$. The sufficiency of this condition is easy to see; for any $\beta \in \mathcal{C}$, we have

$$f(\beta) \overset{(i)}{\geq} f(\beta^*) + \langle \nabla f(\beta^*), \beta - \beta^* \rangle \overset{(ii)}{\geq} f(\beta^*), \tag{5.4}$$

where inequality (i) follows from the convexity of f,[1] and inequality (ii) follows from the optimality condition (5.3). As a special case, when $\mathcal{C} = \mathbb{R}^p$ so that the problem (5.2) is actually unconstrained, then the first-order condition (5.3) reduces to the classical zero-gradient condition $\nabla f(\beta^*) = 0$.

Frequently, it is the case that the constraint set $\mathcal{C}$ can be described in terms of the sublevel sets of some convex constraint functions. For any convex function $g : \mathbb{R}^p \to \mathbb{R}$, it follows from the definition (5.1) that the sublevel set $\{\beta \in \mathbb{R}^p \mid g(\beta) \leq 0\}$ is a convex set. On this basis, the convex optimization problem

$$\underset{\beta \in \mathbb{R}^p}{\text{minimize}} \, f(\beta) \quad \text{such that } g_j(\beta) \leq 0 \text{ for } j = 1, \dots, m, \tag{5.5}$$

---

[1] Inequality (i) is an equivalent definition of convexity for a differentiable function f; the first-order Taylor approximation centered at any point $\tilde{\beta} \in \mathcal{C}$ gives a tangent lower bound to f.

where $g_j, j = 1, \ldots, m$ are convex functions that express constraints to be satisfied, is an instance of the general program (5.2). We let f^* denote the optimal value of the optimization problem (5.5).

An important function associated with the problem (5.5) is the Lagrangian $L : \mathbb{R}^p \times \mathbb{R}^m_+ \to \mathbb{R}$, defined by

$$L(\beta; \lambda) = f(\beta) + \sum_{j=1}^{m} \lambda_j g_j(\beta). \tag{5.6}$$

The nonnegative weights $\lambda \geq 0$ are known as *Lagrange multipliers*; the purpose of the multiplier λ_j is to impose a penalty whenever the constraint $g_j(\beta) \leq 0$ is violated. Indeed, if we allow the multipliers to be chosen optimally, then we recover the original program (5.5), since

$$\sup_{\lambda \geq 0} L(\beta; \lambda) = \begin{cases} f(\beta) & \text{if } g_j(\beta) \leq 0 \text{ for all } j = 1, \ldots, m, \text{ and} \\ +\infty & \text{otherwise,} \end{cases} \tag{5.7}$$

and thus $f^* = \inf_{\beta \in \mathbb{R}^p} \sup_{\lambda \geq 0} L(\beta; \lambda)$. See Exercise 5.2 for further details on this equivalence.

For convex programs, the Lagrangian allows for the constrained problem (5.5) to be solved by reduction to an equivalent unconstrained problem. More specifically, under some technical conditions on f and $\{g_j\}$, the theory of Lagrange duality guarantees that there exists an optimal vector $\lambda^* \geq 0$ of Lagrange multipliers such that $f^* = \min_{\beta \in \mathbb{R}^p} L(\beta; \lambda^*)$. As a result, any optimum β^* of the problem (5.5), in addition to satisfying the feasibility constraints $g_j(\beta^*) \leq 0$, must also be a zero-gradient point of the Lagrangian, and hence satisfy the equation

$$0 = \nabla_\beta L(\beta^*; \lambda^*) = \nabla f(\beta^*) + \sum_{j=1}^{m} \lambda_j^* \nabla g_j(\beta^*). \tag{5.8}$$

When there is only a single constraint function g, this condition reduces to $\nabla f(\beta^*) = -\lambda^* \nabla g(\beta^*)$, and has an intuitive geometric interpretation, as shown in Figure 5.2. In particular, at the optimal solution β^*, the normal vector $\nabla f(\beta^*)$ to the contour line of f points in the opposite direction to the normal vector to the constraint curve $g(\beta) = 0$. Equivalently, the normal vector to the contour f lies at right angles to the tangent vector of the constraint. Consequently, if we start at the optimum β^* and travel along the tangent at $g(\beta) = 0$, we cannot decrease the value of $f(\beta)$ up to first order.

In general, the *Karush–Kuhn–Tucker* conditions relate the optimal Lagrange multiplier vector $\lambda^* \geq 0$, also known as the dual vector, to the optimal primal vector $\beta^* \in \mathbb{R}^p$:

(a) *Primal feasibility:* $g_j(\beta^*) \leq 0$ for all $j = 1, \ldots, m$.

(b) *Complementary slackness:* $\lambda_j^* g_j(\beta^*) = 0$ for all $j = 1, \ldots, m$.

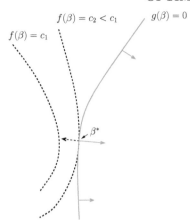

$f(\beta) = c_2 < c_1$ $g(\beta) = 0$

$f(\beta) = c_1$

β^*

Figure 5.2 *Illustration of the method of Lagrange multipliers. We are minimizing a function f subject to a single constraint $g(\beta) \leq 0$. At an optimal solution β^*, the normal vector $\nabla f(\beta^*)$ to the level sets of the cost function f points in the opposite direction to the normal vector $\nabla g(\beta^*)$ of the constraint boundary $g(\beta) = 0$. Consequently, up to first order, the value of $f(\beta^*)$ cannot be decreased by moving along the contour $g(\beta) = 0$.*

(c) *Lagrangian condition:* The pair (β^*, λ^*) satisfies condition (5.8).

These KKT conditions are necessary and sufficient for β^* to be a global optimum whenever the optimization problem satisfies a regularity condition known as *strong duality*. (See Exercise 5.4 for more details.) The complementary slackness condition asserts that the multiplier λ_j^* must be zero if the constraint $g_j(\beta) \leq 0$ is inactive at the optimum—that is, if $g_j(\beta^*) < 0$. Consequently, under complementary slackness, the Lagrangian gradient condition (5.8) guarantees that the normal vector $-\nabla f(\beta^*)$ lies in the positive linear span of the gradient vectors $\{\nabla g_j(\beta^*) \mid \lambda_j^* > 0\}$.

5.2.2 Nondifferentiable Functions and Subgradients

In practice, many optimization problems arising in statistics involve convex but nondifferentiable cost functions. For instance, the ℓ_1-norm $g(\beta) = \sum_{j=1}^{p} |\beta_j|$ is a convex function, but it fails to be differentiable at any point where at least one coordinate β_j is equal to zero. For such problems, the optimality conditions that we have developed—in particular, the first-order condition (5.3) and the Lagrangian condition (5.8)—are not directly applicable, since they involve gradients of the cost and constraint functions. Nonetheless, for convex functions, there is a natural generalization of the notion of gradient that allows for a more general optimality theory.

A basic property of differentiable convex functions is that the first-order tangent approximation always provides a lower bound. The notion of subgra-

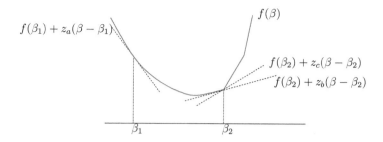

Figure 5.3 *A convex function* $f : \mathbb{R} \to \mathbb{R}$, *along with some examples of subgradients at* β_1 *and* β_2.

dient is based on a natural generalization of this idea. In particular, given a convex function $f : \mathbb{R}^p \to \mathbb{R}$, a vector $z \in \mathbb{R}^p$ is said to be a *subgradient* of f at β if

$$f(\beta') \geq f(\beta) + \langle z, \, \beta' - \beta \rangle \quad \text{for all } \beta' \in \mathbb{R}^p. \tag{5.9}$$

In geometric terms, the subgradient vector z is the normal to a (nonvertical) hyperplane that supports the epigraph of f. The set of all subgradients of f at β is called the *subdifferential*, denoted by $\partial f(\beta)$. Whenever f is differentiable at β, then the sub-differential reduces to a single vector—namely $\partial f(\beta) = \{\nabla f(\beta)\}$. At points of nondifferentiability, the subdifferential is a convex set containing all possible subgradients. For example, for the absolute value function $f(\beta) = |\beta|$, we have

$$\partial f(\beta) = \begin{cases} \{+1\} & \text{if } \beta > 0 \\ \{-1\} & \text{if } \beta < 0 \\ [-1, +1] & \text{if } \beta = 0. \end{cases} \tag{5.10}$$

We frequently write $z \in \text{sign}(\beta)$ to mean that z belongs to sub-differential of the absolute value function at β.

Figure 5.3 shows a function $f : \mathbb{R} \to \mathbb{R}$, and some examples of subgradients at the two points β_1 and β_2. At the point β_1, the function is differentiable and hence there is only one subgradient—namely, $f'(\beta_1)$. At the point β_2, it is not differentiable, and there are multiple subgradients; each one specifies a tangent plane that provides a lower bound on f.

How is this useful? Recall the convex optimization problem (5.5), and assume that one or more of the functions $\{f, g_j\}$ are convex but nondifferentiable. In this case, the zero-gradient Lagrangian condition (5.8) no longer makes sense. Nonetheless, again under mild conditions on the functions, the generalized KKT theory can still be applied using the modified condition

$$0 \in \partial f(\beta^*) + \sum_{j=1}^{m} \lambda_j^* \partial g_j(\beta^*), \tag{5.11}$$

in which we replace the gradients in the KKT condition (5.8) with subdifferentials. Since the subdifferential is a set, Equation (5.11) means that the all-zeros vector belongs to the sum of the subdifferentials.[2]

Example 5.1. Lasso and subgradients. As an example, suppose that we want to solve a minimization problem of the form (5.5) with a convex and differentiable cost function f, and a single constraint specified by $g(\beta) = \sum_{j=1}^{p} |\beta_j| - R$ for some positive constant R. Thus, the constraint $g(\beta) \leq 0$ is equivalent to requiring that β belongs to an ℓ_1-ball of radius R. Recalling the form of the subdifferential (5.10) for the absolute value function, condition (5.11) becomes

$$\nabla f(\beta^*) + \lambda^* z^* = 0, \qquad (5.12)$$

where the subgradient vector satisfies $z_j^* \in \text{sign}(\beta_j^*)$ for each $j = 1, \ldots, p$. When the cost function f is the squared error $f(\beta) = \frac{1}{2N} \|\mathbf{y} - \mathbf{X}\beta\|_2^2$, this condition is equivalent to Equation (2.6) from Chapter 2.   ◇

Example 5.2. Nuclear norm and subgradients. The *nuclear norm* is a convex function on the space of matrices. Given a matrix $\Theta \in \mathbb{R}^{m \times n}$ (where we assume $m \leq n$), it can always be decomposed in the form $\Theta = \sum_{j=1}^{m} \sigma_j u_j v_j^T$. where $\{u_j\}_{j=1}^{m}$ and $\{v_j\}_{j=1}^{m}$ are the (left and right) singular vectors, chosen to be orthonormal in $\mathbb{R}^m$ and $\mathbb{R}^n$, respectively, and the nonnegative numbers $\sigma_j \geq 0$ are the singular values. This is known as the *singular-value decomposition (SVD)* of Θ. The *nuclear norm* is the sum of the singular values—that is, $\|\Theta\|_\star = \sum_{j=1}^{m} \sigma_j(\Theta)$. Note that it is a natural generalization of the vector ℓ_1-norm, since for any (square) diagonal matrix, the nuclear norm reduces to the ℓ_1-norm of its diagonal entries. As we discuss in Chapter 7, the nuclear norm is useful for various types of matrix approximation and decomposition. The subdifferential $\partial \|\Theta\|_\star$ of the nuclear norm at Θ consists of all matrices of the form $\mathbf{Z} = \sum_{j=1}^{m} z_j u_j v_j^T$, where each for $j = 1, \ldots, m$, the scalar $z_j \in \text{sign}(\sigma_j(\Theta))$. We leave it as an exercise for the reader to verify this claim using the definition (5.9).   ◇

5.3   Gradient Descent

Thus far, we have seen various types of optimality conditions for different types of convex programs. We now turn to various classes of iterative algorithms for solving optimization problems. In this section, we focus on first-order algorithms, meaning methods that exploit only gradient (or subgradient) information, as opposed to information from higher-order gradients. First-order methods are particularly attractive for large-scale problems that arise in much of modern statistics.

---

[2]Here we define the sum of two subsets A and B of $\mathbb{R}^p$ as $A + B := \{\alpha + \beta \mid \alpha \in A, \beta \in B\}$.

5.3.1   Unconstrained Gradient Descent

We begin with the simplest case —namely, unconstrained minimization of a convex differentiable function $f : \mathbb{R}^p \to \mathbb{R}$. In this case, assuming that the global minimum is achieved, then a necessary and sufficient condition for optimality of $\beta^* \in \mathbb{R}^p$ is provided by the zero-gradient condition $\nabla f(\beta^*) = 0$. Gradient descent is an iterative algorithm for solving this fixed point equation: it generates a sequence of iterates $\{\beta^t\}_{t=0}^{\infty}$ via the update

$$\beta^{t+1} = \beta^t - s^t \nabla f(\beta^t), \quad \text{for } t = 0, 1, 2, \ldots, \tag{5.13}$$

where $s^t > 0$ is a stepsize parameter. This update has a natural geometric interpretation: by computing the gradient, we determine the direction of steepest descent $-\nabla f(\beta^t)$, and then walk in this direction for a certain amount determined by the stepsize s^t.

More generally, the class of *descent methods* is based on choosing a direction $\Delta^t \in \mathbb{R}^p$ such that $\langle \nabla f(\beta^t), \Delta^t \rangle < 0$, and then performing the update

$$\beta^{t+1} = \beta^t + s^t \Delta^t \quad \text{for } t = 0, 1, 2, \ldots. \tag{5.14}$$

In geometric terms, the inner product condition $\langle \nabla f(\beta^t), \Delta^t \rangle < 0$ means that the chosen direction Δ^t forms an angle of less than 90° with the direction of steepest descent. The gradient descent update (5.13) is a special case with $\Delta^t = -\nabla f(\beta^t)$. Other interesting choices include *diagonally-scaled gradient descent*: given a diagonal matrix $\mathbf{D}^t \succ \mathbf{0}$, it uses the descent direction $\Delta^t = -(\mathbf{D}^t)^{-1} \nabla f(\beta^t)$. This type of diagonal scaling is helpful when the function varies more rapidly along some coordinates than others. More generally, *Newton's method* is applicable to functions that are twice continuously differentiable, and is based on the descent direction

$$\Delta^t = - \left(\nabla^2 f(\beta^t) \right)^{-1} \nabla f(\beta^t), \tag{5.15}$$

where $\nabla^2 f(\beta^t)$ is the Hessian of f, assumed to be invertible. Newton's method is a second-order method, since it involves first and second derivatives. In particular, a Newton step (with stepsize one) amounts to exactly minimizing the second-order Taylor approximation to f at β^t. Under some regularity conditions, it enjoys a quadratic rate of convergence; however, computation of the Newton direction (5.15) is more expensive than first-order methods.

An important issue for all iterative algorithms, among them the gradient descent update (5.13), is how to choose the stepsize s^t. For certain problems with special structure, it can be shown that a constant stepsize (meaning $s^t = s$ for all iterations $t = 0, 1, \ldots$) will guarantee convergence; see Exercise 5.1 for an illustration. In general, it is *not sufficient* to simply choose a stepsize for which $f(\beta^{t+1}) < f(\beta^t)$; without some care, this choice may cause the algorithm to converge to a nonstationary point. Fortunately, there are various kinds of stepsize selection rules that are relatively simple, and have associated convergence guarantees:

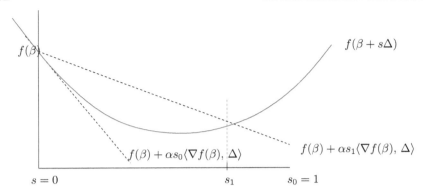

Figure 5.4 *Armijo rule or backtracking line search. Starting with step-size* $s_0 = 1$, *we repeatedly reduce* s *by a fraction* γ *until the condition* $f(\beta + s\Delta) \leq f(\beta) + \alpha s \langle \nabla f(\beta), \Delta \rangle$ *is satisfied. This is achieved here at* s_1.

- *Limited minimization rule:* choose the stepsize $s^t = \arg\min_{s \in [0,1]} f(\beta^t + s\Delta^t)$.
  Although this choice is very intuitive, it does require solving a one-dimensional optimization problem at each step.
- *Armijo or backtracking rule:* Given parameters $\alpha \in (0,1)$ and $\gamma \in (0,1)$ and an initial stepsize $s = 1$, perform the reduction $s \leftarrow \gamma s$ until the descent condition

$$f\left(\beta^t + s\Delta^t\right) \leq f(\beta^t) + \alpha s \left\langle \nabla f(\beta^t), \Delta^t \right\rangle \tag{5.16}$$

is met. In practice, the choices $\alpha = 0.5$ and $\gamma = 0.8$ are reasonable. The condition (5.16) can be interpreted as saying that we will accept a fraction α of the decrease in $f(\beta)$ that is predicted by linear extrapolation (Figure 5.4).

For convex functions, both of these stepsize choices, when combined with suitable choices of the descent directions $\{\Delta^t\}_{t=0}^{\infty}$, yield algorithms that are guaranteed to converge to a global minimum of the convex function f. See the bibliographic section on page 131 for further discussion.

5.3.2 *Projected Gradient Methods*

We now turn to gradient methods for problems that involve additional side constraints. In order to provide some geometric intuition for these methods, it is useful to observe that the gradient step (5.13) has the alternative representation

$$\beta^{t+1} = \arg\min_{\beta \in \mathbb{R}^p} \left\{ f(\beta^t) + \langle \nabla f(\beta^t), \beta - \beta^t \rangle + \frac{1}{2s^t} \|\beta - \beta^t\|_2^2 \right\}. \tag{5.17}$$

Thus, it can be viewed as minimizing the linearization of f around the current iterate, combined with a smoothing penalty that penalizes according to Euclidean distance.

This view of gradient descent—an algorithm tailored specifically for unconstrained minimization—leads naturally to the method of projected gradient descent, suitable for minimization subject to a constraint $\beta \in \mathcal{C}$:

$$\beta^{t+1} = \arg\min_{\beta \in \mathcal{C}} \left\{ f(\beta^t) + \langle \nabla f(\beta^t), \beta - \beta^t \rangle + \frac{1}{2s^t} \|\beta - \beta^t\|_2^2 \right\}. \qquad (5.18)$$

Equivalently, as illustrated in Figure 5.5, this method corresponds to taking a gradient step $\beta^t - s\nabla f(\beta^t)$, and then projecting the result back onto the convex constraint set $\mathcal{C}$. It is an efficient algorithm as long as this projection can be computed relatively easily. For instance, given an ℓ_1-ball constraint $\mathcal{C} = \{\beta \in \mathbb{R}^p \mid \|\beta\|_1 \le R\}$, this projection can be computed easily by a variant of soft thresholding, as we discuss in more detail later.

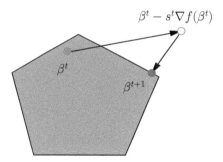

Figure 5.5 *Geometry of projected gradient descent. Starting from the current iterate β^t, it moves in the negative gradient direction to $\beta^t - s^t\nabla f(\beta^t)$, and then performs a Euclidean projection of the result back onto the convex constraint set $\mathcal{C}$ in order to obtain the next iterate β^{t+1}.*

5.3.3 Proximal Gradient Methods

Here we discuss a general class of methods that include projected gradient descent as a special case. As discussed previously, many objective functions f can be decomposed as a sum $f = g + h$, where g is convex and differentiable, and h is convex but nondifferentiable. Suppose that we would like to minimize such an objective function by a gradient-type algorithm. How do we deal with the nondifferentiability of the component h?

In order to see how this difficulty can be finessed, recall that an ordinary gradient step can be viewed as minimizing the combination of a local linear approximation to f combined with a quadratic smoothness term—in particular, see Equation (5.17). This perspective suggests the following strategy: form a local approximation to f by linearizing the differentiable component g, but leaving the nondifferentiable component fixed. This leads to the *generalized gradient update*, defined by

$$\beta^{t+1} = \arg\min_{\beta \in \mathbb{R}^p} \left\{ g(\beta^t) + \langle \nabla g(\beta^t), \beta - \beta^t \rangle + \frac{1}{2s^t} \|\beta - \beta^t\|_2^2 + h(\beta) \right\}, \quad (5.19)$$

where we have approximated the differentiable part g, but retained an exact form of the nondifferentiable component h.

The update (5.19) is closely related to the projected gradient descent update (5.18); in fact, it can be viewed as a Lagrangian analog. In order to make this connection explicit, we define the *proximal map* of a convex function h, a type of generalized projection operator:

$$\mathbf{prox}_h(z) := \arg\min_{\theta \in \mathbb{R}^p} \left\{ \frac{1}{2} \|z - \theta\|_2^2 + h(\theta) \right\}. \quad (5.20)$$

From this definition we immediately have the following relations:

(a) $\mathbf{prox}_{sh}(z) = \arg\min_{\theta \in \mathbb{R}^p} \left\{ \frac{1}{2s} \|z - \theta\|_2^2 + h(\theta) \right\}$.

(b) When

$$h(\theta) = I_{\mathcal{C}}(\theta) = \begin{cases} 0 & \text{if } \theta \in \mathcal{C}, \text{ and} \\ +\infty & \text{otherwise} \end{cases}$$

we have $\mathbf{prox}_h(z) = \arg\min_{\theta \in \mathcal{C}} \|z - \theta\|_2^2$, corresponding to the usual Euclidean projection onto the set $\mathcal{C}$.

(c) If $h(\theta) = \lambda \|\theta\|_1$, then $\mathbf{prox}_h(z) = \mathcal{S}_\lambda(z)$, the element-wise soft-thresholded version of z. See Example 5.3 below.

As we show in Exercise 5.7, it follows that the update (5.19) has the equivalent representation

$$\beta^{t+1} = \mathbf{prox}_{s^t h} \left(\beta^t - s^t \nabla g(\beta^t) \right). \quad (5.21)$$

Similarly, it is easy to see that the proximal-gradient update

$$\beta^{t+1} = \mathbf{prox}_{I_{\mathcal{C}}} \left(\beta^t - s^t \nabla g(\beta^t) \right) \quad (5.22)$$

is exactly the projected gradient step (5.18).

The updates (5.21) will be computationally efficient as long as the proximal map is relatively easy to compute. For many problems that arise in statistics— among them the ℓ_1-norm, group-lasso ℓ_2 norm, and nuclear norms—the proximal map (5.20) can be computed quite cheaply. Typically the update (5.21) is better suited to statistical problems that impose regularization via a penalty, as opposed to a constraint of the form $h(\theta) \leq R$.

Example 5.3. Proximal gradient descent for ℓ_1-penalty. Suppose that the nondifferentiable component is a (scaled) ℓ_1 penalty, say $h(\theta) = \lambda \|\theta\|_1$. With this choice of h, proximal gradient descent with stepsize s^t at iteration t consists of two very simple steps:

1. First, take a gradient step $z = \beta^t - s^t \nabla g(\beta^t)$.

2. Second, perform elementwise soft-thresholding $\beta^{t+1} = \mathcal{S}_{s^t \lambda}(z)$.

In detail, the proximal map (5.21) is given by

$$\begin{aligned}
\mathbf{prox}_{sh}(z) &= \arg\min_{\theta \in \mathbb{R}^p} \left\{ \frac{1}{2s} \|z - \theta\|_2^2 + \lambda \|\theta\|_1 \right\} \\
&= \arg\min_{\theta \in \mathbb{R}^p} \left\{ \frac{1}{2} \|z - \theta\|_2^2 + \lambda s \|\theta\|_1 \right\}.
\end{aligned} \tag{5.23}$$

This optimization problem has an explicit closed-form solution; in particular, since the objective function decouples across coordinates as

$$\frac{1}{2} \|z - \theta\|_2^2 + \lambda s \|\theta\|_1 = \sum_{j=1}^{p} \left\{ \frac{1}{2}(z_j - \theta_j)^2 + \lambda s |\theta_j| \right\}, \tag{5.24}$$

we can solve the p-dimensional problem by solving each of the univariate problems separately. We leave it as an exercise for the reader to verify the solution is obtained by applying the *soft thresholding operator* $\mathcal{S}_\tau : \mathbb{R}^p \to \mathbb{R}^p$ with coordinates

$$[\mathcal{S}_\tau(z)]_j = \text{sign}(z_j)(|z_j| - \tau)_+, \tag{5.25}$$

with the threshold choice $\tau = s\lambda$. (Here we use $(x)_+$ as a shorthand for $\max\{x, 0\}$.) $\diamond$

Example 5.4. Proximal gradient descent for nuclear norm penalty. As a second illustration, suppose that h is λ times the nuclear norm. As previously introduced in Example 5.2, the nuclear norm is a real-valued function on the space of $m \times n$ matrices, given by $\|\Theta\|_\star = \sum_{j=1}^{m} \sigma_j(\Theta)$, where $\{\sigma_j(\Theta)\}$ are the singular values of Θ. With this choice of h, the generalized projection operator (5.20) takes the form

$$\mathbf{prox}_{sh}(\mathbf{Z}) = \arg\min_{\Theta \in \mathbb{R}^{m \times n}} \left\{ \frac{1}{2s} \|\mathbf{Z} - \Theta\|_F^2 + \lambda \|\Theta\|_\star \right\}. \tag{5.26}$$

Here the Frobenius norm $\|\mathbf{Z} - \Theta\|_F^2 = \sum_{j=1}^{m} \sum_{k=1}^{n} (Z_{jk} - \Theta_{jk})^2$ is simply the usual Euclidean norm applied to the entries of the matrices. Although this proximal map (5.26) is no longer separable, it still has a relatively simple solution. Indeed, as we explore in Exercise 5.8, the update $\Pi_{s,h}(\mathbf{Z})$ is obtained by computing the singular value decomposition of $\mathbf{Z}$, and then soft-thresholding its singular values. $\diamond$

Nesterov (2007) provides sufficient conditions for the convergence of the updates (5.21) when applied to a composite objective function $f = g + h$. Suppose that the component g is continuously differentiable with a Lipschitz gradient, meaning that there is some constant L such that

$$\|\nabla g(\beta) - \nabla g(\beta')\|_2 \leq L \|\beta - \beta'\|_2 \quad \text{for all } \beta, \beta' \in \mathbb{R}^p. \tag{5.27}$$

Under this condition and with a constant stepsize $s^t = s \in (0, 1/L]$, it can be shown that there is a constant C, independent of the iteration number, such that the updates (5.21) satisfy

$$f(\beta^t) - f(\beta^*) \leq \frac{C}{t+1} \, \|\beta^t - \beta^*\|_2 \quad \text{for all } t = 1, 2, \ldots, \tag{5.28}$$

where β^* is an optimal solution. In words, the difference between the value $f(\beta^t)$ of the tth iterate and the optimal value $f(\beta^*)$ decreases at the rate $\mathcal{O}(1/t)$. This rate is known as *sublinear convergence*, and is guaranteed for any fixed stepsize in the interval $(0, 1/L]$. Such a choice requires an upper bound on the Lipschitz constant L, which may or may not be available. In practice, the Armijo rule also yields the same rate (5.28). (See Figure 5.6.)

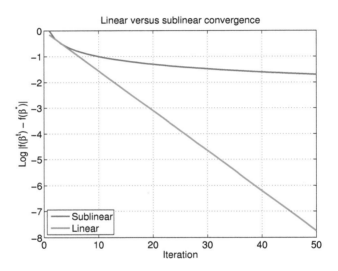

Figure 5.6 *Plot of* $\log |f(\beta^t) - f(\beta^*)|$ *versus the iteration number* t, *comparing a sublinear convergence rate* (5.28) *with a linear or geometric convergence rate* (5.30). *For an algorithm with geometric convergence, the error decay on this logarithmic scale is linear with a negative slope.*

Significantly faster rates are possible if the objective function has additional structure. For instance, suppose that in addition to having a Lipschitz continuous gradient (5.27), the differentiable component g is *strongly convex*, meaning that there exists some $\gamma > 0$ such that

$$g(\beta + \Delta) - g(\beta) - \langle \nabla g(\beta), \Delta \rangle \geq \gamma^2 \|\Delta\|_2^2 \quad \text{for all } \beta, \Delta \in \mathbb{R}^p. \tag{5.29}$$

This condition guarantees that g has at least as much curvature as the quadratic function $\beta \mapsto \gamma^2 \|\beta\|_2^2$ in all directions. Under conditions (5.27) and (5.29), it can be shown that with a constant stepsize $s \in (0, 1/L]$, the updates (5.21) will achieve a *linear or geometric rate* of convergence, meaning that there exists a positive constant C and contraction factor $\kappa \in (0, 1)$ such that

$$f(\beta^t) - f(\beta^*) \leq C \kappa^t \|\beta^0 - \beta^*\|_2 \quad \text{for all } t = 1, 2, \ldots, \tag{5.30}$$

Thus, under the additional strong convexity condition, the error $f(\beta^t) - f(\beta^*)$ is guaranteed to contract at a geometric rate specified by $\kappa \in (0, 1)$. See Figure 5.6 for an illustration of the difference between this linear rate and the earlier sublinear rate (5.28).

Example 5.5. Proximal gradient for lasso. For the lasso, we have

$$g(\boldsymbol{\beta}) = \frac{1}{2N} \|\mathbf{y} - \mathbf{X}\beta\|_2^2 \text{ and } h(\beta) = \lambda \|\beta\|_1,$$

so that the proximal gradient update (5.21) takes the form

$$\beta^{t+1} = \mathcal{S}_{s^t \lambda} \left(\beta^t + s^t \frac{1}{N} \mathbf{X}^T (\mathbf{y} - \mathbf{X}\beta^t) \right). \tag{5.31}$$

Note that this has a very similar form to the coordinate descent update (see Section 5.4), especially if we take the stepsize $s = 1$ and assume that the predictors are standardized. Then both procedures operate on the same quantities, one in a cyclical manner and the other (proximal gradients) in a simultaneous manner on all coordinates. It is not clear which is a more effective approach. The coordinate descent procedure can exploit sparsity of the coefficient vector and doesn't need to worry about step-size optimization, while the proximal gradient may gain efficiency by moving all parameters at the same time. It may also have speed advantages in problems where the multiplication of a vector by both $\mathbf{X}$ and $\mathbf{X}^T$ can be done quickly, for example by a fast Fourier transform. The Lipschitz constant L here is the maximum eigenvalue of $\mathbf{X}^T\mathbf{X}/N$; one can use a fixed stepsize in $(0, 1/L]$ or a form of backtracking step selection. We compare these in a numerical example in Section 5.5. ◇

5.3.4   Accelerated Gradient Methods

In this section, we discuss a class of accelerated gradient methods due to Nesterov (2007). Suppose that we have a convex differentiable function f, and recall the standard gradient step (5.13). For certain objective functions, this update may exhibit an undesirable type of "zig-zagging" behavior from step to step, which could conceivably slow down convergence. With the motivation of alleviating this drawback, Nesterov (2007) proposed the class of accelerated gradient methods that use weighted combinations of the current and previous gradient directions.

   In more detail, the accelerated gradient method involves a pair of sequences $\{\beta^t\}_{t=0}^\infty$ and $\{\theta^t\}_{t=0}^\infty$, and some initialization $\beta^0 = \theta^0$. For iterations

$t = 0, 1, 2, \ldots$, the pair is then updated according to the recursions

$$\beta^{t+1} = \theta^t - s^t \nabla f(\theta^t), \quad \text{and} \tag{5.32a}$$

$$\theta^{t+1} = \beta^{t+1} + \frac{t}{t+3}(\beta^{t+1} - \beta^t). \tag{5.32b}$$

For non-smooth functions f that have the "smooth plus non-smooth" decomposition $g + h$, Nesterov's acceleration scheme can be combined with the proximal gradient update: in particular, we replace the ordinary gradient step (5.32a) with the update

$$\beta^{t+1} = \mathbf{prox}_{s^t h}\left(\theta^t - s^t \nabla g(\theta^t)\right). \tag{5.33}$$

In either case, the stepsize s^t is either fixed to some value, or chosen according to some type of backtracking line search.

Example 5.6. Proximal gradient descent with momentum. Let us consider the combination of proximal gradient steps with the acceleration scheme in application to the ℓ_1-regularized lasso program. Recalling the form (5.31) of the composite gradient update, we see that the accelerated scheme consists of the updates

$$\beta^{t+1} = \mathcal{S}_{s^t \lambda}\left(\theta^t + s^t \tfrac{1}{N}\mathbf{X}^T(\mathbf{y} - \mathbf{X}\theta^t)\right)$$

$$\theta^{t+1} = \beta^{t+1} + \frac{t}{t+3}\left(\beta^{t+1} - \beta^t\right). \tag{5.34a}$$

This algorithm for the lasso is essentially equivalent, modulo some minor differences in the acceleration weights, to the Fast Iterative Soft-thresholding Algorithm (FISTA) of Beck and Teboulle (2009).

To investigate how well this works, we generated data from a regression problem with $N = 1000$ observations and $p = 500$ features. The features x_{ij} are standard Gaussian having pairwise correlation 0.5. Twenty of the 500 coefficients β_j were nonzero, each distributed as standard Gaussian variates and chosen at random locations between 1 and p. Figure 5.7 shows the performance of the generalized gradient and Nesterov's method, for two different values of the regularization parameter λ. We tried the algorithms using a fixed value of the step-size s^t (equal to the reciprocal of the largest eigenvalue of $\frac{1}{N}\mathbf{X}^T\mathbf{X}$). We also tried the approximate backtracking line search for $s^t \in [0, 0.5]$. We see that Nesterov's momentum method yields substantial speedups, over the generalized gradient, and backtracking is faster than the fixed stepsize choice. In the latter comparison this does not even take into account the cost of computing the largest eigenvalue of $\frac{1}{N}\mathbf{X}^T\mathbf{X}$: backtracking can speed up the computation by allowing a larger stepsize to be used when it is appropriate. Note that we are simply counting the number of iterations, rather than measuring the total elapsed time; however the Nesterov momentum steps are only slightly more costly than the generalized gradient steps. We also note that the relative error and hence the iterates $f(\beta^t)$ are not strictly monotone decreasing for Nesterov's momentum method.

$\diamondsuit$

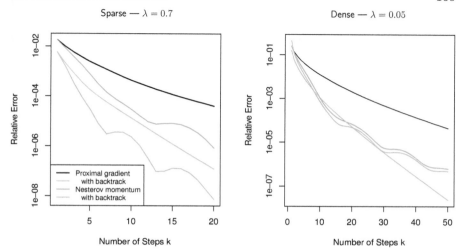

Figure 5.7 *Performance of the generalized gradient and Nesterov momentum methods for a sample lasso problem described in the text. The vertical axis shows the error measure* $[f(\beta^t) - f(\beta^*)]/f(\beta^*)$, *where* β^* *is the minimizer, and* β^t *is the solution after* t *steps. On the left, the solution* β^* *is sparse with just* 20 *of the* 500 *coefficients not equal to zero; on the right,* 237 *of the coefficients are nonzero.*

In computational terms, the momentum updates (5.32) and (5.33) only involve slightly more work than an ordinary gradient update. Nonetheless, Nesterov (2007) proves that the change yields a significant improvement in convergence rates: in particular, whenever g satisfies the Lipschitz condition (5.27), then there is a constant $C > 0$ such that the iterates satisfy

$$f(\beta^t) - f(\beta^*) \leq \frac{C}{(t+1)^2} \|\beta^0 - \beta^*\|_2. \tag{5.35}$$

Consequently, the error $f(\beta^t) - f(\beta^*)$ decreases at the rate $\mathcal{O}(1/t^2)$, as opposed to the slower $\mathcal{O}(1/t)$ rate of a nonaccelerated method (see Equation (5.28)). When g is strongly convex (5.29), the accelerated gradient method again enjoys a geometric rate of convergence (5.30), although with a smaller contraction factor κ. More precisely, the nonaccelerated method converges with a contraction factor determined by the condition number of g, whereas the accelerated variant converges according to the *square root* of this condition number.

5.4 Coordinate Descent

Certain classes of problems, among them the lasso and variants, have an additional separability property that lends itself naturally to a coordinate minimization algorithm. *Coordinate descent* is an iterative algorithm that updates from β^t to β^{t+1} by choosing a single coordinate to update, and then performing a univariate minimization over this coordinate. More precisely, if coordinate k is chosen at iteration t, then the update is given by

$$\beta_k^{t+1} = \arg\min_{\beta_k} f\left(\beta_1^t, \beta_2^t, \ldots, \beta_{k-1}^t, \beta_k, \beta_{k+1}^t, \ldots, \beta_p^t\right), \qquad (5.36)$$

and $\beta_j^{t+1} = \beta_j^t$ for $j \neq k$. A typical choice would be to cycle through the coordinates in some fixed order. This approach can also be generalized to *block coordinate descent*, in which the variables are partitioned into non-overlapping blocks (as in the group lasso), and we perform minimization over a single block at each round.

5.4.1 Separability and Coordinate Descent

When does this procedure converge to the global minimum of a convex function? One sufficient (but somewhat restrictive) condition is that f be continuously differentiable and strictly convex in each coordinate. However, the use of various statistical regularizers leads to optimization problems that need not be differentiable. For such cases, more care is required when using coordinate minimization, because, as we discuss below, it can become "stuck" at nonoptimal points. One form of problem structure that ensures good behavior of coordinate minimization is a type of separability condition. In particular, suppose that the cost function f has the additive decomposition

$$f(\beta_1, \ldots \beta_p) = g(\beta_1, \ldots \beta_p) + \sum_{j=1}^{p} h_j(\beta_j), \qquad (5.37)$$

where $g : \mathbb{R}^p \to \mathbb{R}$ is differentiable and convex, and the univariate functions $h_j : \mathbb{R} \to \mathbb{R}$ are convex (but not necessarily differentiable). An important example of this problem structure is the standard lasso program (2.5), with $g(\beta) = \frac{1}{2N}\|\mathbf{y} - \mathbf{X}\beta\|_2^2$ and $h_j(\beta_j) = \lambda \cdot |\beta_j|$. Tseng (1988, 2001) shows that for any convex cost function f with the separable structure (5.37), the coordinate descent Algorithm (5.36) is guaranteed to converge to the global minimizer. The key property underlying this result is the separability of the nondifferentiable component $h(\beta) = \sum_{j=1}^{p} h_j(\beta_j)$, as a sum of functions of each individual parameter. This result implies that coordinate descent is a suitable algorithm for the lasso as well as certain other problems discussed in this book. In contrast, when the nondifferentiable component h is *not* separable, coordinate descent is no longer guaranteed to converge. Instead, it is possible to create problems for which it will become "stuck," and fail to reach the global minimum.

Example 5.7. Failure of coordinate descent. As an illustration, we consider an instance of a problem that violates (5.37)—the fused lasso, discussed in Section 4.5. Here the nondifferentiable component takes the form $h(\beta) = \sum_{j=1}^{p} |\beta_j - \beta_{j-1}|$. Figure 5.8 illustrates the difficulty. We created a fused lasso problem with 100 parameters, with the solutions for two of the parameters, $\beta_{63} = \beta_{64} \approx -1$. The left and middle panels show slices of the function f varying β_{63} and β_{64}, with the other parameters set to the global minimizers. We see that the coordinate-wise descent algorithm has got stuck in a corner of the response surface, and is stationary under single-coordinate moves. In order to advance to the minimum, we have to move both β_{63} and β_{64} together.

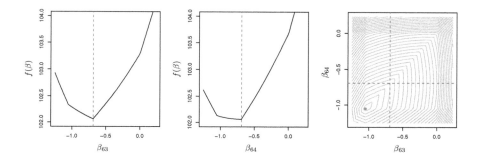

Figure 5.8 *Failure of coordinate-wise descent in a fused lasso problem with 100 parameters. The optimal values for two of the parameters, β_{63} and β_{64}, are both -1.05, as shown by the dot in the right panel. The left and middle panels show slices of the objective function f as a function of β_{63} and β_{64}, with the other parameters set to the global minimizers. The coordinate-wise minimizer over both β_{63} and β_{64} (separately) is -0.69, rather than -1.05. The right panel shows contours of the two-dimensional surface. The coordinate-descent algorithm is stuck at the point $(-0.69, -0.69)$. Despite being strictly convex, the surface has corners, in which the coordinate-wise procedure can get stuck. In order to travel to the minimum we have to move both β_{63} and β_{64} together.*

$\diamondsuit$

Tseng (2001) gives a more general and intuitive condition for convergence of coordinate descent, one which depends on the behavior of the directional derivatives of the cost function f. For a given direction $\Delta \in \mathbb{R}^p$, the lower directional derivative at β is given by

$$f'(\beta; \Delta) := \liminf_{s \downarrow 0} \frac{f(\beta + s\Delta) - f(\beta)}{s}. \tag{5.38}$$

In rough terms, a coordinate descent algorithm only gains information about directions of the form $e^j = (0, 0, \ldots 0, e_j, 0, \ldots, 0)$ for some $e_j \in \mathbb{R}$. Therefore, suppose that the coordinate descent algorithm reaches a point β for which

$$f'(\beta; e^j) \geq 0 \quad \text{for all } j = 1, \ldots, p, \text{ and coordinate vectors } e^j. \tag{5.39}$$

At any such point, there are no coordinate directions that will further reduce the function value. Therefore, we require that any β satisfying the condition (5.39) also satisfies $f'(\beta; \Delta) \geq 0$ for all directions $\Delta \in \mathbb{R}^p$. Tseng (2001) calls this condition *regularity*. It rules out a situation like that of Figure 5.8, in which moves along all coordinate directions fail to decrease the criterion, but an oblique move does improve the criterion. As a side-note, it is worth observing that separability of the nondifferentiable component of the objective function implies regularity, but that there are nondifferentiable and nonseparable functions that are still regular. An example is the function

$$h(\beta_1, \ldots, \beta_p) = |\beta|^T \mathbf{P} |\beta| = \sum_{j,k=1}^{p} |\beta_j| P_{jk} |\beta_k|, \qquad (5.40)$$

where $\mathbf{P}$ is a symmetric positive definite matrix.

5.4.2 Linear Regression and the Lasso

Recall the optimization problem (2.5) that underlies the lasso estimator. As discussed in Chapter 2, the optimality conditions for this problem are

$$-\frac{1}{N} \sum_{i=1}^{N} (y_i - \beta_0 - \sum_{k=1}^{p} x_{ik} \beta_k) x_{ij} + \lambda s_j = 0, \qquad (5.41)$$

where $s_j \in \text{sign}(\beta_j)$ for $j = 1, 2, \ldots, p$. The coordinate descent procedure simply solves these equations in a cyclical fashion, iterating over $j = 1, 2, \ldots, p, 1, 2, \ldots$.

Since the intercept β_0 is typically not penalized, we can center both the response y_i and the covariate vectors x_i by their means, and then omit the intercept in the calculations of the other β_j. (Of course, as in OLS, the intercept is calculated at the end using the $\widehat{\beta}_0 = \bar{y} - \sum_{k=1}^{p} \bar{x}_k \widehat{\beta}_k$.) To simplify matters, we define the *partial residual* $r_i^{(j)} = y_i - \sum_{k \neq j} x_{ik} \widehat{\beta}_k$, which removes from the outcome the current fit from all but the jth predictor. Then the solution for $\widehat{\beta}_j$ satisfies

$$\widehat{\beta}_j = \frac{\mathcal{S}_\lambda \left(\frac{1}{N} \sum_{i=1}^{N} r_i^{(j)} x_{ij} \right)}{\frac{1}{N} \sum_{i=1}^{N} x_{ij}^2}, \qquad (5.42)$$

whereas before $\mathcal{S}_\lambda(\theta) = \text{sign}(\theta)(|\theta| - \lambda)_+$ is the soft-thresholding operator. If in addition to centering, the variables are standardized to have unit variance (typically a good idea, especially if the variables are in different units), then the update has the particularly succinct form

$$\widehat{\beta}_j = \mathcal{S}_\lambda(\tilde{\beta}_j), \qquad (5.43)$$

where $\tilde{\beta}_j$ is the simple linear regression coefficient of the partial residual on variable j. If instead we have an elastic net penalty $(1 - \alpha)\beta_j^2/2 + \alpha|\beta_j|$, the update (5.42) becomes

$$\widehat{\beta}_j = \frac{\mathcal{S}_{\alpha\lambda}\left(\frac{1}{N}\sum_{i=1}^N r_i^{(j)}x_{ij}\right)}{\frac{1}{N}\sum_{i=1}^N x_{ij}^2 + (1 - \alpha)\lambda}, \tag{5.44}$$

or in the standardized case

$$\widehat{\beta}_j = \frac{\mathcal{S}_{\alpha\lambda}(\tilde{\beta}_j)}{1 + (1 - \alpha)\lambda}. \tag{5.45}$$

There are a number of strategies for making these operations efficient. For ease of notation we assume that the predictors are standardized to have mean zero and variance one; for nonstandardized data, the steps are similar.

Partial residuals. Note that we can write $r_i^{(j)} = y_i - \sum_{k \neq j} x_{ik}\widehat{\beta}_k = r_i + x_{ij}\widehat{\beta}_j$, where r_i denotes the current residual for observation i. Since the vectors $\{\mathbf{x}_j\}_{j=1}^p$ are standardized, we can write

$$\frac{1}{N}\sum_{i=1}^N x_{ij}r_i^{(j)} = \frac{1}{N}\sum_{i=1}^N x_{ij}r_i + \widehat{\beta}_j, \tag{5.46}$$

a representation that reveals the computational efficiency of coordinate descent. Many coefficients are zero and remain so after thresholding, and so nothing needs to be changed. The primary cost arises from computing the sum in Equation (5.46), which requires $\mathcal{O}(N)$ operations. On the other hand, if a coefficient does change after the thresholding, r_i is changed in $\mathcal{O}(N)$ and the step costs $\mathcal{O}(2N)$. A full cycle through all p variables costs $\mathcal{O}(pN)$ operations. Friedman et al. (2010b) refer to this as *naive updating*, since it works directly with the inner products of the data.

Covariance updating. Naive updating is generally less efficient than *covariance updating* when $N \gg p$ and N is large. Up to a factor $1/N$, we can write the first term on the right of expression (5.46)

$$\sum_{i=1}^N x_{ij}r_i = \langle \mathbf{x}_j, \mathbf{y} \rangle - \sum_{k \,|\, |\widehat{\beta}_k| > 0} \langle \mathbf{x}_j, \mathbf{x}_k \rangle \widehat{\beta}_k. \tag{5.47}$$

In this approach, we compute inner products of each feature with $\mathbf{y}$ initially, and then each time a new feature $\mathbf{x}_k$ enters the model for the first time, we compute and store its inner product with all the rest of the features, requiring $\mathcal{O}(Np)$ operations. We also store the p gradient components (5.47). If one of the coefficients currently in the model changes, we can update each gradient in $\mathcal{O}(p)$ operations. Hence with k nonzero terms in the model, a complete cycle costs $\mathcal{O}(pk)$ operations if no new variables become nonzero, and costs $\mathcal{O}(Np)$ for each new variable entered. Importantly, each step does not require making $\mathcal{O}(N)$ calculations.

Warm starts. Typically one wants a sequence of lasso solutions, say for a decreasing sequence of values $\{\lambda_\ell\}_0^L$. It is easy to see that the largest value that we need consider is

$$\lambda_0 = \tfrac{1}{N} \max_j |\langle \mathbf{x}_j, \mathbf{y} \rangle|, \tag{5.48}$$

since any value larger would yield an empty model. One strategy, as employed by the R package `glmnet`, is to create a sequence of values $\{\lambda_\ell\}_{\ell=0}^L$ decreasing from λ_0 down to $\lambda_L = \epsilon\lambda_0 \approx 0$ on a log scale. The solution $\widehat{\beta}(\lambda_\ell)$ is typically a very good warm start for the solution $\widehat{\beta}(\lambda_{\ell+1})$. Likewise the number of nonzero elements tends to increase slowly with ℓ, starting at zero at $\ell = 0$. Doubling the number $L = 100$ to say $2L$ does not double the compute time, since the warm starts are much better, and fewer iterations are needed each time.

Active-set convergence. After a single iteration through the set of p variables at a new value λ_ℓ, starting from the warm start $\widehat{\beta}(\lambda_{\ell-1})$, we can define the active set $\mathcal{A}$ to index those variables with nonzero coefficients at present. The idea is to iterate the algorithm using only the variables in $\mathcal{A}$. Upon convergence, we do a pass through all the omitted variables. If they all pass the simple exclusion test $\frac{1}{N}|\langle \mathbf{x}_j, \mathbf{r} \rangle| < \lambda_\ell$, where $\mathbf{r}$ is the current residual, we have the solution for the entire set of p variables. Those that fail are included in $\mathcal{A}$ and the process is repeated. In practice we maintain an *ever-active* set— any variable that had a nonzero coefficient somewhere along the path until present is kept in $\mathcal{A}$.

Strong-set convergence. Similar to the above, we identify a subset of variables likely to be candidates for the active set. Let $\mathbf{r}$ be the residual at $\widehat{\beta}(\lambda_{\ell-1})$, and we wish to compute the solution at λ_ℓ. Define the strong set $\mathcal{S}$ as

$$\mathcal{S} = \{j \mid |\tfrac{1}{N}\langle \mathbf{x}_j, \mathbf{r} \rangle| > \lambda_\ell - (\lambda_{\ell-1} - \lambda_\ell)\}. \tag{5.49}$$

We now compute the solution restricting attention to *only* the variables in $\mathcal{S}$. Apart from rare exceptions, the strong set will cover the optimal active set. Strong rules are extremely useful, especially when p is very large (in the 100Ks or millions). We discuss them in some detail in Section 5.10.

Sparsity. The main computational operation in all the above is an inner-product of a pair of N-vectors, at least one of which is a column of the design matrix $\mathbf{X}$. If $\mathbf{X}$ is sparse, we can compute these inner products efficiently. An example is document classification, where often the feature vector follows the so-called "bag-of-words" model. Each document is scored for the presence/absence of each of the words in the entire dictionary under consideration (sometimes counts are used, or some transformation of counts). Since most words are absent, the feature vector for each document is mostly zero, and so the entire matrix is mostly zero. Such matrices can be stored efficiently in *sparse-column format*, where we store only the nonzero entries and the coordinates where they occur. Now when we compute inner products, we sum only over the nonzero entries.

Penalty strength. The default formulation applies the same penalty parameter λ to each term in the model. It is a simple matter to include a relative penalty strength $\gamma_j \geq 0$ per variable, making the overall penalty

$$\lambda \sum_{j=1}^{p} \gamma_j P_\alpha(\beta_j). \tag{5.50}$$

This allows for some γ_j to be zero, which means those variables are always in the model, unpenalized.

Parameter bounds. Coordinate descent also makes it easy to set upper and lower bounds on each parameter:

$$L_j \leq \beta_j \leq U_j, \tag{5.51}$$

where typically $-\infty \leq L_j \leq 0 \leq U_j \leq \infty$. For example, we sometimes want to constrain all coefficients to be nonnegative. One simply computes the coordinate update, and if the parameter violates the bound, it is set to the closest boundary.

5.4.3 Logistic Regression and Generalized Linear Models

Here we move from squared-error loss to other members of the exponential family—the so-called *generalized linear models*. For simplicity, we focus on the most prominent (nonlinear) member of this class—namely, logistic regression. In logistic regression, the response is binary, and can be modeled as a class label G taking the values -1 or 1. The standard logistic model represents the class probabilities as a linear model in the log-odds

$$\log \frac{\Pr(G = -1 \mid x)}{\Pr(G = 1 \mid x)} = \beta_0 + x^T \beta. \tag{5.52}$$

See Section 3.2 for more detail.

We consider fitting this model by regularized maximum (binomial) likelihood. Introducing the shorthand notation $p(x_i; \beta_0, \beta) = \Pr(G = 1 | x_i)$ for the probability (5.52) of observation i, we maximize the penalized log-likelihood

$$\frac{1}{N} \sum_{i=1}^{N} \{ I(g_i = 1) \log p(x_i; \beta_0, \beta) + I(g_i = -1) \log(1 - p(x_i; \beta_0, \beta)) \} - \lambda P_\alpha(\beta). \tag{5.53}$$

Denoting $y_i = I(g_i = -1)$, the log-likelihood part of (5.53) can be written in the more explicit form

$$\ell(\beta_0, \beta) = \frac{1}{N} \sum_{i=1}^{N} \left[y_i \cdot (\beta_0 + x_i^T \beta) - \log(1 + e^{\beta_0 + x_i^T \beta}) \right], \tag{5.54}$$

which corresponds to a concave function of the parameters. By way of background, the Newton algorithm for maximizing the (unpenalized) log-likelihood (5.54) amounts to iteratively reweighted least squares. Hence, if the current estimates of the parameters are $(\widetilde{\beta}_0, \widetilde{\beta})$, we form a second-order Taylor expansion about current estimates. In terms of the shorthand $\widetilde{p}(x_i) = p(x_i; \widetilde{\beta}_0, \widetilde{\beta})$, and $w_i = \widetilde{p}(x_i)(1 - \widetilde{p}(x_i))$, this Taylor expansion leads to the quadratic objective function

$$\ell_Q(\beta_0, \beta) = -\frac{1}{2N} \sum_{i=1}^{N} w_i(z_i - \beta_0 - x_i^T \beta)^2 + C(\widetilde{\beta}_0, \widetilde{\beta})^2. \tag{5.55}$$

where $z_i = \widetilde{\beta}_0 + x_i^T \widetilde{\beta} + \frac{y_i - \widetilde{p}(x_i)}{\widetilde{p}(x_i)(1 - \widetilde{p}(x_i))}$ is the current working response. The Newton update is obtained by minimizing ℓ_Q, which is a simple weighted-least-squares problem. In order to solve the regularized problem, one could apply coordinate descent directly to the criterion (5.53). A disadvantage of this approach is that the optimizing values along each coordinate are not explicitly available and require a line search. In our experience, it is better to apply coordinate descent to the quadratic approximation, resulting in a nested algorithm. For each value of λ, we create an outer loop which computes the quadratic approximation ℓ_Q about the current parameters $(\widetilde{\beta}_0, \widetilde{\beta})$. Then we use coordinate descent to solve the penalized weighted least-squares problem

$$\underset{(\beta_0, \beta) \in \mathbb{R}^{p+1}}{\text{minimize}} \left\{ -\ell_Q(\beta_0, \beta) + \lambda P_\alpha(\beta) \right\}. \tag{5.56}$$

By analogy with Section 5.3.3, this is known as a generalized Newton algorithm, and the solution to the minimization problem (5.56)) defines a *proximal Newton map* (see the paper (Lee et al. 2014) for details). Overall, the procedure consists of a sequence of nested loops:

OUTER LOOP: Decrement λ.

MIDDLE LOOP: Update the quadratic approximation ℓ_Q using the current parameters $(\widetilde{\beta}_0, \widetilde{\beta})$.

INNER LOOP: Run the coordinate descent algorithm on the penalized weighted-least-squares problem (5.56).

When $p \gg N$, one cannot run λ all the way to zero, because the saturated logistic regression fit is undefined (parameters wander off to $\pm\infty$ in order to achieve probabilities of 0 or 1). Also, the Newton algorithm is not guaranteed to converge without step-size optimization (Lee, Lee, Abneel and Ng 2006). The `glmnet` program does not implement any checks for divergence; this would slow it down, and when used as recommended, it does not seem to be necessary. We have a closed form expression for the starting solutions, and each subsequent solution is warm-started from the previous close-by solution, which generally makes the quadratic approximations very accurate. We have not encountered any divergence problems so far.

The `glmnet` package generalizes this procedure to other GLMs, such as multiclass logistic regression, the Poisson log-linear model and Cox's proportional hazards model for survival data. More details are given in Chapter 3. The speed of this procedure is studied in Section 5.5.

5.5   A Simulation Study

Both the coordinate descent algorithm and Nesterov's composite gradient method are simple and computationally efficient approaches for solving the lasso. How do they compare in terms of computational cost per iteration? If (at a given iteration) the current iterate β^t has k nonzero coefficients, each pass of coordinate descent over all p predictors (using naive updating) takes $\mathcal{O}(pN + kN)$ operations. On the other hand, the generalized gradient update (5.31) requires $\mathcal{O}(kN)$ operations to compute the matrix-vector product $\mathbf{X}\beta$, and then $\mathcal{O}(pN)$ to compute the product $\mathbf{X}^T(\mathbf{y} - \mathbf{X}\beta)$, again a total of $\mathcal{O}(pN + kN)$ operations.

In order to examine more closely the relative efficiency of coordinate descent, proximal gradient descent, and Nesterov's momentum method, we carried out a small simulation study.[3] We generated an $N \times p$ predictor matrix $\mathbf{X}$ with standard Gaussian entries and pairwise correlation 0 or 0.5 between the features. Coefficients β_j were defined by $|\beta_j| = \exp[-.5(u(j - 1))^2]$ with $u = \sqrt{\pi/20}$ and alternating signs $+1, -1, +1, \ldots$. Then the outcome y_i was generated as

$$y_i = \sum_{j=1}^p x_{ij}\beta_j + \sigma\varepsilon_i \tag{5.57}$$

with σ chosen so that the signal to noise ratio $\mathrm{Sd}[\mathrm{E}(y_i)]/\sigma$ equals 3. Table 5.1 shows the average (standard error) of CPU times for coordinate descent, generalized gradient and Nesterov's momentum methods, for a scenario with $N > p$ and another with $N < p$. Shown is the total time over a path of 20 values of the regularization parameterλ. Warm starts were used in each case, with

Table 5.1 *Lasso for linear regression: Average (standard error) of CPU times over ten realizations, for coordinate descent, generalized gradient, and Nesterov's momentum methods. In each case, time shown is the total time over a path of 20 λ values.*

	$N = 10000, p = 100$		$N = 200, p = 10000$	
Correlation	0	0.5	0	0.5
Coordinate descent	0.110 (0.001)	0.127 (0.002)	0.298 (0.003)	0.513 (0.014)
Proximal gradient	0.218 (0.008)	0.671 (0.007)	1.207 (0.026)	2.912 (0.167)
Nesterov	0.251 (0.007)	0.604 (0.011)	1.555 (0.049)	2.914 (0.119)

convergence defined as the maximum change in the parameter vector being less than 10^{-4}. An approximate backtracking line search was used for the

---

[3] We thank Jerome Friedman for the programs used in this section.

Table 5.2 *Lasso for logistic regression: average (standard error) of CPU times over ten realizations, for coordinate descent, generalized gradient, and Nesterov's momentum methods. In each case, time shown is the total time over a path of 20 λ values.*

| | $N = 10000, p = 100$ | | $N = 200, p = 10000$ | |
Correlation	0	0.5	0	0.5
Coordinate descent	0.309 (0.086)	0.306 (0.086)	0.646 (0.006)	0.882 (0.026)
Proximal gradient	2.023 (0.018)	6.955 (0.090)	2.434 (0.095)	4.350 (0.133)
Nesterov	1.482 (0.020)	2.867 (0.045)	2.910 (0.106)	8.292 (0.480)

latter two methods. We see that coordinate descent is 2–6 times faster than the other methods, with a greater speedup in the $p > N$ case. Interestingly, momentum does not provide a consistent speedup over proximal gradient descent, as the aforementioned theory would suggest. Our investigation into this suggests that the warm starts are the reason: by starting close to the solution, the "zig-zagging", that is ameliorated by the momentum term, is not nearly as much of a problem as it is when starting far from the solution.

Table 5.2 shows the corresponding results for logistic regression. The predictors were generated as before, but now there are 15 nonzero β_j with alternating signs, and $|\beta_j| = 15 - j + 1$. Then defining $p_i = 1/(1 + \exp(-\sum x_{ij}\beta_j))$ we generate 0/1 y_i with $\text{Prob}(y_i = 1) = p_i$.

We see that coordinate descent is 5–10 times faster than the other methods, with a greater speedup in the $p > N$ case. Again, momentum does not provide a consistent speedup over proximal gradient descent.

The reader should take comparisons like those above with a grain of salt, as the performance of a method will depend on the details of its implementation. Further suspicion should arise, since two of the authors of this text are coauthors of the method (coordinate descent) that performs best. For our part, we can only say that we have tried to be fair to all methods and have coded all methods as efficiently as we could. More importantly, we have made available all of the scripts and programs to generate these results on the book website, so that the reader can investigate the comparisons further.

5.6   Least Angle Regression

Least angle regression, also known as the *homotopy* approach, is a procedure for solving the lasso with squared-error loss that delivers the entire solution path as a function of the regularization parameter λ. It is a fairly efficient algorithm, but does not scale up to large problems as well as some of the other methods in this chapter. However it has an interesting statistical motivation and can be viewed as a kind of "democratic" version of forward stepwise regression.

Forward stepwise regression builds a model sequentially, adding one variable at a time. At each step, it identifies the best variable to include in the *active set*, and then updates the least-squares fit to include all the active variables. Least angle regression (LAR) uses a similar strategy, but only enters "as much" of a predictor as it deserves. At the first step it identifies the variable most correlated with the response. Rather than fit this variable completely, the LAR method moves the coefficient of this variable continuously toward its least-squares value (causing its correlation with the evolving residual to decrease in absolute value). As soon as another variable catches up in terms of correlation with the residual, the process is paused. The second variable then joins the active set, and their coefficients are moved together in a way that keeps their correlations tied and decreasing. This process is continued until all the variables are in the model, and ends at the full least-squares fit. The details are given in Algorithm 5.1. Although the LAR algorithm is stated in terms of correlations, since the input features are standardized, it is equivalent and easier to work with inner products. The number of terms K at step 3 requires some explanation. If $p > N - 1$, the LAR algorithm reaches a zero residual solution after $N - 1$ steps (the -1 is because there is an intercept in the model, and we have centered the data to take care of this).

Algorithm 5.1 LEAST ANGLE REGRESSION.

---

1. Standardize the predictors to have mean zero and unit ℓ_2 norm. Start with the residual $r_0 = y - \bar{y}$, $\beta^0 = (\beta_1, \beta_2, \ldots, \beta_p) = 0$.

2. Find the predictor x_j most correlated with r_0; i.e., with largest value for $|\langle x_j, r_0 \rangle|$. Call this value λ_0, define the active set $\mathcal{A} = \{j\}$, and $X_{\mathcal{A}}$, the matrix consisting of this single variable.

3. For $k = 1, 2, \ldots, K = \min(N - 1, p)$ do:

   (a) Define the least-squares direction $\delta = \frac{1}{\lambda_{k-1}}(X_{\mathcal{A}}^T X_{\mathcal{A}})^{-1} X_{\mathcal{A}}^T r_{k-1}$, and define the p-vector Δ such that $\Delta_{\mathcal{A}} = \delta$, and the remaining elements are zero.

   (b) Move the coefficients β from β^{k-1} in the direction Δ toward their least-squares solution on $X_{\mathcal{A}}$: $\beta(\lambda) = \beta^{k-1} + (\lambda_{k-1} - \lambda)\Delta$ for $0 < \lambda \leq \lambda_{k-1}$, keeping track of the evolving residuals $r(\lambda) = y - X\beta(\lambda) = r_{k-1} - (\lambda_{k-1} - \lambda)X\Delta$.

   (c) Keeping track of $|\langle x_\ell, r(\lambda) \rangle|$ for $\ell \notin \mathcal{A}$, identify the largest value of λ at which a variable "catches up" with the active set; if the variable has index j, that means $|\langle x_j, r(\lambda) \rangle| = \lambda$. This defines the next "knot" λ_k.

   (d) Set $\mathcal{A} = \mathcal{A} \cup \{j\}$, $\beta^k = \beta(\lambda_k) = \beta^{k-1} + (\lambda_{k-1} - \lambda_k)\Delta$, and $r_k = y - X\beta^k$.

4. Return the sequence $\{\lambda_k, \beta^k\}_0^K$.

We make a few observations to clarify the steps in the algorithm. In step 3b, it is easy to check that $|\langle x_j, r(\lambda) \rangle| = \lambda$, $\forall j \in \mathcal{A}$—that is, the correlations remain tied along this path, and decrease to zero with λ. In fact $\beta^0 = \beta^{k-1} + \lambda_{k-1}\Delta$ is the least-squares coefficient vector corresponding to the subset $\mathcal{A}$.

By construction the coefficients in LAR change in a piecewise linear fashion. Figure 5.9 [left panel] shows the LAR coefficient profile evolving as a

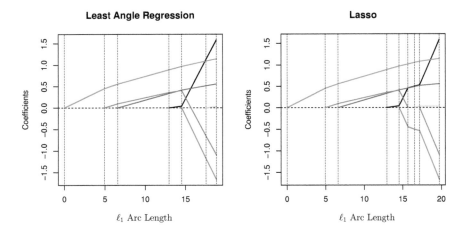

Figure 5.9 *Left panel shows the LAR coefficient profiles on the simulated data, as a function of the L^1 arc length. The right panel shows the lasso profile. They are identical until the red coefficient crosses zero at an arc length of about 16.*

function of their ℓ_1 arc length. [4] Note that we do not need to take small steps and recheck the correlations in step 3c. Variable ℓ "catching up" means that $|\langle \mathbf{x}_\ell, \mathbf{r}(\lambda) \rangle| = \lambda$, a pair of equations linear in λ. We solve for λ for each $\ell \notin \mathcal{A}$ and pick the largest (Exercise 5.9).

The right panel of Figure 5.9 shows the lasso coefficient profiles on the same data. They are almost identical to those in the left panel, and differ for the first time when the pink coefficient passes back through zero. These observations lead to a simple modification in step 3c of the LAR algorithm that gives the entire lasso path, which is also piecewise-linear:

3(c)+ *lasso modification*: If a nonzero coefficient crosses zero before the next variable enters, drop it from $\mathcal{A}$ and recompute the current joint least-squares direction.

Notice in the figure that the pink coefficient remains zero for a while, and then it becomes active again, but this time negative.

We can give a heuristic argument for why these procedures are so similar. As observed, we have at any stage of the algorithm

$$\mathbf{x}_j^T(\mathbf{y} - \mathbf{X}\beta(\lambda)) = \lambda \cdot s_j, \ \forall j \in \mathcal{A}, \tag{5.58}$$

where $s_j \in \{-1, 1\}$ indicates the sign of the inner-product, and λ is the com-

---

[4]The ℓ_1 arc-length of a differentiable curve $\{s \mapsto \beta(s) \mid s \in [0, S]\}$ is given by $\mathrm{TV}(\beta, S) = \int_0^S \|\dot{\beta}(s)\|_1 ds$, where $\dot{\beta}(s) = \partial\beta(s)/\partial s$. For the piecewise-linear LAR coefficient profile, this amounts to summing the ℓ_1-norms of the changes in coefficients from step to step.

mon value. Also by definition of the LAR active set, $|\mathbf{x}_k^T(\mathbf{y}-\mathbf{X}\beta(\lambda))| \leq \lambda \; \forall k \notin \mathcal{A}$. Now consider the lasso criterion[5]

$$R(\beta) = \tfrac{1}{2}\|\mathbf{y} - \mathbf{X}\beta\|_2^2 + \lambda\|\beta\|_1. \tag{5.59}$$

Let $\mathcal{B}$ be the active set of variables in the solution for a given value of λ. For these variables $R(\beta)$ is differentiable, and the stationarity conditions give

$$\mathbf{x}_j^T(\mathbf{y} - \mathbf{X}\beta) = \lambda \cdot \text{sign}(\beta_j), \; \forall j \in \mathcal{B}. \tag{5.60}$$

Comparing (5.60) with (5.58), we see that they are identical only if the sign of β_j matches the sign of the inner product. That is why the LAR algorithm and lasso start to differ when an active coefficient passes through zero; condition (5.60) is violated for that variable, and it is removed from the active set $\mathcal{B}$ in step 3(c)+. Exercise 5.9 shows that these equations imply a piecewise-linear coefficient profile as λ decreases, as was imposed in the LAR update. The stationarity conditions for the nonactive variables require that

$$|\mathbf{x}_k^T(\mathbf{y} - \mathbf{X}\beta)| \leq \lambda, \; \forall k \notin \mathcal{B}, \tag{5.61}$$

which again agrees with the LAR algorithm.

The LAR algorithm exploits the fact that the coefficient paths for the lasso are piecewise linear. This property holds for a more general class of problems; see Rosset and Zhu (2007) for details.

5.7 Alternating Direction Method of Multipliers

The *alternating direction method of multipliers* (ADMM) is a Lagrangian-based approach that has some attractive features for large-scale applications. It is based on a marriage of different ideas that developed over a long period of time. Here we provide a brief overview, referring the reader to Boyd et al. (2011) for a comprehensive discussion.

Consider a problem of the form

$$\underset{\beta\in\mathbb{R}^m, \theta\in\mathbb{R}^n}{\text{minimize}} \; f(\beta) + g(\theta) \quad \text{subject to } \mathbf{A}\beta + \mathbf{B}\theta = c, \tag{5.62}$$

where $f : \mathbb{R}^m \to \mathbb{R}$ and $g : \mathbb{R}^n \to \mathbb{R}$ are convex functions, and $\mathbf{A} \in \mathbb{R}^{d\times m}$ and $\mathbf{B} \in \mathbb{R}^{d\times n}$ are (known) matrices of constraints, and $c \in \mathbb{R}^d$ is a constraint vector. To solve this problem we introduce a vector $\mu \in \mathbb{R}^d$ of Lagrange multipliers associated with the constraint, and then consider the augmented Lagrangian

$$L_\rho(\beta, \theta, \mu) := f(\beta) + g(\theta) + \langle \mu, \mathbf{A}\beta + \mathbf{B}\theta - c \rangle + \frac{\rho}{2}\|\mathbf{A}\beta + \mathbf{B}\theta - c\|_2^2, \tag{5.63}$$

---

[5]We have omitted the factor $\frac{1}{N}$, to stay faithful to the original LAR procedure; all values of λ are hence larger by a factor of N.

where $\rho > 0$ is a small fixed parameter. The quadratic term involving ρ is an augmented Lagrangian that enforces the constraint in a smoother fashion. The ADMM algorithm is based on minimizing the augmented Lagrangian (5.63) successively over β and θ, and then applying a dual variable update to μ. Doing so yields the updates

$$\beta^{t+1} = \underset{\beta \in \mathbb{R}^m}{\arg\min} \, L_\rho(\beta, \theta^t, \mu^t) \tag{5.64a}$$

$$\theta^{t+1} = \underset{\theta \in \mathbb{R}^n}{\arg\min} \, L_\rho(\beta^{t+1}, \theta, \mu^t) \tag{5.64b}$$

$$\mu^{t+1} = \mu^t + \rho \left(\mathbf{A}\beta^{t+1} + \mathbf{B}\theta^{t+1} - c \right), \tag{5.64c}$$

for iterations $t = 0, 1, 2, \ldots$. The update (5.64c) can be shown to be a dual ascent step for the Lagrange multiplier vector μ. Under relatively mild conditions, one can show that this procedure converges to an optimal solution to Problem (5.62).

The ADMM framework has several advantages. First, convex problems with nondifferentiable constraints can be easily handled by the separation of parameters into β and θ. We illustrate this procedure via application to the lasso, as discussed in the example to follow. A second advantage of ADMM is its ability to break up a large problem into smaller pieces. For datasets with large number of observations we break up the data into blocks, and carry out the optimization over each block. As discussed in more detail in Exercise 5.12, constraints are included to ensure that the solution vectors delivered by the optimization over each data block agree with one another at convergence. In a similar way, the problem can be split up into feature blocks, and solved in a coordinated blockwise fashion.

Example 5.8. ADMM for the lasso. The Lagrange form of the lasso can be expressed in equivalent form as

$$\underset{\beta \in \mathbb{R}^p, \theta \in \mathbb{R}^p}{\text{minimize}} \left\{ \frac{1}{2} \|\mathbf{y} - \mathbf{X}\beta\|_2^2 + \lambda \|\theta\|_1 \right\} \quad \text{such that } \beta - \theta = 0. \tag{5.65}$$

When applied to this problem, the ADMM updates take the form

$$\begin{aligned} \beta^{t+1} &= (\mathbf{X}^T\mathbf{X} + \rho\mathbf{I})^{-1}(\mathbf{X}^T\mathbf{y} + \rho\theta^t - \mu^t) \\ \theta^{t+1} &= S_{\lambda/\rho}(\beta^{t+1} + \mu^t/\rho) \\ \mu^{t+1} &= \mu^t + \rho(\beta^{t+1} - \theta^{t+1}). \end{aligned} \tag{5.66}$$

Thus, the algorithm involves a ridge regression update for β, a soft-thresholding step for θ and then a simple linear update for μ. The first step is the main work, and after an initial singular value decomposition of $\mathbf{X}$, subsequent iterations can be done quickly. The initial SVD requires $\mathcal{O}(p^3)$ operations, but can be done off-line, whereas subsequent iterations have cost $\mathcal{O}(Np)$. Consequently, after the start-up phase, the cost per iteration is similar to coordinate descent or the composite gradient method.   ◇

5.8 Minorization-Maximization Algorithms

In this section, we turn to a class of methods, known either as minorization-maximization or majorization-minimization (MM) algorithms, that are especially useful for optimization of nonconvex functions. These belong to the class of auxiliary-variable methods, in that they are based on introducing extra variables and using them to majorize (or upper bound) the objective function to be minimized. Although these methods apply more generally to constrained problems, here we describe them in application to a simple unconstrained problem of the form $\text{minimize}_{\beta \in \mathbb{R}^p} f(\beta)$, where $f : \mathbb{R}^p \to \mathbb{R}$ is a (possibly) nonconvex function.

A function $\Psi : \mathbb{R}^p \times \mathbb{R}^p \mapsto \mathbb{R}^1$ majorizes the function f at a point $\beta \in \mathbb{R}^p$ if

$$f(\beta) \leq \Psi(\beta, \theta) \quad \text{for all } \theta \in \mathbb{R}^p \tag{5.67}$$

with *equality* holding when $\beta = \theta$. (Naturally, there is a corresponding definition of minorization, with the inequality reversed in direction.) Figure 5.10 shows a schematic of a majorizing function.

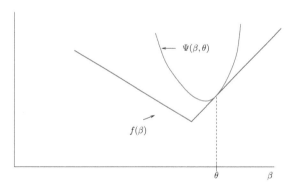

Figure 5.10 *Illustration of a majorizing function for use in an MM algorithm. The function $\Psi(\beta, \theta)$ lies on or above $f(\beta)$ for all β and is equal to $f(\beta)$ when $\beta = \theta$. The MM algorithm seeks to minimize the target function f by solving a sequence of subproblems involving the majorizing function Ψ and the current iterate.*

An MM algorithm for performing an unconstrained minimization of f involves initializing β^0, and then updating via the recursion

$$\beta^{t+1} = \arg\min_{\beta \in \mathbb{R}^p} \Psi\left(\beta, \beta^t\right) \quad \text{for } t = 0, 1, 2, \dots. \tag{5.68}$$

By the majorization property (5.67), this scheme generates a sequence for which the cost $f(\beta^t)$ is nonincreasing. In particular, we have

$$f(\beta^t) = \Psi(\beta^t, \beta^t) \overset{(i)}{\geq} \Psi(\beta^{t+1}, \beta^t) \overset{(ii)}{\geq} f(\beta^{t+1}), \tag{5.69}$$

where inequality (i) uses the fact that β^{t+1} is a minimizer of the function $\beta \mapsto \Psi(\beta, \beta^t)$, and inequality (ii) uses the majorization property (5.67). If the original function f is strictly convex, it can be shown the MM algorithm converges to the global minimizer.

There are different classes of majorizing functions that are useful for different problems. In general, a good majorization function is one for which the update (5.68) is relatively easy to compute, at least relative to direct minimization of f. See Lange (2004) for more details.

Example 5.9. Proximal gradient as an MM algorithm. Recall from Section 5.3.3 the proximal gradient algorithm that can be applied to cost functions that decompose as a sum $f = g + h$, where g is convex and differentiable, and h is convex and (potentially) nondifferentiable. By applying a second-order Taylor series expansion (with remainder) to g, we obtain

$$f(\beta) = g(\beta) + h(\beta)$$
$$= g(\theta) + \langle \nabla g(\theta), \theta - \beta \rangle + \frac{1}{2}\langle \theta - \beta, \nabla^2 g(\beta')(\theta - \beta) \rangle + h(\beta),$$

where $\beta' = s\beta + (1 - s)\theta$ for some $s \in [0, 1]$. It can be verified that Lipschitz condition (5.27) on the gradient ∇g implies a uniform upper bound on the Hessian, namely $\nabla^2 g(\beta') \preceq L\,\mathbf{I}_{p \times p}$, from which we obtain the inequality

$$f(\beta) \leq \underbrace{g(\theta) + \langle \nabla g(\theta), \theta - \beta \rangle + \frac{L}{2}\|\theta - \beta\|_2^2 + h(\beta)}_{\Psi(\beta, \theta)},$$

with equality holding when $\theta = \beta$. Thus, we see that the proximal gradient method can be viewed as an MM algorithm with a particular choice of majorizing function. $\Diamond$

Apart from a direct bound on the Hessian, there are other ways of deriving majorizing functions, For example, Jensen's inequality can be used to derive the usual EM algorithm as an instance of an MM algorithm (Hunter and Lange 2004, Wu and Lange 2010). As we discuss in Chapter 8, MM algorithms turn out to be useful in procedures for sparse multivariate analysis.

5.9 Biconvexity and Alternating Minimization

Recall the class of coordinate descent algorithms discussed in Section 5.4. Algorithms of this form are also useful for optimizing a class of (potentially) nonconvex functions known as biconvex functions. A function $f : \mathbb{R}^m \times \mathbb{R}^n \to \mathbb{R}$ is *biconvex* if for each $\beta \in \mathbb{R}^n$, the function $\alpha \mapsto f(\alpha, \beta)$ is convex, and for each $\alpha \in \mathbb{R}^m$, the function $\beta \mapsto f(\alpha, \beta)$ is convex. Of course, any function that is jointly convex in the pair (α, β) is also biconvex. But a function can be biconvex without being jointly convex. For instance, consider the biconvex function

$$f(\alpha, \beta) = (1 - \alpha\beta)^2 \text{ for } |\alpha| \leq 2, |\beta| \leq 2. \tag{5.70}$$

As illustrated in Figure 5.11, it is convex when sliced along lines parallel to the axes, as required by the definition of biconvexity, but other slices can lead to nonconvex functions.

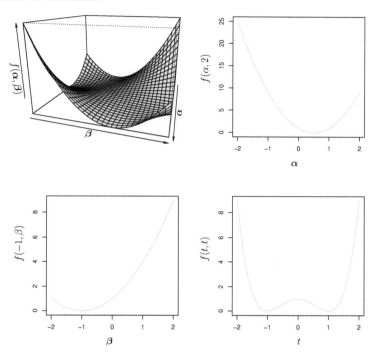

Figure 5.11 *Example of a biconvex function. Shown in the top left is the function* $f(\alpha, \beta) = (1 - \alpha\beta)^2$ *over the range* $-2 \le \alpha, \beta \le 2$. *The top right and bottom left panels show slices of the function for* $\beta = 2$ *and* $\alpha = -1$. *The bottom right panel shows the function along the line* $\alpha = \beta = t$.

More generally, let $\mathcal{A} \subseteq \mathbb{R}^m$ and $\mathcal{B} \subseteq \mathbb{R}^n$ be two nonempty and convex sets, and let $\mathcal{C} \subseteq \mathcal{A} \times \mathcal{B}$. For each fixed $\alpha \in \mathcal{A}$ and $\beta \in \mathcal{B}$, the sets

$$\mathcal{C}_\alpha := \{\beta \in \mathcal{B} \mid (\alpha, \beta) \in \mathcal{C}\}, \quad \text{and} \quad \mathcal{C}_\beta := \{\alpha \in \mathcal{A} \mid (\alpha, \beta) \in \mathcal{C}\} \quad (5.71)$$

are called the α and β sections of $\mathcal{C}$. The set $\mathcal{C} \subseteq \mathcal{A} \times \mathcal{B}$ is called a *biconvex set* if the section $\mathcal{C}_\alpha$ is convex for each $\alpha \in \mathcal{A}$, and the section $\mathcal{C}_\beta$ is convex for every $\beta \in \mathcal{B}$. Given a biconvex set $\mathcal{C}$, a function $f : \mathcal{C} \to \mathbb{R}$ is a *biconvex function* if the function $\alpha \mapsto f(\alpha, \beta)$ is convex in α for each fixed $\beta \in \mathcal{B}$, and the function $\beta \mapsto f(\alpha, \beta)$ is convex in β for each fixed $\alpha \in \mathcal{A}$.

Given these ingredients, a biconvex optimization problem has the form minimize$_{(\alpha, \beta) \in \mathcal{C}} f(\alpha, \beta)$, where the set $\mathcal{C}$ is biconvex on $\mathcal{A} \times \mathcal{B}$, and the objective function is biconvex on $\mathcal{C}$.

The most obvious method for solving a biconvex optimization problem is based on *Alternate Convex Search* (ACS), which is simply block coordinate descent applied to the α and β blocks:

(a) Initialize (α^0, β^0) at some point in $\mathcal{C}$.

(b) For iterations $t = 0, 1, 2, \ldots$:

    1. Fix $\beta = \beta^t$, and perform the update $\alpha^{t+1} \in \arg\min_{\alpha \in \mathcal{C}_{\beta^t}} f(\alpha, \beta^t)$.

    2. Fix $\alpha = \alpha^{t+1}$, and perform the update $\beta^{t+1} \in \arg\min_{\beta \in \mathcal{C}_{\alpha^{t+1}}} f(\alpha^{t+1}, \beta)$.

Given the biconvex structure, each of the two updates involve solving a convex optimization problem. The ACS procedure will be efficient as long as these convex sub-problems can be solved relatively quickly.

By construction, the sequence of function values $\{f(\alpha^t, \beta^t)\}_{t=0}^{\infty}$ is nonincreasing. Consequently, if f is bounded from below over $\mathcal{C}$, then the function values converge to some limiting value. We note that this form of convergence is relatively weak, and only ensures that the function values converge. The solution sequence $\{(\alpha^t, \beta^t\}$ may not converge, and in some cases may diverge to infinity. Assuming convergence, to what does the solution sequence converge? Since a biconvex function f need not be convex in general, we cannot expect it to converge to the global minimum. All we can say in general is that if it converges, it converges to a partial optimum.

More specifically, we say that $(\alpha^*, \beta^*) \in \mathcal{C}$ is a *partial optimum* if

$$f(\alpha^*, \beta^*) \le f(\alpha^*, \beta) \quad \text{for all } \beta \in \mathcal{C}_{\alpha^*}, \text{ and}$$
$$f(\alpha^*, \beta^*) \le f(\alpha, \beta^*) \quad \text{for all } \alpha \in \mathcal{C}_{\beta^*}.$$

Example 5.10. Alternating subspace algorithm. One biconvex problem in which convergence of ACS can be fully characterized is the alternating subspace algorithm for computing the maximal singular vectors/value of a matrix. Given a matrix $\mathbf{X} \in \mathbb{R}^{m \times n}$, consider the problem of finding the best rank-one approximation in the Frobenius norm.[6] This approximation problem can be formulated in terms of minimizing the objective function

$$f(\alpha, \beta, s) = \|\mathbf{X} - s\,\alpha\beta^T\|_{\mathrm{F}}^2 \tag{5.72}$$

over vectors $\alpha \in \mathbb{R}^m$ and $\beta \in \mathbb{R}^n$, with $\|\alpha\|_2 = \|\beta\|_2 = 1$, and a scalar $s > 0$. The ACS procedure for this problem starts with any random unit-norm initialization for β^0, and then for iterations $t = 1, 2, \ldots$, it performs the updates

$$\alpha^t = \frac{\mathbf{X}\beta^{t-1}}{\|\mathbf{X}\beta^{t-1}\|_2}, \quad \text{and} \quad \beta^t = \frac{\mathbf{X}^T\alpha^t}{\|\mathbf{X}^T\alpha^t\|_2}. \tag{5.73}$$

The scalar s can be computed as $s = \|\mathbf{X}\beta^t\|_2$ at convergence. It can be shown (see Exercise 5.13 that as long as β^0 is not orthogonal to the largest right

---

[6]The Frobenius norm of a matrix is the Euclidean norm applied to its vectorized version.

singular vector, the iterates (α^t, β^t) converge to the left and right singular vectors of $\mathbf{X}$ corresponding to the largest singular value of $\mathbf{X}$.

The procedure is related to the *power method* for finding the largest eigenvector of a symmetric positive semi-definite matrix. The β^t iterates for the right singular vector have the form

$$\beta^{t+1} = \frac{\mathbf{X}^T\mathbf{X}\beta^t}{\|\mathbf{X}^T\mathbf{X}\beta^t\|_2}, \tag{5.74}$$

with similar updates for α^t in terms of $\mathbf{X}\mathbf{X}^T$. Consequently, the procedure simply "powers up" the operator $\mathbf{X}^T\mathbf{X}$, with the normalization driving all but the largest eigenvalue to zero. See De Leeuw (1994) and Golub and Loan (1996, §7.3) for further details on the power method.   ◇

In Chapter 7, we present Algorithm 7.2 on page 189 as another example of an ACS procedure.

5.10   Screening Rules

As seen in Section 5.6, inner products play an important role in the lasso problem. For simplicity we assume all variables are mean centered (so we can ignore the intercept), and we consider solving the lasso problem[7]

$$\underset{\beta\in\mathbb{R}^p}{\text{minimize}} \; \frac{1}{2}\|\mathbf{y} - \mathbf{X}\beta\|_2^2 + \lambda\|\beta\|_1 \tag{5.75}$$

with a decreasing sequence of values for λ. The first variable to enter the model has largest absolute inner-product $\lambda_{\max} = \max_j |\langle \mathbf{x}_j, \mathbf{y}\rangle|$, which also defines the entry value for λ. Also, at any stage, all variables $\mathbf{x}_j$ in the active set have $|\langle \mathbf{x}_j, \mathbf{y}-\hat{\mathbf{y}}_\lambda\rangle| = \lambda$, and all those out have smaller inner-products with the residuals. Hence one might expect a priori that predictors having small inner products with the response are not as likely to have a nonzero coefficient as compared to those with larger inner products. Based on this intuition, one might be able to eliminate predictors from the problem, and thereby reduce the computational load. For example, in some genomic applications we might have millions of variables (SNPs), and anticipate fitting models with only a handful of terms. In this section, we discuss screening rules that exploit this intuition, and have the potential to speed up the computation substantially while still delivering the exact numerical solution.

We begin our discussion with the "dual polytope projection" (DPP) rule (Wang, Lin, Gong, Wonka and Ye 2013). Suppose we wish to compute a lasso solution at $\lambda < \lambda_{\max}$. The DPP rule discards the jth variable if

$$|\mathbf{x}_j^T\mathbf{y}| < \lambda_{\max} - \|\mathbf{x}_j\|_2\|\mathbf{y}\|_2\frac{\lambda_{\max} - \lambda}{\lambda} \tag{5.76}$$

---

[7]In this section we have omitted the $\frac{1}{N}$ in the first part of the objective (to match the referenced formulas); this increases the scale of λ by a factor N.

It may come as a surprise that such a rule can work, as it surprised us when we first saw it. We know that in a linear regression, a predictor can be insignificant on its own, but can become significant when included in the model with other predictors. It seems that the same phenomenon should occur with the lasso.

In fact, there is no contradiction, and a similar rule applies at any stage of the regularization path (not just the start). Suppose we have the lasso solution $\hat{\beta}(\lambda')$ at λ', and we wish to screen variables for the solution at $\lambda < \lambda'$. Then if

$$\left|\mathbf{x}_j^T\left(\mathbf{y} - \mathbf{X}\hat{\beta}(\lambda')\right)\right| < \lambda' - \|\mathbf{x}_j\|_2 \|\mathbf{y}\|_2 \frac{\lambda' - \lambda}{\lambda}, \qquad (5.77)$$

variable j is not part of the active set at λ. We refer to this as the sequential DPP rule.

Figure 5.12 shows the performance of this rule on a simulated example with 5000 predictors (details in caption). The global DPP applies rule (5.76) for all values of λ, and we can see it quickly runs out of steam. By the time λ is small enough to admit 8 predictors into the model, all 5000 predictors survive the screen. But the sequential DPP rule is much more aggressive, and even with 250 predictors in the model, only 1200 need to be considered. So the sequential screening rule (5.77) works much better if λ' and λ are close together. We derive the lasso dual and the DPP rules in Appendix B on Page 132.

In order to achieve even better performance, it is natural to consider screening rules that are less conservative, and allow for occasional failures. Such rules can be incorporated as part of an overall strategy that still yields the exact solution upon termination. A variant of the global DPP rule (5.76) is the *global strong rule*, which discards predictor j whenever

$$\left|\mathbf{x}_j^T \mathbf{y}\right| < \lambda - (\lambda_{\max} - \lambda) = 2\lambda - \lambda_{\max}. \qquad (5.78)$$

This tends to discard more predictors than the global DPP rule (compare blue with orange points in Figure 5.12.) Similarly the *sequential strong rule* discards the jth predictor from the optimization problem at λ if

$$\left|\mathbf{x}_j^T\left(\mathbf{y} - \mathbf{X}\hat{\beta}(\lambda')\right)\right| < 2\lambda - \lambda'. \qquad (5.79)$$

Intuitively, the active set will include predictors that can achieve inner-product λ with the residuals. So we include all those that achieve inner product close to λ using the *current residuals at* $\lambda' > \lambda$, where close is defined by the gap $\lambda' - \lambda$.

As with the sequential DPP rule, the sequential strong rule is based on solving the lasso over a grid of decreasing λ values. Figure 5.12 includes the global and sequential strong rules. In both cases they dominate the DPP counterparts. Neither of the strong rules make any errors in this example, where an error means that it discards some predictor with a nonzero coefficient in the

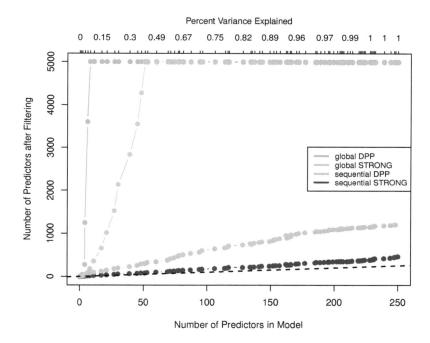

Figure 5.12 *Lasso regression: Results of different rules applied to a simulated dataset. There are $N = 200$ observations and $p = 5000$ uncorrelated Gaussian predictors; one-quarter of the true coefficients are nonzero. Shown are the number of predictors left after screening at each stage, plotted against the number of predictors in the model for a given value of λ. The value of λ is decreasing as we move from left to right. In the plots, we are fitting along a path of 100 decreasing λ values equally spaced on the log-scale, A broken line with unit slope is added for reference. The proportion of variance explained by the model is shown along the top of the plot. There were no violations for either of the strong rules.*

actual solution. The sequential strong rule (5.79) has remarkable performance, discarding almost all of the redundant predictors.

We now give further motivation for the strong rules (Tibshirani, Bien, Friedman, Hastie, Simon, Taylor and Tibshirani$_2$ 2012). Suppose that predictor j is not in the model at $\lambda = \lambda_{\max}$. The KKT conditions for the lasso then ensure that $|\mathbf{x}_j^T \mathbf{y}| < \lambda_{\max}$, so that the global rule (5.78) can be interpreted as dictating that as we move from $\lambda_{\max}$ to λ, the inner product $|\mathbf{x}_j^T (\mathbf{y} - \mathbf{X}\hat{\beta}(\lambda))|$ can increase by at most $\lambda_{\max} - \lambda$. Consequently, if the inner product is below the strong bound $\lambda - (\lambda_{\max} - \lambda)$, it cannot reach the necessary level λ for inclusion in the model, where we have again used the KKT conditions in our reasoning. If we define $c_j(\lambda) := \mathbf{x}_j^T (\mathbf{y} - \mathbf{X}\hat{\beta}(\lambda))$, then for either the global or sequential strong rules to hold, it would be sufficient to have

$$\left| \frac{dc_j(\lambda)}{d\lambda} \right| \leq 1, \tag{5.80}$$

assuming that this derivative exists.[8] Now the KKT conditions at λ are

$$c_j(\lambda) = \lambda\, s_j(\lambda), \quad \text{for } j = 1, 2, \ldots, p, \tag{5.81}$$

where $s_j(\lambda) = \text{sign}(\hat{\beta}_j(\lambda))$ if $\hat{\beta}_j(\lambda) \neq 0$ and $s_j(\lambda) \in [-1, 1]$ if $\hat{\beta}_j(\lambda) = 0$. By the chain rule

$$\frac{dc_j(\lambda)}{d\lambda} = s_j(\lambda) + \lambda \cdot \frac{ds_j(\lambda)}{d\lambda}.$$

If we ignore the second term, then we have $|\frac{dc_j(\lambda)}{d\lambda}| \leq 1$. Now the second term equals zero when a variable has a nonzero coefficient in an interval of λ values, for then $s_j(\lambda)$ is constant (equaling ± 1). In addition, the slope condition (5.80) always holds if $(\mathbf{X}^T \mathbf{X})^{-1}$ is diagonally dominant (Tibshirani et al. 2012), a condition meaning that the predictors are nearly uncorrelated. In general, however, the slope condition can fail over short stretches of λ, and in these instances, the strong rules can fail (i.e., discard predictors in error). However these failures are rare, and are virtually nonexistent when $p \gg N$.

In summary, we have found empirically that the strong rules, and especially the sequential strong rule (5.79) seem to be very good heuristics for discarding variables. This is the case in the lasso, lasso-penalized logistic regression, and the elastic net.

One can use the sequential strong rule to save computation time, without sacrificing the exact solution, as follows. We compute the solution along a fine grid of decreasing λ values. For each value of λ, the screening rule is applied, yielding a subset of the predictors. Then the problem is solved using only this subset. The KKT conditions (5.81) for all predictors are then checked. If they are satisfied, we are done. Otherwise the predictors that violate the conditions

---

[8]The arguments here are only heuristic, because $dc_j(\lambda)/d\lambda$ and $ds_j(\lambda)/d\lambda$ discussed below it do not exist at $\hat{\beta}_j(\lambda) = 0$.

are added to the active set and the problem is solved again. In principle this must be iterated until no violations occur.

This approach is effective computationally because violations of the strong rule turn out to be rare, especially when $p \gg N$. Tibshirani et al. (2012) implement these rules for the coordinate descent approach in `glmnet` and the generalized gradient and Nesterov first-order methods. They report speedup factors in the range from 2 to 80, depending on the setting.

Finally, suppose instead that we are interested in a more general convex problem of the form

$$\underset{\beta}{\text{minimize}} \left\{ f(\beta) + \lambda \sum_{j=1}^{r} c_j \|\beta_j\|_{p_j} \right\}. \qquad (5.82)$$

Here f is a convex and differentiable function, and $\beta = (\beta_1, \beta_2, \dots \beta_r)$ with each β_j being a scalar or a vector. Also $\lambda \geq 0$, and $c_j \geq 0$, $p_j \geq 1$ for each $j = 1, \dots r$. Then given $\lambda' > \lambda$, the sequential strong rule for discarding predictor j takes the form

$$\left\| \nabla_j f\big(\hat{\beta}(\lambda')\big) \right\|_{q_j} < c_j(2\lambda - \lambda'), \qquad (5.83)$$

where $\nabla_j f(\hat{\beta}) = (\partial f(\hat{\beta})/\partial \beta_{j_1}, \dots \partial f(\hat{\beta})/\partial \beta_{j_m})$ where $1/p_j + 1/q_j = 1$ (i.e., $\| \cdot \|_{p_j}$ and $\| \cdot \|_{q_j}$ are dual norms). The rule (5.83) can be applied to a wide variety of problems, including logistic regression and other generalized linear models, the group lasso and the graphical lasso.

Bibliographic Notes

The behavior of descent algorithms, including convergence proofs for methods based on appropriate stepsize selection rules, such as limited minimization or the Armijo rule, is a classical subject in optimization; see Chapters 1 and 2 of Bertsekas (1999) for more details. Further background on Lagrangian methods and duality can be found in Bertsekas (1999), as well as Boyd and Vandenberghe (2004). Rockafellar (1996) provides a more advanced treatment of convex duality and convex analysis. Nesterov (2007) derives and analyzes the generalized gradient method (5.21) for composite objectives; see also Nesterov's book (2004) for related analysis of projected gradient methods. Minorization-maximization procedures, also known as auxiliary function methods, are discussed in Lange (2004) and Hunter and Lange (2004).

Gorski, Pfeuffer and Klamroth (2007) provide an overview of biconvex functions, and alternating algorithms for optimizing them. El Ghaoui, Viallon and Rabbani (2010) introduced the use of screening rules such as (5.76); inspired by this work, we derived a very similar formula, and which led to our development of the strong rules in Section 5.10. However, the more recent DPP rules of Wang, Lin, Gong, Wonka and Ye (2013) dominate these earlier safe rules, and provide a simple sequential formula. Fu (1998) was an early proponent of coordinate descent for the lasso.

Appendix A: The Lasso Dual

In this appendix, we derive a useful dual of the lasso primal problem (2.5), which we write in a slightly more convenient form

$$\text{Lasso Primal: } \underset{\beta \in \mathbb{R}^p}{\text{minimize}} \frac{1}{2}\|\mathbf{y} - \mathbf{X}\beta\|_2^2 + \lambda\|\beta\|_1. \tag{5.84}$$

Introducing the residual vector $\boldsymbol{r} = \mathbf{y} - \mathbf{X}\beta$, we can rewrite the primal Equation (5.84) as

$$\underset{\beta \in \mathbb{R}^p}{\text{minimize}} \frac{1}{2}\|\boldsymbol{r}\|_2^2 + \lambda\|\beta\|_1 \text{ subject to } \boldsymbol{r} = \mathbf{y} - \mathbf{X}\beta. \tag{5.85}$$

Letting $\boldsymbol{\theta} \in \mathbb{R}^N$ denote a Lagrange multiplier vector, the Lagrangian of this problem can be written as

$$L(\beta, \boldsymbol{r}, \boldsymbol{\theta}) := \frac{1}{2}\|\boldsymbol{r}\|_2^2 + \lambda\|\beta\|_1 - \boldsymbol{\theta}^T(\boldsymbol{r} - \mathbf{y} + \mathbf{X}\beta). \tag{5.86}$$

The dual objective is derived by minimizing this expression (5.86) with respect to β and $\boldsymbol{r}$. Isolating those terms involving β, we find

$$\min_{\beta \in \mathbb{R}^p} -\boldsymbol{\theta}^T\mathbf{X}\beta + \lambda\|\beta\|_1 = \begin{cases} 0 & \text{if } \|\mathbf{X}^T\boldsymbol{\theta}\|_\infty \leq \lambda \\ -\infty & \text{otherwise} \end{cases} \tag{5.87}$$

where $\|\mathbf{X}^T\boldsymbol{\theta}\|_\infty = \max_j |\mathbf{x}_j^T\boldsymbol{\theta}|$. Next we isolate terms involving $\boldsymbol{r}$ and find

$$\min_{\boldsymbol{r}} \frac{1}{2}\|\boldsymbol{r}\|_2^2 - \boldsymbol{\theta}^T\boldsymbol{r} = -\frac{1}{2}\boldsymbol{\theta}^T\boldsymbol{\theta}, \tag{5.88}$$

with $\boldsymbol{r} = \boldsymbol{\theta}$. Substituting relations (5.87) and (5.88) into the Lagrangian representation (5.86), we obtain

$$\text{Lasso Dual: } \underset{\boldsymbol{\theta}}{\text{maximize}} \frac{1}{2}\{\|\mathbf{y}\|_2^2 - \|\mathbf{y} - \boldsymbol{\theta}\|_2^2\} \text{ subject to } \|\mathbf{X}^T\boldsymbol{\theta}\|_\infty \leq \lambda. \tag{5.89}$$

Overall, this form of the lasso dual amounts to projecting $\mathbf{y}$ onto the feasible set $\mathcal{F}_\lambda = \{\boldsymbol{\theta} \in \mathbb{R}^N \mid \|\mathbf{X}^T\boldsymbol{\theta}\|_\infty \leq \lambda\}$. $\mathcal{F}_\lambda$ is the intersection of the $2p$ half-spaces defined by $\{|\mathbf{x}_j^T\boldsymbol{\theta}| \leq \lambda\}_{j=1}^p$, a convex-polytope in $\mathbb{R}^N$. In the language of Section 5.3.3, the solution is given by the proximal map $\boldsymbol{\theta}^* = \mathbf{prox}_{I(\mathcal{F}_\lambda)}(\mathbf{y})$. Figure 5.13 provides an illustration of this geometric interpretation.

Appendix B: Derivation of the DPP Rule

Here we derive the sequential DPP screening rule (5.77); our proof follows that in Wang, Lin, Gong, Wonka and Ye (2013). We first modify the lasso dual via a change of variables $\boldsymbol{\phi} = \boldsymbol{\theta}/\lambda$, leading to

$$\underset{\boldsymbol{\theta}}{\text{maximize}} \frac{1}{2}\{\|\mathbf{y}\|_2^2 - \lambda^2\|\mathbf{y}/\lambda - \boldsymbol{\phi}\|_2^2\} \text{ subject to } \|\mathbf{X}^T\boldsymbol{\phi}\|_\infty \leq 1. \tag{5.90}$$

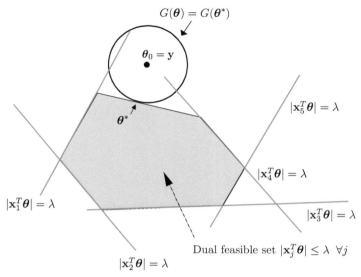

Figure 5.13 *The Lagrange dual of the lasso, with $G(\boldsymbol{\theta}) = \frac{1}{2}(\|\mathbf{y}\|_2^2 - \|\mathbf{y} - \boldsymbol{\theta}\|_2^2)$. The blue shaded region is the feasible set $\mathcal{F}_\lambda$. The unconstrained dual solution is $\boldsymbol{\theta}_0 = \mathbf{y}$, the null residual. The dual solution $\boldsymbol{\theta}^* = \mathbf{prox}_{I(\mathcal{F}_\lambda)}(\mathbf{y})$, the projection of $\mathbf{y}$ onto the convex set $\mathcal{F}_\lambda$.*

Theorem 5.1. Suppose we are given a solution $\hat{\phi}(\lambda')$ to the lasso dual (5.90) for a specific $\lambda_{\max} \geq \lambda' > 0$. Let λ be a nonnegative value different from λ'. If the following holds:

$$|\mathbf{x}_j^T \hat{\phi}(\lambda')| < 1 - \|\mathbf{x}_j\|_2 \|\mathbf{y}\|_2 \left| \frac{1}{\lambda'} - \frac{1}{\lambda} \right|, \qquad (5.91)$$

then $\hat{\beta}_j(\lambda) = 0$.

Since $\hat{\phi}(\lambda') = (\mathbf{y} - \mathbf{X}\hat{\beta}(\lambda'))/\lambda'$, simple algebra leads to (5.77).

Proof: We know from the stationarity conditions for the lasso that

$$|\mathbf{x}_j^T \hat{\phi}(\lambda)| < 1 \implies \hat{\beta}_j(\lambda) = 0. \qquad (5.92)$$

From the dual (5.90), $\hat{\phi}(\lambda)$ is the projection of $\mathbf{y}/\lambda$ into the feasible set $\mathcal{F}_\lambda$. By the projection theorem (Bertsekas 2003) for closed convex sets, $\hat{\phi}(\lambda)$ is continuous and nonexpansive, which implies

$$\|\hat{\phi}(\lambda) - \hat{\phi}(\lambda')\|_2 \leq \left\| \frac{\mathbf{y}}{\lambda} - \frac{\mathbf{y}}{\lambda'} \right\|_2 \qquad (5.93)$$

$$= \|\mathbf{y}\|_2 \left| \frac{1}{\lambda} - \frac{1}{\lambda'} \right|.$$

Then

$$|\mathbf{x}_j^T \hat{\boldsymbol{\phi}}(\lambda)| \leq |\mathbf{x}_j^T \hat{\boldsymbol{\phi}}(\lambda) - \mathbf{x}_j^T \hat{\boldsymbol{\phi}}(\lambda')| + |\mathbf{x}_j^T \hat{\boldsymbol{\phi}}(\lambda')| \qquad (5.94)$$

$$< \|\mathbf{x}_j\|_2 \|\hat{\boldsymbol{\phi}}(\lambda) - \hat{\boldsymbol{\phi}}(\lambda')\|_2 + 1 - \|\mathbf{x}_j\|_2 \|\mathbf{y}\|_2 \left|\frac{1}{\lambda'} - \frac{1}{\lambda}\right|$$

$$\leq \|\mathbf{x}_j\|_2 \|\mathbf{y}\|_2 \left|\frac{1}{\lambda'} - \frac{1}{\lambda}\right| + 1 - \|\mathbf{x}_j\|_2 \|\mathbf{y}\|_2 \left|\frac{1}{\lambda'} - \frac{1}{\lambda}\right| \qquad = 1.$$

$\square$

Exercises

Ex. 5.1 Consider the unconstrained minimization of the quadratic function $f(\beta) = \frac{1}{2}\beta^T \mathbf{Q}\beta - \langle \beta, b \rangle$, where $\mathbf{Q} \succ 0$ is a symmetric positive definite matrix, and $b \in \mathbb{R}^p$.

(a) Show that the optimal solution β^* exists and is unique, and specify its form in terms of $(\mathbf{Q}, b)$.

(b) Write out the gradient descent updates with constant stepsize s for this problem.

(c) Show that there exists some constant $c > 0$, depending only on $\mathbf{Q}$, such that gradient descent converges for any fixed stepsize $s \in (0, c)$.

Ex. 5.2 Consider the constrained program minimize $f(\beta)$ subject to $g_j(\beta) \leq 0$ for $j = 1, \ldots, m$, and let f^* be its optimal value.
Define the Lagrangian function

$$L(\beta; \lambda) = f(\beta) + \sum_{j=1}^{m} \lambda_j g_j(\beta). \qquad (5.95)$$

(a) Show that

$$\sup_{\lambda \geq 0} L(\beta; \lambda) = \begin{cases} f(\beta) & \text{if } g_j(\beta) \leq 0 \text{ for } j = 1, \ldots, m \\ +\infty & \text{otherwise.} \end{cases}$$

(b) Use part (a) to show that $f^* = \inf_\beta \sup_{\lambda \geq 0} L(\beta; \lambda)$.

(c) How is f^* related to the quantity $\sup_{\lambda \geq 0} \inf_\beta L(\beta, \lambda)$?

Ex. 5.3 Let $f : \mathbb{R}^p \to \mathbb{R}$ be a convex and differentiable function, and consider a subspace constraint of the form $\mathcal{C} = \{\beta \in \mathbb{R}^p \mid M\beta = c\}$, where $M \in \mathbb{R}^{m \times p}$ is a fixed matrix, and $c \in \mathbb{R}^m$ is a fixed vector.

(a) Suppose that $\beta^* \in \mathcal{C}$ satisfies the first-order optimality condition (5.4). Show that there must exist a vector $\lambda^* \in \mathbb{R}^m$ such that

$$\nabla f(\beta^*) + M^T \lambda^* = 0 \qquad (5.96)$$

(b) Conversely, suppose that condition (5.96) holds for some $\lambda^* \in \mathbb{R}^m$. Show that the first-order optimality condition (5.4) must be satisfied.

Ex. 5.4 Consider the Lagrangian $L(\beta, \lambda) = f(\beta) + \sum_{j=1}^{m} \lambda_j g_j(\beta)$ associated with the constrained problem (5.5), and assume that the optimal value f^* is finite. Suppose that there exist vectors $\beta^* \in \mathbb{R}^p$ and $\lambda^* \in \mathbb{R}_+^m$ such that

$$L(\beta^*, \lambda) \overset{(i)}{\leq} L(\beta^*, \lambda^*) \overset{(ii)}{\leq} L(\beta, \lambda^*) \qquad (5.97)$$

for all $\beta \in \mathbb{R}^p$ and $\lambda \in \mathbb{R}_+^m$. Show that β^* is optimal for the constrained program.

Ex. 5.5 *Subgradient of Euclidean norm.* Consider the Euclidean or ℓ_2 norm $\|\beta\|_2 = \sqrt{\sum_{j=1}^{p} \beta_j^2}$, which is used in the group lasso. Show that:

(a) For any $\beta \neq 0$, the norm $g(\beta) := \|\beta\|_2$ is differentiable with $\nabla g(\beta) = \frac{\beta}{\|\beta\|_2}$.

(b) For $\beta = 0$, any vector $\widehat{s} \in \mathbb{R}^p$ with $\|\widehat{s}\|_2 \leq 1$ is an element of the subdifferential of g at 0.

Ex. 5.6 Show that the function

$$h(\beta_1, \ldots, \beta_p) = |\beta|^T \mathbf{P} |\beta|$$

in Equation (5.40) satisfies the regularity conditions below Equation (5.39) on page 111. (As a consequence, coordinate descent will still work even though this function is not separable).

Ex. 5.7 Show that the proximal-gradient update step (5.21) is equal to the step (5.19)

Ex. 5.8 Show that when h is given by the nuclear norm, the composite gradient update (5.26) can be obtained by the following procedure:

(a) Compute the singular value decomposition of the input matrix $\mathbf{Z}$, that is $\mathbf{Z} = \mathbf{U}\mathbf{D}\mathbf{V}^T$ where $\mathbf{D} = \text{diag}\{\sigma_j(\mathbf{Z})\}$ is a diagonal matrix of the singular values.

(b) Apply the soft-thresholding operator (5.25) to compute the "shrunken" singular values

$$\gamma_j := \mathcal{S}_{s\lambda}(\sigma_j(\mathbf{Z})), \quad \text{for } j = 1, \ldots, p.$$

(c) Return the matrix $\widehat{\mathbf{Z}} = \mathbf{U} \, \text{diag}\{\gamma_1, \ldots, \gamma_p\}\mathbf{V}^T$.

Ex. 5.9 Consider a regression problem with all variables and response having mean zero and standard deviation one in the dataset. Suppose also that each variable has identical absolute correlation with the response—that is

$$\frac{1}{N}|\langle \mathbf{x}_j, \mathbf{y}\rangle| = \lambda, \qquad \text{for all } j = 1, \ldots, p.$$

Let $\widehat{\beta}$ be the least-squares coefficient vector of $\mathbf{y}$ on $\mathbf{X}$, assumed to be unique for this exercise. Let $\mathbf{u}(\alpha) = \alpha \mathbf{X}\widehat{\beta}$ for $\alpha \in [0,1]$ be the vector that moves a fraction α toward the least-squares fit $\mathbf{u}$. Let $RSS = \|y - \mathbf{X}\widehat{\beta}\|_2^2$, the residual sum-of-squares from the full least-squares fit.

(a) Show that

$$\frac{1}{N}|\langle \mathbf{x}_j, \mathbf{y} - \mathbf{u}(\alpha)\rangle| = (1 - \alpha)\lambda \quad \text{for } j = 1, \ldots, p,$$

and hence the correlations of each $\mathbf{x}_j$ with the residuals remain equal in magnitude as we progress toward $\mathbf{u}$.

(b) Show that these correlations are all equal to

$$\lambda(\alpha) = \frac{(1 - \alpha)}{\sqrt{(1 - \alpha)^2 + \frac{\alpha(2 - \alpha)}{N} \cdot RSS}} \cdot \lambda,$$

and hence they decrease monotonically to zero.

(c) Use these results to show that the LAR algorithm in Section 5.6 keeps the correlations tied and monotonically decreasing.

Ex. 5.10 Consider step 3c of the LAR Algorithm 5.1. Define $c_\ell = \langle \mathbf{x}_\ell, \mathbf{r}_{k-1}\rangle$ and $a_\ell = \langle \mathbf{x}_\ell, \mathbf{X}_\mathcal{A}\delta\rangle$, $\ell \notin \mathcal{A}$. Define

$$\alpha_\ell = \min_+ \left\{ \frac{\lambda_{k-1} - c_\ell}{1 - a_\ell}, \frac{\lambda_{k-1} + c_\ell}{1 + a_\ell} \right\},$$

where $\min_+$ only considers positive entries. Show that the variable to enter at step k has index $j = \arg\min_{\ell \notin \mathcal{A}} \alpha_\ell$, with value $\lambda_k = \lambda_{k-1} - \alpha_j$.

Ex. 5.11 *Strong rules*

(a) Show that if the slope condition (5.80) holds, then the global and sequential strong rules (5.78) and (5.79) are guaranteed to work.

(b) In the case of orthogonal design $\mathbf{X}^T\mathbf{X} = \mathbf{I}$, show that the slope condition (5.80) always holds.

(c) Design a simulation study to investigate the accuracy of the DPP and strong rules for the lasso, in the cases $(N, p) = (100, 20)$, $(N, p) = (100, 100)$, and $(N, p) = (100, 1000)$.

Ex. 5.12 *ADMM for consensus optimization:* Suppose that we have a dataset $\{x_i, y_i\}_{i=1}^N$, and that our goal is to minimize an objective function $L(\mathbf{X}\beta - \mathbf{y})$ that decomposes additively as a sum of N terms, one for each sample. A natural approach is to divide the dataset into B blocks, and denote by $L_b(\mathbf{X}_b\beta_b - \mathbf{y}_b)$ the objective function over the bth block of data, where $\mathbf{X}_b$ and $\mathbf{y}_b$ are the corresponding blocks of $\mathbf{X}$ and $\mathbf{y}$. We thus arrive at the problem

$$\underset{\beta \in \mathbb{R}^p}{\text{minimize}} \left\{ \sum_{b=1}^B L_b(\mathbf{X}_b\beta_b - \mathbf{y}_b) + r(\theta) \right\} \text{ such that } \beta_b = \theta \text{ for all } b = 1, \ldots B.$$

(5.98)

(a) Show that the ADMM algorithm for this problem takes the form

$$\beta_b^{t+1} \leftarrow \underset{\beta_b}{\arg\min} \left(L_b(\mathbf{X}_b\beta_b - \mathbf{y}_b) + (\rho/2)\|\beta_b - \theta^t + \mu_b^t\|_2^2 \right) \tag{5.99a}$$

$$\theta^{t+1} \leftarrow \underset{\theta}{\arg\min} \left(r(z) + (N\rho/2)\|\theta - \bar{\beta}^{t+1} - \bar{\mu}^t\|_2^2 \right) \tag{5.99b}$$

$$\mu_b^{t+1} \leftarrow \mu_b^t + (\beta_b^{t+1} - \theta^{t+1}) \tag{5.99c}$$

where the $\bar{\mu}^k$ and $\bar{\beta}^{k+1}$ denote averages over blocks. Interpret it as consensus optimization.

(b) Now consider the lasso, which uses the regularizer $r(\theta) = \lambda\|\theta\|_1$. Show that the algorithm has the form

$$\beta_b^{t+1} \leftarrow (\mathbf{X}_b^T\mathbf{X}_b + \rho\mathbf{I})^{-1}(\mathbf{X}_b\mathbf{y}_b + \rho(\theta^t - \mu_b^t)) \tag{5.100a}$$

$$\theta^{t+1} \leftarrow \mathcal{S}_{\lambda/(\rho N)}(\bar{\beta}^{t+1} + \bar{\mu}^t) \tag{5.100b}$$

$$\mu_b^{t+1} \leftarrow \mu_b^t + (\beta_b^{t+1} - \theta^{t+1}) \tag{5.100c}$$

(c) Implement the updates (5.100) in software and demonstrate it on a numerical example.

Ex. 5.13

(a) Derive the alternating convex minimization (ACS) for problem (5.72), and show that it has the form of a power iteration (Equations (5.73) and (5.74)).

(b) Show that it converges to the eigenvector corresponding to the largest eigenvalue of $\mathbf{X}^T\mathbf{X}$, provided that the starting vector $\mathbf{v}_0$ is not orthogonal to this largest eigenvector.

Chapter 6

Statistical Inference

An attractive feature of ℓ_1-regularized procedures is their ability to combine variable selection with parameter fitting. We often select a model based on cross-validation—as an estimate for prediction or generalization error—and then do further validation on a held-out test set.

It is sometimes of interest to determine the statistical strength of the included variables, as in "p-values" in traditional models. The adaptive nature of the estimation procedure makes this problem difficult—both conceptually and analytically. We describe some useful approaches to the inference problem in this chapter. We begin by discussing two "traditional" approaches—Bayesian methods and the bootstrap, and then present some newer approaches to this problem.

6.1   The Bayesian Lasso

The Bayesian paradigm treats the parameters as random quantities, along with a prior distribution that characterizes our belief in what their values might be. Here we adopt the approach of Park and Casella (2008), involving a model of the form

$$\mathbf{y} \mid \beta, \lambda, \sigma \sim N(\mathbf{X}\beta, \sigma^2 \mathbf{I}_{N \times N}) \tag{6.1a}$$

$$\beta \mid \lambda, \sigma \sim \prod_{j=1}^{p} \frac{\lambda}{2\sigma} e^{-\frac{\lambda}{\sigma}|\beta_j|}, \tag{6.1b}$$

using the i.i.d. Laplacian prior (6.1b). Under this model, it is easy to show that the negative log posterior density for $\beta \mid \mathbf{y}, \lambda, \sigma$ is given by

$$\frac{1}{2\sigma^2} \|\mathbf{y} - \mathbf{X}\beta\|_2^2 + \frac{\lambda}{\sigma} \|\beta\|_1, \tag{6.2}$$

where we have dropped an additive constant independent of β. Consequently, for any fixed values of σ and λ, the posterior mode coincides with the lasso estimate (with regularization parameter $\sigma\lambda$). Park and Casella (2008) include σ^2 in the prior specification (6.1b) for technical reasons. Here we have assumed there is no constant in the model, and that the columns of $\mathbf{X}$ are

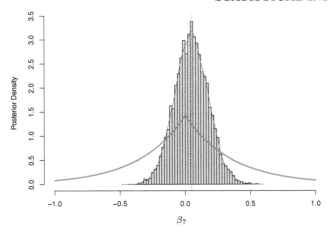

Figure 6.1 *Prior and posterior distribution for the seventh variable in the diabetes example, with λ held fixed. The prior in the figure is a double exponential (Laplace) distribution with density proportional to $\exp(-.0065|\beta_7|)$. The prior rate .0065 is a representative value just for illustration.*

mean-centered, as is $\mathbf{y}$.[1] The posterior distribution provides more than point estimates: it provides an entire joint distribution.

The red curve in Figure 6.1 is the Laplace prior used in the Bayesian lasso, applied to variable β_7 in the "diabetes data." These data consist of observations on 442 patients, with the response of interest being a quantitative measure of disease progression one year after baseline. There are ten baseline variables—age, sex, body-mass index, average blood pressure, and six blood serum measurements—plus quadratic terms, giving a total of 64 features. The prior has a sharp peak at zero, which captures our belief that some parameters are zero. Given a probability distribution (likelihood) for the observed data given the parameters, we update our prior by conditioning on the observed data, yielding the posterior distribution of the parameters. The histogram in Figure 6.1 characterizes the posterior distribution for β_7 for the diabetes data. The prior distribution has a variance parameter that characterizes the strength of our belief in zero as a special value. The posterior mode is slightly away from zero, although a 95% posterior credible interval comfortably covers zero. Exact Bayesian calculations are typically intractable, except for the simplest of models. Fortunately, modern computation allows us to use Markov chain Monte Carlo (MCMC) to efficiently sample realizations from the posterior distributions of the parameters of interest. Figure 6.2 [left panel] shows a summary of MCMC samples from the posterior distribution of $\beta \mid \lambda$; the median of 10,000 posterior samples is shown at each of 100 values of λ. Here

---

[1]This is not a real restriction on the model, and is equivalent to assuming an improper flat prior on β_0, which is rarely of interest.

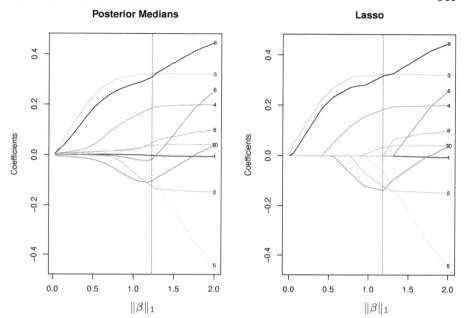

Figure 6.2 *Bayesian lasso on the diabetes data. The left plot shows the posterior medians from MCMC runs (conditional on λ). The right plot shows the lasso profile. In the left plot, the vertical line is at the posterior median of $\|\beta\|_1$ (from an unconditional model), while for the right plot the vertical line was found by N-fold cross-validation.*

σ^2 is allowed to vary (with $\pi(\sigma^2) \sim \frac{1}{\sigma^2}$). This, and the fact that we have displayed medians, accounts for the slight discrepancies with the right plot (the lasso), which shows the posterior mode for fixed values of $\sigma\lambda$. A complete Bayesian model will also specify a prior distribution for λ; in this case, a diffuse Gamma distribution is conjugate and hence convenient for the MCMC sampling. This is where the Bayesian approach can be worth the considerable extra effort and leap of faith. The full posterior distribution includes λ as well as β, so that model selection is performed automatically. Furthermore, the posterior credible intervals for β take into account the posterior variability in λ. Figure 6.3 shows a summary of 10,000 MCMC samples from the posterior distribution for the diabetes data. While the posterior mode has nine nonzero coefficients, the posterior distributions suggest that only 5–8 of these are well separated from zero.

Specifying the Bayesian model is technically challenging, and there are several choices to be made along the way. These include priors for λ and σ^2, which themselves have hyperparameters that have to be set. Our examples were fit in R using the function blasso in the package monomvn (Gramacy 2011), and for the most part we used the default parameter settings. For

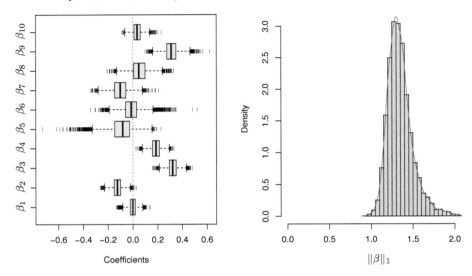

Figure 6.3 *Posterior distributions for the β_j and $\|\beta\|_1$ for the diabetes data. Summary of* 10,000 *MCMC samples, with the first* 1000 *"burn-in" samples discarded.*

this 442×10 problem it took 5 seconds on a 2.3 GHz Macbook Pro. However, Bayesian computations do not scale well; experiments in the next section show that the computational cost scales roughly as $\mathcal{O}(p^2)$.

6.2   The Bootstrap

The bootstrap is a popular nonparametric tool for assessing the statistical properties of complex estimators (Efron 1979, Efron and Tibshirani 1993). To motivate its use, suppose that we have obtained an estimate $\widehat{\beta}(\hat{\lambda}_{CV})$ for a lasso problem according to the following procedure:

1. Fit a lasso path to $(\mathbf{X}, \mathbf{y})$ over a dense grid of values $\Lambda = \{\lambda_\ell\}_{\ell=1}^L$.

2. Divide the training samples into 10 groups at random.

3. With the kth group left out, fit a lasso path to the remaining 9/10ths, using the same grid Λ.

4. For each $\lambda \in \Lambda$ compute the mean-squared prediction error for the left-out group.

5. Average these errors to obtain a prediction error curve over the grid Λ.

6. Find the value $\hat{\lambda}_{CV}$ that minimizes this curve, and then return the coefficient vector from our original fit in step (1) at that value of λ.

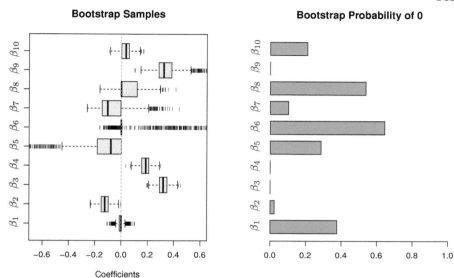

Figure 6.4 *[Left] Boxplots of 1000 bootstrap realizations of $\widehat{\beta}^*(\widehat{\lambda}_{CV})$ obtained by the nonparametric bootstrap, which corresponds to re-sampling from the empirical CDF $\widehat{F}_N$. Comparing with the corresponding Bayesian posterior distribution in Figure 6.3, we see a close correspondence in this case. [Right] Proportion of times each coefficient is zero in the bootstrap distribution.*

How do we assess the sampling distribution of $\widehat{\beta}(\widehat{\lambda}_{CV})$? That is, we are interested in the distribution of the random estimate $\widehat{\beta}(\widehat{\lambda}_{CV})$ as a function of the N i.i.d. samples $\{(x_i, y_i)\}_{i=1}^N$. The nonparametric bootstrap is one method for approximating this sampling distribution: in order to do so, it approximates the cumulative distribution function F of the random pair (X, Y) by the empirical CDF $\widehat{F}_N$ defined by the N samples. We then draw N samples from $\widehat{F}_N$, which amounts to drawing N samples with replacement from the given dataset. Figure 6.4[left] shows boxplots of 1000 bootstrap realizations $\widehat{\beta}^*(\widehat{\lambda}_{CV})$ obtained in this way, by repeating steps 1–6 on each bootstrap sample.[2] There is a reasonable correspondence between this figure, and the corresponding Bayesian results in Figure 6.3. The right plot shows the proportion of times that each variable was exactly zero in the bootstrap distribution. None of the Bayesian posterior realizations are exactly zero, although often some are close to zero. (The `blasso` function has an argument that allows for variable selection via "reversible jump" MCMC, but this was not used here.) Similar to the right-hand plot, Meinshausen and Bühlmann (2010) pro-

---

[2]On a technical note, we implement the bootstrap with observation weights $w_i^* = k/N$, with $k = 0, 1, 2, \dots$. In cross-validation, the units are again the original N observations, which carry along with them their weights w_i^*.

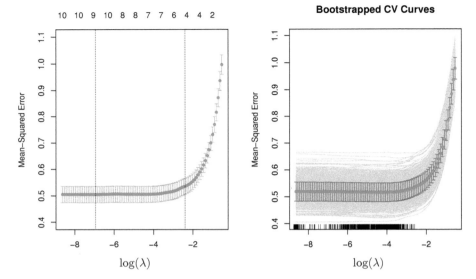

Figure 6.5 *[Left] Cross-validation curve for lasso on the diabetes data, with one-standard-error bands computed from the 10 realizations. The vertical line on the left corresponds to the minimizing value for* λ. *The line on the right corresponds to the* one-standard-error *rule; the biggest value of* λ *for which the CV error is within one standard error of the minimizing value. [Right] 1000 bootstrap CV curves, with the average in red, and one-standard-error bands in blue. The rug-plot at the base shows the locations of the minima.*

duce a *stability* plot for lasso under bootstrap resampling; as a function of λ they display what fraction of times a variable is nonzero in the bootstrapped coefficient paths.

Figure 6.5 shows the bootstrapped cross-validation curves, and their minima. Not surprisingly, the bootstrapped minima have a wide spread, since the original CV curve is flat over a broad region. Interestingly, the bootstrap standard-error bands bear a close correspondence to those computed from the original CV fit in the left plot. Figure 6.6 shows pairwise plots of the bootstrapped coefficients. From such plots we can see, for example, how correlated variables can trade off with each other, both in value and their propensity for being zero.

In Table 6.1, we show comparative timings in seconds for problems with $N = 400$ and different numbers of predictors. We generated 1000 bootstrap samples; for the Bayesian lasso we generated 2000 posterior samples, with the idea of discarding the first 1000 samples as a burn-in. While such comparisons depend on implementation details, the relative growth with p is informative. The Bayesian lasso is perhaps faster for small problems, but its complexity

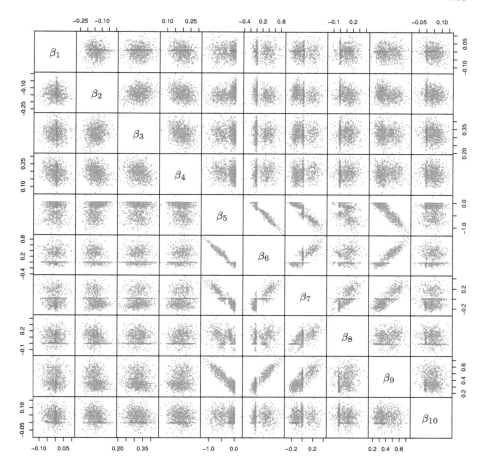

Figure 6.6 *Pairwise plots of the bootstrapped coefficients $\widehat{\beta}^*(\hat{\lambda}_{CV})$. The red points correspond to values that are zero on at least one coordinate for that plot. The samples $\mathbf{x}_5$ and $\mathbf{x}_6$ have high correlation (0.9); we see the corresponding negative correlation in their coefficients, with zero playing a prominent role.*

Table 6.1 *Timings for Bayesian lasso and bootstrapped lasso, for four different problem sizes. The sample size is $N = 400$.*

p	Bayesian Lasso	Lasso/Bootstrap
10	3.3 secs	163.8 secs
50	184.8 secs	374.6 secs
100	28.6 mins	14.7 mins
200	4.5 hours	18.1 mins

seems to scale as $\mathcal{O}(p^2)$. In contrast, the scaling of the bootstrap seems to be closer to $\mathcal{O}(p)$, because it exploits the sparseness and convexity of the lasso.

The above procedure used the *nonparametric bootstrap*, in which we estimate the unknown population F by the empirical distribution function $\widehat{F}_N$, the nonparametric maximum likelihood estimate of F. Sampling from $\widehat{F}_N$ corresponds to sampling with replacement from the data. In contrast, the *parametric bootstrap* samples from a parametric estimate of F, or its corresponding density function f. In this example, we would fix $\mathbf{X}$ and obtain estimates $\widehat{\beta}$ and $\widehat{\sigma}^2$ either from the full least-squares fit, or from the fitted lasso with parameter λ. We would then sample $\mathbf{y}$ values from the Gaussian model (6.1a), with β and σ^2 replaced by $\widehat{\beta}$ and $\widehat{\sigma}^2$.

Using the full least-squares estimates for $\widehat{\beta}$ and $\widehat{\sigma}^2$, the parametric bootstrap results for our example are shown in Figure 6.7. They are similar to both the nonparametric bootstrap results and those from the Bayesian lasso. In general, we might expect that the parametric bootstrap would likely produce results even closer to the Bayesian lasso as compared to the nonparametric bootstrap, since the parametric bootstrap and Bayesian lasso both use the assumed parametric form for data distribution (6.1a). Note also that the use of the full least squares estimates for $\widehat{\beta}$ and $\widehat{\sigma}^2$ would not work when $p \gg N$, and we would need to generate a different dataset for each value of λ. This would slow down the computations considerably.

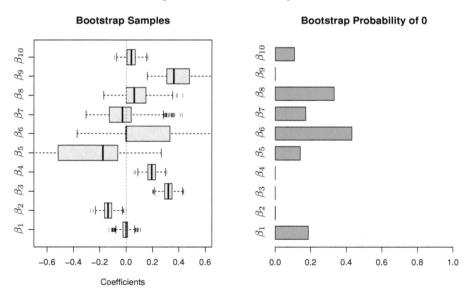

Figure 6.7 *[Left] Boxplots of 1000 parametric bootstrap realizations of $\widehat{\beta}^*(\hat{\lambda}_{CV})$. Comparing with the corresponding Bayesian posterior distribution in Figure 6.3, we again see a close correspondence. [Right] Proportion of times each coefficient is zero in the bootstrap distribution.*

In summary, in this section we have compared the Bayesian and bootstrap approach on a Gaussian linear-regression problem, for which Bayesian software was available at the time of writing. As we move to GLMs and other models, the Bayesian technical complexities grow. The bootstrap, on the other hand, can be applied seamlessly in many situations. In a general sense, the similar results for the Bayesian lasso and lasso/bootstrap are not surprising. The histogram of values from the nonparametric bootstrap can be viewed as a kind of posterior-Bayes estimate under a noninformative prior in the multinomial model (Rubin 1981, Efron 1982).

Which approach is better? Both the Bayesian and bootstrap approaches provide a way to assess variability of lasso estimates. The Bayesian approach is more principled but leans more heavily on parametric assumptions, as compared to the nonparametric bootstrap. The bootstrap procedure scales better computationally for large problems. Some further discussion of the relationship between Bayesian and bootstrap approaches is given in Efron (2011).

6.3   Post-Selection Inference for the Lasso

In this section we present some relatively recent ideas on making inference after selection by adaptive methods such as the lasso and forward-stepwise regression. The first method we discuss in Section 6.3.1 pioneered a particular line of research, and has been followed in rapid succession by a series of generalizations and improvements discussed in Section 6.3.2.

6.3.1   The Covariance Test

In this section we describe a method proposed for assigning p-values to predictors as they are successively entered by the lasso. This method is based on the LAR algorithm and its piecewise construction of the path of lasso solutions (Section 5.6).

Suppose that we are in the usual linear regression setup, with an outcome vector $\mathbf{y} \in \mathbb{R}^N$ and matrix of predictor variables $\mathbf{X} \in \mathbb{R}^{N \times p}$ related by

$$\mathbf{y} = \mathbf{X}\beta + \epsilon, \quad \epsilon \sim N(\mathbf{0}, \sigma^2 \mathbf{I}_{N \times N}), \tag{6.3}$$

where $\beta \in \mathbb{R}^p$ are unknown coefficients to be estimated.

To understand the motivation for the covariance test, let's first consider forward-stepwise regression. This procedure enters predictors one at a time, choosing the predictor that most decreases the residual sum of squares at each stage. Defining RSS_k to be the residual sum of squares for the model containing k predictors, we can use this change in residual sum-of-squares to form a test statistic

$$R_k = \frac{1}{\sigma^2}(\mathrm{RSS}_{k-1} - \mathrm{RSS}_k) \tag{6.4}$$

(with σ assumed known for now), and compare it to a χ_1^2 distribution.

Figure 6.8(a) shows the quantiles of R_1 from forward stepwise regression (the chi-squared statistic for the first predictor to enter) versus those of a χ_1^2 variate, in the fully null case ($\beta = 0$). The observed quantiles are much larger than those of the χ_1^2 distribution. A test at the 5% level, for example, using the χ_1^2 cutoff of 3.84, would have an actual type I error of about 39%.

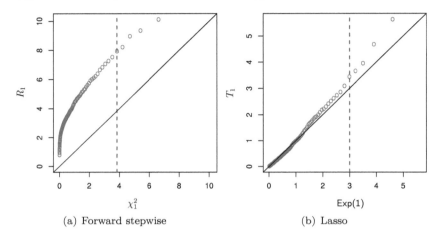

(a) Forward stepwise   (b) Lasso

Figure 6.8 *A simulation example with $N = 100$ observations and $p = 10$ orthogonal predictors and $\beta = 0$. (a) a quantile-quantile plot, constructed over 1000 simulations, of the standard chi-squared statistic R_1 in (6.4), measuring the drop in residual sum-of-squares for the first predictor to enter in forward stepwise regression, versus the χ_1^2 distribution. The dashed vertical line marks the 95% quantile of the χ_1^2 distribution. (b) a quantile-quantile plot of the covariance test statistic T_1 in (6.5) for the first predictor to enter in the lasso path, versus its asymptotic null distribution Exp(1). The covariance test explicitly accounts for the adaptive nature of lasso modeling, whereas the chi-squared test is not appropriate for adaptively selected models as in forward-stepwise regression.*

The reason for this is clear: the chi-squared test assumes that the models being compared are pre-specified, not chosen on the basis of the data. But the forward stepwise procedure has deliberately chosen the strongest predictor among all of the available choices, so it is not surprising that it yields a larger drop in training error than expected.

It seems difficult to derive an appropriate p-value for forward stepwise regression, one that properly accounts for the adaptive nature of the fitting. For the first step and the test of the global null hypothesis, one can use a permutation distribution. For subsequent steps, it is not clear how to correctly carry out permutations. One can resort to sample splitting: we divide the data in half, compute the sequence of models on one half and then evaluate their significance on the other half. But this can lead to a significant loss of power, unless the sample size is large.

Surprisingly, it turns out that for the lasso, a simple test can be derived that properly accounts for the adaptivity. Denote the knots returned by the LAR algorithm (Algorithm 5.1 on page 119) by $\lambda_1 > \lambda_2 \ldots > \lambda_K$. These are the values of the regularization parameter λ where there is a change in the set of active predictors. Suppose that we wish to test significance of the predictor entered by LAR at λ_k. Let $\mathcal{A}_{k-1}$ be the active set (the predictors with nonzero coefficients) before this predictor was added and let the estimate at the end of this step be $\hat{\beta}(\lambda_{k+1})$. We refit the lasso, keeping $\lambda = \lambda_{k+1}$ but using just the variables in $\mathcal{A}_{k-1}$. This yields the estimate $\hat{\beta}_{\mathcal{A}_{k-1}}(\lambda_{k+1})$. The *covariance test statistic* is defined by

$$T_k = \frac{1}{\sigma^2} \cdot \left(\langle \mathbf{y}, \mathbf{X}\hat{\beta}(\lambda_{k+1}) \rangle - \langle \mathbf{y}, \mathbf{X}\hat{\beta}_{\mathcal{A}_{k-1}}(\lambda_{k+1}) \rangle \right). \tag{6.5}$$

This statistic measures how much of the covariance between the outcome and the fitted model can be attributed to the predictor that has just entered the model; i.e., how much improvement there was over the interval $(\lambda_k, \lambda_{k+1})$ in this measure. Interestingly, for forward-stepwise regression, the corresponding covariance statistic is equal to R_k (6.4); however, for the lasso this is not the case (Exercise 6.2).

Remarkably, under the null hypothesis that all $k - 1$ signal variables are in the model, and under general conditions on the model matrix $\mathbf{X}$, for the predictor entered at the next step we have

$$T_k \overset{d}{\to} \text{Exp}(1) \tag{6.6}$$

as $N, p \to \infty$. Figure 6.8(b) shows the quantile-quantile plot for T_1 versus $\text{Exp}(1)$. When σ^2 is unknown, we estimate it using the full model: $\hat{\sigma}^2 = \frac{1}{N-p}\text{RSS}_p$. We then plug this into (6.5), and the exponential test becomes an $F_{2,N-p}$ test.

Table 6.2 shows the results of forward stepwise regression and LAR/lasso applied to the diabetes data. Only the first ten steps are shown in each case. We see that forward stepwise regression enters eight terms at level 0.05, while the covariance test enters only four. However as we argued above, the forward stepwise p-values are biased downward, and hence they are not trustworthy. In Exercise 6.3 we discuss a method for combining a set of sequential p-values to control the false discovery rate of the list of selected predictors. When applied to the covariance test at an FDR of 5%, it yields a model containing the first four predictors. For comparison, cross-validation estimated the optimal model size for prediction to be in the range of 7 to 14 predictors.

Why is the mean of the forward-stepwise statistic R_1 much larger than one, while the mean of T_1 is approximately equal to one? The reason is *shrinkage*: the lasso picks the best predictor available at each stage, but does not fit it fully by least squares. It uses shrunken estimates of the coefficients, and this shrinkage compensates exactly for the inflation due to the selection. This test is also the natural analogue of the degrees of freedom result for the lasso and

Table 6.2 *Results of forward stepwise regression and LAR/lasso applied to the di-*
abetes data introduced in Chapter 2. Only the first ten steps are shown in each case.
The p-values are based on (6.4), (6.5), and (6.11), respectively. Values marked as 0
are < 0.01.

| | Forward Stepwise | | LAR/lasso | | |
Step	Term	p-value	Term	p-value Covariance	Spacing
1	bmi	0	bmi	0	0
2	ltg	0	ltg	0	0
3	map	0	map	0	0.01
4	age:sex	0	hdl	0.02	0.02
5	bmi:map	0	bmi:map	0.27	0.26
6	hdl	0	age:sex	0.72	0.67
7	sex	0	glu^2	0.48	0.13
8	glu^2	0.02	bmi^2	0.97	0.86
9	age^2	0.11	age:map	0.88	0.27
10	tc:tch	0.21	age:glu	0.95	0.44

LAR, discussed in Section 2.5. The lasso with k nonzero coefficients has k
degrees of freedom in expectation, and LAR spends one degree of freedom in
each segment $(\lambda_{k+1}, \lambda_k)$ along the path. The covariance test has mean equal
to one, the degrees of freedom per step. In a sense, the Exp(1) distribution is
the analogue of the χ_1^2 distribution, for adaptive fitting.

The exponential limiting distribution for the covariance test (6.5) requires
certain conditions on the data matrix $\mathbf{X}$, namely that the signal variables
(having nonzero true coefficients) are not too correlated with the noise variables. These conditions are similar to those needed for support recovery for
the lasso (Chapter 11). In the next section we discuss a more general scheme
that gives the *spacing test*, whose null distribution holds exactly for finite N
and p, and works for any $\mathbf{X}$.

6.3.2 A General Scheme for Post-Selection Inference

Here we discuss a general scheme for inference after selection—one that yields
exact p-values and confidence intervals in the Gaussian case. It can deal with
any procedure for which the selection events can be characterized by a set of
linear inequalities in $\mathbf{y}$. In other words, the selection event can be written as
$\{\mathbf{A}\mathbf{y} \leq b\}$ for some matrix $\mathbf{A}$ and vector b. In particular, it can be applied to
successive steps of the LAR algorithm, where it gives an exact (finite sample)
form of the covariance test. Similarly, it can be applied to forward stepwise
regression, and to the lasso at a fixed choice of the regularization parameter λ.

Why can the selection events for these procedures be written in the form
$\{\mathbf{A}\mathbf{y} \leq b\}$? This is easiest to see for forward-stepwise regression. In this case we

take $b = 0$. At the first step, forward-stepwise regression chooses the predictor whose absolute inner product with $\mathbf{y}$ is the largest (see Figure 6.10 for an illustration). This can be expressed by forming $2(p - 1)$ rows in the matrix $\mathbf{A}$, each computing a difference of inner products, once each for the positive and negative directions. Similarly, at the next step we add $2(p - 2)$ rows contrasting the inner product between the selected predictor and the other $p - 2$ predictors, and so on.

The lasso solution at a fixed value of λ is characterized by an active set of variables $\mathcal{A}$, along with the signs of their coefficients. Again, it turns out that the selection event that led to this particular combination can be written in the form $\{\mathbf{Ay} \leq b\}$ for some $\mathbf{A}$ and b. That is, the set $\{\mathbf{y}|\mathbf{Ay} \leq b\}$ corresponds to the values of the outcome vector $\mathbf{y}$ that would yield this same collection of active variables and signs (with $\mathbf{X}$ fixed) (see Lee, Sun, Sun and Taylor (2016), and Exercise 6.10). The same is true for the LAR algorithm after its kth step.

Now suppose that $\mathbf{y} \sim N(\boldsymbol{\mu}, \sigma^2 \mathbf{I}_{N \times N})$, and that we want to make inferences conditional on the event $\{\mathbf{Ay} \leq b\}$. In particular, we wish to make inferences about $\boldsymbol{\eta}^T \boldsymbol{\mu}$, where $\boldsymbol{\eta}$ might depend on the selection event. With lasso, LAR, or forward-stepwise regression having selected this set, we can now make inference statements about the selected variables. For example, we could be interested in the (ordinary) regression coefficients of $\mathbf{y}$ on $\mathbf{X}_{\mathcal{A}}$, namely $\hat{\theta} = (\mathbf{X}_{\mathcal{A}}^T \mathbf{X}_{\mathcal{A}})^{-1} \mathbf{X}_{\mathcal{A}}^T \mathbf{y}$. These correspond to the population parameters $\theta = (\mathbf{X}_{\mathcal{A}}^T \mathbf{X}_{\mathcal{A}})^{-1} \mathbf{X}_{\mathcal{A}}^T \boldsymbol{\mu}$, the coefficients in the projection of $\boldsymbol{\mu}$ on $\mathbf{X}_{\mathcal{A}}$. So here $\boldsymbol{\eta}^T \boldsymbol{\mu}$ could correspond to one of these coefficients, and hence $\boldsymbol{\eta}$ is one of the columns of $\mathbf{X}_{\mathcal{A}}(\mathbf{X}_{\mathcal{A}}^T \mathbf{X}_{\mathcal{A}})^{-1}$. We pursue this example in Section 6.3.2.1.

Lee et al. (2016) and Taylor, Lockhart, Tibshirani[2] and Tibshirani (2014) show that

$$\{\mathbf{Ay} \leq b\} = \{\mathcal{V}^-(\mathbf{y}) \leq \boldsymbol{\eta}^T \mathbf{y} \leq \mathcal{V}^+(\mathbf{y}), \ \mathcal{V}^0(\mathbf{y}) \geq 0\}, \tag{6.7}$$

and furthermore, $\boldsymbol{\eta}^T \mathbf{y}$ and $(\mathcal{V}^-(\mathbf{y}), \mathcal{V}^+(\mathbf{y}), \mathcal{V}^0(\mathbf{y}))$ are statistically independent. See Figure 6.9 for a geometric view of this surprising result, known as the *polyhedral lemma*. The three values on the right in (6.7) are computed via

$$\alpha = \frac{\mathbf{A}\boldsymbol{\eta}}{\|\boldsymbol{\eta}\|_2^2}$$

$$\mathcal{V}^-(\mathbf{y}) = \max_{j:\alpha_j < 0} \frac{b_j - (\mathbf{Ay})_j + \alpha_j \boldsymbol{\eta}^T \mathbf{y}}{\alpha_j}$$

$$\mathcal{V}^+(\mathbf{y}) = \min_{j:\alpha_j > 0} \frac{b_j - (\mathbf{Ay})_j + \alpha_j \boldsymbol{\eta}^T \mathbf{y}}{\alpha_j} \tag{6.8}$$

$$\mathcal{V}^0(\mathbf{y}) = \min_{j:\alpha_j = 0} (b_j - (\mathbf{Ay})_j)$$

(Exercise 6.7). Hence the selection event $\{\mathbf{Ay} \leq b\}$ is equivalent to the event that $\boldsymbol{\eta}^T \mathbf{y}$ falls into a certain range, a range depending on $\mathbf{A}$ and b. This equivalence and the independence means that the conditional inference on

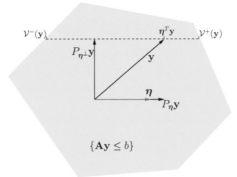

Figure 6.9 *Schematic illustrating the polyhedral lemma (6.7), for the case $N = 2$ and $\|\boldsymbol{\eta}\|_2 = 1$. The yellow region is the selection event $\{\mathbf{A}\mathbf{y} \leq b\}$. We decompose $\mathbf{y}$ as the sum of two terms: its projection $P_{\boldsymbol{\eta}}\mathbf{y}$ onto $\boldsymbol{\eta}$ (with coordinate $\boldsymbol{\eta}^T\mathbf{y}$) and its projection onto the $(N-1)$-dimensional subspace orthogonal to $\boldsymbol{\eta}$: $\mathbf{y} = P_{\boldsymbol{\eta}}\mathbf{y} + P_{\boldsymbol{\eta}^\perp}\mathbf{y}$. Conditioning on $P_{\boldsymbol{\eta}^\perp}\mathbf{y}$, we see that the event $\{\mathbf{A}\mathbf{y} \leq b\}$ is equivalent to the event $\{\mathcal{V}^-(\mathbf{y}) \leq \boldsymbol{\eta}^T\mathbf{y} \leq \mathcal{V}^+(\mathbf{y})\}$. Furthermore $\mathcal{V}^+(\mathbf{y})$ and $\mathcal{V}^-(\mathbf{y})$ are independent of $\boldsymbol{\eta}^T\mathbf{y}$ since they are functions of $P_{\boldsymbol{\eta}^\perp}\mathbf{y}$ only, which is independent of $\mathbf{y}$.*

$\boldsymbol{\eta}^T\boldsymbol{\mu}$ can be made using the truncated distribution of $\boldsymbol{\eta}^T\mathbf{y}$, a truncated normal distribution.

To use this fact, we define the cumulative distribution function (CDF) of a truncated normal distribution with support confined to $[c, d]$:

$$F_{\mu,\sigma^2}^{c,d}(x) = \frac{\Phi((x-\mu)/\sigma) - \Phi((c-\mu)/\sigma)}{\Phi((d-\mu)/\sigma) - \Phi((c-\mu)/\sigma)}, \tag{6.9}$$

with Φ the CDF of the standard Gaussian. Now the CDF of a random variable, evaluated at the value of that random variable, has a uniform distribution. Hence we can write

$$F_{\boldsymbol{\eta}^T\boldsymbol{\mu},\ \sigma^2\|\boldsymbol{\eta}\|_2^2}^{\mathcal{V}^-,\mathcal{V}^+}\left(\boldsymbol{\eta}^T\mathbf{y}\right) \mid \{\mathbf{A}\mathbf{y} \leq b\} \sim \mathrm{U}(0,1). \tag{6.10}$$

This result is used to make conditional inferences about any linear functional $\boldsymbol{\eta}^T\boldsymbol{\mu}$. For example, we can compute a p-value for testing $\boldsymbol{\eta}^T\boldsymbol{\mu} = 0$. We can also construct a $1 - \alpha$ level selection interval for $\theta = \boldsymbol{\eta}^T\boldsymbol{\mu}$ by inverting this test, as follows. Let $P(\theta) = F_{\theta,\ \sigma^2\|\boldsymbol{\eta}\|_2^2}^{\mathcal{V}^-,\mathcal{V}^+}\left(\boldsymbol{\eta}^T\mathbf{y}\right) \mid \{\mathbf{A}\mathbf{y} \leq b\}$. The lower boundary of the interval is the largest value of θ such that $1 - P(\theta) \leq \alpha/2$, and the upper boundary is the smallest value of θ such that $P(\theta) \leq \alpha/2$.

Example 6.1. To help understand these results, we present an example. We simulated $N = 60$ observations from the model $Y = \sum_{j=1}^p X_j\beta_j + Z$, with $X_1, X_2, \ldots, X_p$, $Z \sim N(0,1)$, and each standardized to have sample mean zero and unit ℓ_2 norm. We considered the global null case with all $\beta_j = 0$, and

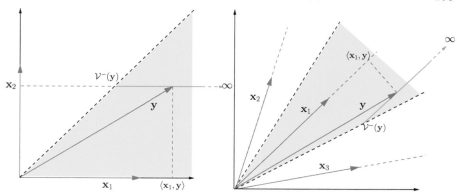

Figure 6.10 *Selection regions in Example 6.1 for which* $\lambda_1 = \langle \mathbf{x}_1, \mathbf{y} \rangle$. *Left panel: two orthogonal predictors; right panel: three correlated predictors. The red line indicates the part of the set* $P_{\boldsymbol{\eta}^\perp}\mathbf{y} + t\boldsymbol{\eta}$ *inside the selection region. In the left panel,* $\mathcal{V}^-(\mathbf{y}) = \langle \mathbf{x}_2, \mathbf{y} \rangle$, *while in the right it is* λ_2.

found the predictor j_1 having largest absolute inner product with $\mathbf{y}$. This is the first variable to enter the LAR or lasso path. We wish to make inference on λ_1, the value of the largest knot in LAR, under the global null hypothesis. Thus $\boldsymbol{\eta} = \mathbf{x}_{j_1}$ and $\boldsymbol{\eta}^T\mathbf{y}$ is the attained inner product (for simplicity we condition on a positive sign for the inner-product). Note that with our standardization, $\boldsymbol{\eta}^T\mathbf{y} = \mathbf{x}_{j_1}^T\mathbf{y}$ is also the simple least-squares coefficient of $\mathbf{y}$ on the chosen $\mathbf{x}_{j_1}$, and so we are also making (conditional) inference on the population coefficient in the simple regression of $\mathbf{y}$ on $\mathbf{x}_{j1}$. We chose five scenarios with number of predictors $p \in \{2, 5, 10, 20, 50\}$. We also considered two correlation patterns for the predictors: uncorrelated and pairwise correlation 0.5. Figure 6.10 illustrates the corresponding version of Figure 6.9 for the two situations. The upper bound in all cases is $\mathcal{V}^+ = \infty$, and the lower bound $\mathcal{V}^-$ depends on $\mathbf{y}$ in each simulation. In the orthogonal case (left panel), conditioning on $P_{\boldsymbol{\eta}^\perp}\mathbf{y}$ reduces to conditioning on the values of $|\mathbf{x}_k^T\mathbf{y}|$, for all predictors k not achieving the maximum absolute inner product. Hence the lower bound on $\boldsymbol{\eta}^T\mathbf{y}$ is the second-largest among these. The right panel shows the nonorthogonal case, with correlations between the X_j. Here the situation is slightly more complex, but nevertheless a simple formula can be used to derive $\mathcal{V}^-(\mathbf{y})$: it turns out to be λ_2, the second knot in the LAR sequence (Exercise 6.11). Figure 6.11 shows the resulting truncated normal densities from (6.10), averaged over 100 simulations. We plotted the density for the average value of $\mathcal{V}^-$ over the simulations. The colored squares along the bottom show the average largest inner product $\lambda_1 = \boldsymbol{\eta}^T\mathbf{y}$ in each setting. In the lower panel, with larger p, the effective number of variables is smaller due to the correlation, so the maximum is smaller as well. We pursue this example further in Section 6.3.2.2.   ◇

This general mechanism (6.10) allows one to make inferences about any

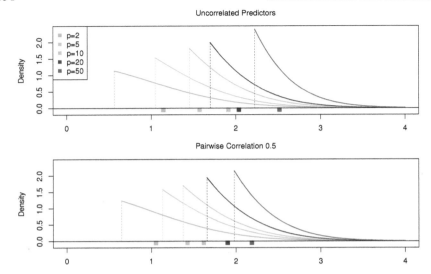

Figure 6.11 *Simulation: $N = 60$ observations from the model with $X_1, X_2, \ldots X_p \sim N(0, 1)$, $Y = \sum_j X_{ij}\beta_j + Z$ with $Z \sim N(0, 1)$, with all $\beta_j = 0$; two different predictor-correlation settings. The selection chooses the predictor j_1 having largest absolute inner product with $\mathbf{y}$. Shown is the truncated density on the left-hand side of (6.10) for $p = 2, 5, 10, 20, 50$. The colored squares along the bottom show the average largest inner product in each setting.*

linear functional $\boldsymbol{\eta}^T \boldsymbol{\mu}$; for example, inferences about any parameter $\boldsymbol{\eta}^T \boldsymbol{\mu}$ at a given step of the LAR algorithm, or at a lasso solution computed at λ. The form of $\mathbf{A}$ and b is different depending on the setting, but otherwise the construction is the same. We illustrate two applications in the next two sections.

6.3.2.1   Fixed-λ Inference for the Lasso

Consider the solution to the lasso, at some fixed value of λ. We can apply result (6.10) by constructing $\mathbf{A}$ and b so that the event $\{\mathbf{Ay} \leq b\}$ represents the set of outcome vectors $\mathbf{y}$ that yield the observed active set and signs of the predictors selected by the lasso at λ. These inequalities derive from the sub-gradient conditions that characterize the solution (Exercise 6.10). This yields an active set $\mathcal{A}$ of variables, and we can now make conditional inference on the population regression coefficients of $\mathbf{y}$ on $\mathbf{X}_{\mathcal{A}}$, for example. This means we will perform a separate conditional analysis for $\boldsymbol{\eta}$ equal to each of the columns of $\mathbf{X}_{\mathcal{A}}(\mathbf{X}_{\mathcal{A}}^T \mathbf{X}_{\mathcal{A}})^{-1}$. Hence we can obtain exact p-values and confidence intervals for the parameters of the active set in the lasso solution at λ. These quantities have the correct type-I error and coverage *conditional on the membership and signs of the active set.*[3]

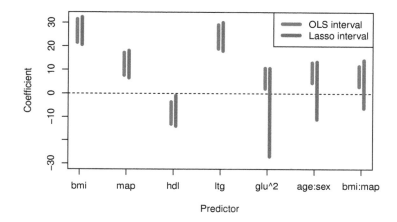

Figure 6.12 *Lasso model fit to the diabetes data. The solution at $\lambda = 7$ yields a model with seven nonzero coefficients. Shown are the 95% confidence intervals for the least-squares fit using the chosen variables. The OLS intervals ignore the selection, while the lasso intervals are exact under a Gaussian assumption, and condition on the selection event. Disclosure: λ was selected by cross-validation (1 SE rule), and σ^2 in (6.10) was estimated using the residuals from the full regression on all 64 variables.*

Figure 6.12 shows the result of fixed-λ lasso inference for the diabetes data with $\lambda = 7$; seven variables were selected. Notice that we now focus attention on the OLS regression coefficients using the reduced model containing only those seven predictors. The blue intervals are based on the usual multiple regression normal theory, ignoring the fact that we used the data to select the seven variables from the full set of 64. The red post-selection intervals were constructed by inverting relationship (6.10), and take the selection into account. We see that these two sets of intervals are similar for the larger coefficients, but the selection-adjusted ones are (appropriately) longer for the smaller coefficients.

How did we choose $\lambda = 7$? Here we cheated a bit, and used ten-fold cross-validation (using the one-standard-error rule). In practice one would need to condition on this selection event as well, which would add considerably more complexity to the selection set. Simulations suggest this does not widen the confidence intervals substantially. In the next section, we discuss conditional inference at the LAR sequence $\{\lambda_k\}$, which limits the set of λs to the knots in

---

[3]Lee et al. (2016) also discuss inference without conditioning on the signs, by considering the union of all regions with the same active set.

the lasso path. We also needed to estimate σ in (6.10); since $N > p$, we used the root mean-squared error from the full regression on all 64 predictors for this purpose.

6.3.2.2   The Spacing Test for LAR

Here we apply the inference procedure (6.10) to successive steps of the LAR algorithm. We already visited the first step in Example 6.1 on page 152, for testing the global null hypothesis. There we set $\boldsymbol{\eta}_1^T \mathbf{y} = \lambda_1 = \max_j |\langle \mathbf{x}_j, \mathbf{y} \rangle|$, and the test amounts to testing if this maximum covariance exceeds what we expect by chance. We saw that $\mathcal{V}^- = \lambda_2$, $\mathcal{V}^+ = +\infty$, and hence the resulting test can be written in a very simple form:

$$R_1 = 1 - F_{0,\sigma^2}^{\mathcal{V}^-,\mathcal{V}^+} \left(\lambda_1 | \{\mathbf{A}\mathbf{y} \leq b\} \right) = \frac{1 - \Phi(\lambda_1/\sigma)}{1 - \Phi(\lambda_2/\sigma)} \sim U(0,1). \qquad (6.11)$$

Remarkably, the uniform distribution above holds *exactly* for finite N and p, and for any $\mathbf{X}$. This is known as the *spacing* test (Taylor et al. 2014) for the global null hypothesis: it is a nonasymptotic version of the covariance test, and is asymptotically equivalent to it (Exercise 6.5). The spacing test is a monotone function of $\lambda_1 - \lambda_2$: the larger this spacing, the smaller the p-value.

Similarly, there is a more general form of the spacing test for testing that the partial regression coefficient of the variable added at any given LAR step is zero. These tests are based on the successive values for λ_k, and result in expressions more complex than Equation (6.11).

In detail, if variable $\mathbf{x}_{j_k}$ is chosen at the kth step in the LAR algorithm, one can show that the corresponding knot λ_k is given by $\lambda_k = \boldsymbol{\eta}_k^T \mathbf{y}$, with

$$\boldsymbol{\eta}_k = \frac{\mathbf{P}_{\mathcal{A}_{k-1}}^{\perp} \mathbf{x}_{j_k}}{s_k - \mathbf{x}_{j_k}^T \mathbf{X}_{\mathcal{A}_{k-1}} (\mathbf{X}_{\mathcal{A}_{k-1}}^T \mathbf{X}_{\mathcal{A}_{k-1}})^{-1} s_{\mathcal{A}_{k-1}}} \qquad (6.12)$$

(Exercise 6.8). Here $\mathcal{A}_{k-1}$ indexes the active set after $k-1$ steps, and

$$\mathbf{P}_{\mathcal{A}_{k-1}}^{\perp} = \mathbf{I}_N - \mathbf{X}_{\mathcal{A}_{k-1}} (\mathbf{X}_{\mathcal{A}_{k-1}}^T \mathbf{X}_{\mathcal{A}_{k-1}})^{-1} \mathbf{X}_{\mathcal{A}_{k-1}}^T$$

is the residual projection operator that "adjusts" $\mathbf{x}_{j_k}$ for $\mathbf{X}_{\mathcal{A}_{k-1}}$. Finally, s_k and $s_{\mathcal{A}_{k-1}}$ are the signs for the coefficients for variables k and those indexed by $\mathcal{A}_{k-1}$ (the latter being a $(k-1)$-vector). Using this value of $\boldsymbol{\eta}$, the spacing test follows from the general inference procedure outlined above, culminating in (6.10) on page 152. The matrix $\mathbf{A}$ at knot λ_k has considerably more rows than in the fixed-λ case, since we are conditioning on the entire sequence $\{\lambda_\ell\}_1^k$. Nevertheless the computations are quite manageable, and one can compute exact p-values as well as confidence intervals for the chosen variables, as in the fixed-λ case.

Taylor et al. (2014) give some simplified versions of the general spacing test—approximations to the exact case, that empirically are very close, and asymptotically are equivalent (also with the covariance test). The most appealing of these has the form

$$
R_k = \frac{\Phi\left(\frac{\lambda_{k-1}}{\sigma\|\boldsymbol{\eta}_k\|_2}\right) - \Phi\left(\frac{\lambda_k}{\sigma\|\boldsymbol{\eta}_k\|_2}\right)}{\Phi\left(\frac{\lambda_{k-1}}{\sigma\|\boldsymbol{\eta}_k\|_2}\right) - \Phi\left(\frac{\lambda_{k+1}}{\sigma\|\boldsymbol{\eta}_k\|_2}\right)},
\tag{6.13}
$$

which is an exact generalization of (6.11), using $\mathcal{V}^- = \lambda_{k-1}$ and $\mathcal{V}^+ = \lambda_{k+1}$. It is easy to see that the term of interest (top-right in (6.13)) is

$$
\tilde{\theta}_k = \frac{\lambda_k}{\sigma\|\boldsymbol{\eta}\|_2} = \frac{\boldsymbol{\eta}_k^T \mathbf{y}}{\sigma\|\boldsymbol{\eta}\|_2}
\tag{6.14}
$$

is the (absolute) standardized partial regression coefficient for $\mathbf{x}_{j_k}$ in the presence of $\mathbf{X}_{\mathcal{A}_{k-1}}$ (Exercise 6.9); this view shows that testing for λ_k amounts to testing for this partial regression coefficient.

The rightmost column of Table 6.2 shows the result of this more general spacing test applied to the diabetes data. Qualitatively the results look similar to those from the covariance test.

Although the spacing test and fixed-λ approaches are similar in their construction, and are both exact, they are different in an important way. In particular, the spacing test applies to each step of the sequential LAR procedure, and uses specific λ values (the knots). In contrast, the fixed-λ inference can be applied at any value of λ, but then treats this value as fixed. Hence it ignores any additional variability caused by choosing λ from the data.

6.3.3 What Hypothesis Is Being Tested?

In adaptive testing, this question is tricky. The covariance test uses a set of conditional hypotheses: at each stage of LAR, we are testing whether the coefficients of all other predictors not yet in the model are zero. This is sometimes called the *complete null* hypothesis

It turns out that the spacing test has a different focus. At the first step, it tests the global null hypothesis, as does the covariance test. But at subsequent steps, it tests whether the partial correlation of the given predictor entered at that step is zero, adjusting for other variables that are currently in the model. This is sometimes called the *incremental null* hypothesis. Unlike the covariance test, it does not try to assess the overall correctness of the current model. The fixed-λ test is similar; it conditions on the current active set of predictors and tests whether the coefficient of *any* given predictor is zero in the projected model. In contrast, Section 6.4 below discusses a procedure which forms confidence intervals for the *population* regression parameters in the full model.

6.3.4   Back to Forward Stepwise Regression

At the beginning of this section, we complained that naïve inference for forward-stepwise regression ignores the effects of selection, as in Figure 6.8(a) and the left side of Table 6.2. Coming full circle, we note that the general inference procedure outlined in Section 6.3.2 can in fact be applied to forward stepwise regression, providing proper selective inference for that procedure as well. In that case, the constraint matrix $\mathbf{A}$ is somewhat complicated, containing approximately $2pk$ rows at step k. However the resulting procedure is computationally tractable: details are in Taylor et al. (2014) and Loftus and Taylor (2014).

6.4   Inference via a Debiased Lasso

The aim of the method that we describe here is quite different from those discussed in Section 6.3. It does not attempt to make inferences about the partial regression coefficients in models derived by LAR or the lasso. Instead it directly estimates confidence intervals for the full set of population regression parameters, under an assumed linear model. To do so, it uses the lasso estimate[4] as a starting point and applies a debiasing operation to yield an estimate that can be used for constructing confidence intervals.

Suppose we assume that the linear model $\mathbf{y} = \mathbf{X}\beta + \boldsymbol{\epsilon}$ is correct, and we want confidence intervals for the components $\{\beta_j\}_1^p$. Then if $N > p$, we can simply fit the full model by least squares and use standard intervals from least-squares theory

$$\widehat{\beta}_j \pm z^{(\alpha)} v_j \hat{\sigma}, \qquad (6.15)$$

where $\widehat{\beta}$ is the OLS estimate, $v_j^2 = \left(\mathbf{X}^T\mathbf{X}\right)_{jj}^{-1}$, $\hat{\sigma}^2 = \sum_i (y_i - \hat{y}_i)^2/(N-p)$, and z^α is the α-percentile of the standard normal distribution. However this approach does not work when $N < p$.

One proposal that has been suggested (Zhang and Zhang 2014, Bühlmann 2013, van de Geer, Bühlmann, Ritov and Dezeure 2013, Javanmard and Montanari 2014), is to use a debiased version of the lasso estimator, namely

$$\widehat{\beta}^d = \widehat{\beta}_\lambda + \frac{1}{N}\boldsymbol{\Theta}\mathbf{X}^T(\mathbf{y} - \mathbf{X}\widehat{\beta}_\lambda), \qquad (6.16)$$

where $\widehat{\beta}_\lambda$ is the lasso estimate at λ, and $\boldsymbol{\Theta}$ is an approximate inverse of $\widehat{\boldsymbol{\Sigma}} = \frac{1}{N}\mathbf{X}^T\mathbf{X}$.[5] From this, we can write

$$\widehat{\beta}^d = \beta + \frac{1}{N}\boldsymbol{\Theta}\mathbf{X}^T\boldsymbol{\epsilon} + \underbrace{(\mathbf{I}_p - \frac{1}{N}\boldsymbol{\Theta}\mathbf{X}^T\mathbf{X})(\widehat{\beta}_\lambda - \beta)}_{\hat{\boldsymbol{\Delta}}} \qquad (6.17)$$

---

[4]Fit using a value of λ based on consistency considerations.

[5]If $N \geq p$, then $\boldsymbol{\Theta}^{-1} = \frac{1}{N}\mathbf{X}^T\mathbf{X}$ and (6.16) would be exactly unbiased for β. However when $N < p$, $\mathbf{X}^T\mathbf{X}/N$ is not invertible and we try to find an approximate inverse.

with $\epsilon \sim N(\mathbf{0}, \sigma^2 \mathbf{I}_p)$. These authors provide (different) estimates of $\boldsymbol{\Theta}$ so that $\|\hat{\boldsymbol{\Delta}}\|_\infty \to 0$. From Equation (6.17), one can use the approximation $\hat{\beta}^d \sim N(\beta, \frac{\sigma^2}{N}\boldsymbol{\Theta}\hat{\boldsymbol{\Sigma}}\boldsymbol{\Theta}^T)$ to form confidence intervals for the components β_j. The debiasing operation (6.16) can be viewed as an approximate Newton step for optimizing the residual sum of squares, starting at the lasso estimate β (Exercise 6.6). There have been different proposals for estimating $\boldsymbol{\Theta}$:

- van de Geer et al. (2013) estimate $\boldsymbol{\Theta}$ using neighborhood-based methods to impose sparsity on the components (see Chapter 9 for details on sparse graph estimation).

- Javanmard and Montanari (2014) use a different approach: for each j they define m_j to be the solution to the convex program

$$\underset{m \in \mathbb{R}^p}{\text{minimize}} \qquad m^T \hat{\boldsymbol{\Sigma}} m \tag{6.18}$$

$$\text{subject to} \qquad \|\hat{\boldsymbol{\Sigma}} m - e_j\|_\infty \leq \gamma, \tag{6.19}$$

with e_j being the jth unit vector. Then they set

$$\hat{\boldsymbol{\Theta}} := (m_1, m_2, \ldots, m_p). \tag{6.20}$$

This tries to make both $\hat{\boldsymbol{\Sigma}}\hat{\boldsymbol{\Theta}} \approx \mathbf{I}$ and the variances of $\hat{\beta}_j^d$ small.

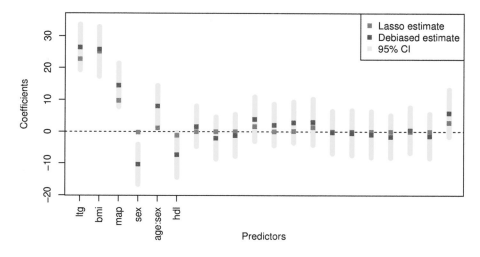

Figure 6.13 *Diabetes data: Lasso estimates, debiased lasso estimates, and confidence intervals from the debiasing approach. These intervals have not been adjusted for multiple comparisons. The first 6 predictors have intervals not containing zero; when Bonferroni-adjusted, this number drops to three.*

Figure 6.13 shows the result of applying the debiasing procedure of Javanmard and Montanari (2014) to the diabetes data. For six of the predictors, the

95% confidence intervals do not contain zero. However these intervals are not corrected for multiple comparisons; if we instead use a Bonferroni-adjusted level of 0.05/64, then the number of significant predictors drops to three. The top three predictors agree with those from the covariance and spacing tests of Table 6.2; the fourth predictor (sex) is not entered until step seven by the forward stepwise algorithm of Table 6.2, and not in the first ten steps by the other two procedures.

6.5   Other Proposals for Post-Selection Inference

The PoSI method (Berk, Brown, Buja, Zhang and Zhao 2013, "Post-Selection Inference") fits the selected submodel, and then adjusts the standard (non-adaptive) confidence intervals by accounting for all possible models that might have been delivered by the selection procedure. The adjustment is not a function of the particular search method used to find the given model. This can be both an advantage and a disadvantage. On the positive side, one can apply the method to published results for which the search procedure is not specified by the authors, or there is doubt as to whether the reported procedure is an accurate account of what was actually done. On the negative side, it can produce very wide (conservative) confidence intervals in order to achieve its robustness property.

In detail, consider again the linear model $\mathbf{y} = \mathbf{X}\beta + \epsilon$, and suppose that a model-selection procedure $\mathcal{M}$ chooses a submodel M, with estimate $\widehat{\beta}_M$. The authors of PoSI argue that inferences should most naturally be made not about the true underlying parameter vector β, but rather the parameters in the projection of $\mathbf{X}\beta$ onto $\mathbf{X}_M$:

$$\beta_M = (\mathbf{X}_M^T \mathbf{X}_M)^{-1} \mathbf{X}_M^T \mathbf{X}\beta. \qquad (6.21)$$

This approach was also adopted with the conditional inference discussed in Section 6.3.2. Consider a confidence interval for the jth component of β_M of the form

$$\text{CI}_{j \cdot M} = \widehat{\beta}_{j \cdot M} \pm K \hat{\sigma} v_{j \cdot M}, \qquad (6.22)$$

with $v_{j \cdot M}^2 = (\mathbf{X}_M^T \mathbf{X}_M)_{jj}^{-1}$. Then the PoSI procedure delivers a constant K so that

$$\Pr(\beta_{j \cdot M} \in \text{CI}_{j \cdot M}) \geq 1 - 2\alpha \qquad (6.23)$$

over all possible model selection procedures $\mathcal{M}$. The value of K is a function of the data matrix $\mathbf{X}$ and the maximum number of nonzero components allowed in β_M, but not the outcome vector $\mathbf{y}$. The authors show that K grows like $\sqrt{2 \log(p)}$ for orthogonal designs but can grow as quickly as $\sqrt{p}$ for nonorthogonal designs.

Note that any individual parameter in any projected submodel of the form (6.21) can be written as $a^T \beta$, with least-squares estimate $a^T \widehat{\beta}$, where $\widehat{\beta}$ is the least-squares estimate for the full model. Scheffé (1953) provides a way to obtain simultaneous inference for *all* such linear combinations:

$$\Pr\left[\sup_a \frac{[a^T(\hat{\beta} - \beta)]^2}{a^T(\mathbf{X}^T\mathbf{X})^{-1}a \cdot \hat{\sigma}^2} \leq K_{Sch}^2\right] = 1 - 2\alpha. \tag{6.24}$$

Assuming that the full model is correct, with Gaussian errors, it can be shown that $K_{Sch} = \sqrt{pF_{p,N-p,1-2\alpha}}$, which provides the $\sqrt{p}$ upper bound referred to above. The PoSI authors show that using numerical methods and direct search, smaller values of K can be found for actual model matrices $\mathbf{X}$, in particular when the user limits the search to all models within a certain maximum size.

For the diabetes data, Andreas Buja computed for us the value of K for submodels of size 5 (this took a little less than 2 hours of computing time). The resulting values of K were 4.21 (90%), 4.42 (95%), and 4.85 (99%). At the 95% level, this yielded four significant predictors bmi, map, hdl, and ltg. This is one more predictor than we obtain from the lasso method of Figure 6.12, if the latter intervals were adjusted for multiplicity.

The PoSI intervals have advantages compared to the lasso method of Figure 6.12 in that they don't require σ to be known or λ to be fixed. On the other hand, the confidence intervals from PoSI can be very wide. In the diabetes dataset there are four very strong predictors: their lasso intervals are essentially unaffected by the selection and look much like the standard least-squares intervals. Even with a Bonferroni adjustment from 0.05 to 0.01, the intervals have approximate length $\pm 2.33 \cdot \sigma v_{j \cdot M}$ compared to $\pm 4.42 \cdot \sigma v_{j \cdot M}$ for PoSI. However the authors of PoSI make the point that their method provides much stronger protection against all kinds of (unreported) things that people actually do with their data, like fishing for models that have lots of significant predictors.

A major limitation of PoSI at this time is computation. According to the authors, with parallel computation the current problem could probably be solved for models up to size 7 or 8 out of 64, but no larger.

Bibliographic Notes

Our discussion of the Bayesian Lasso is based on Park and Casella (2008). The bootstrap is due to Efron (1979); the book by Efron and Tibshirani (1993) is a comprehensive reference. The connection between Bayesian methods and the bootstrap is explored in various papers (Rubin 1981, Efron 1982, Efron 2011).

The covariance test was introduced in Lockhart, Taylor, Tibshirani[2] and Tibshirani (2014); the discussion following that paper is a valuable resource on model selection. This work was extended to general models and exact tests in Taylor, Loftus and Tibshirani[2] (2016). The spacing test is proposed in Taylor et al. (2014), while Lee et al. (2016) derive the fixed λ inference procedure for the lasso. Taylor et al. (2014) and Loftus and Taylor (2014) propose tests for forward stepwise regression, the latter including categorical variables via the group lasso penalty. Grazier G'Sell, Wager, Chouldechova and Tibshirani

(2015) propose FDR-controlling procedures for sequential testing and apply them to the model selection p-values described here. Grazier G'Sell, Taylor and Tibshirani (2013) develop a covariance test for the graphical lasso while Choi, Taylor and Tibshirani (2014) do the same for principal components. Fithian, Sun and Taylor (2014) provide a general theoretical framework for conditional inference after model selection, with particular emphasis on exponential families.

The "debiasing approach" (Section 6.4) was proposed by a number of authors. For example, Zhang and Zhang (2014) derive confidence intervals for contrasts of high-dimensional regression coefficients, by replacing the usual score vector with the residual from a relaxed projection (i.e., the residual from sparse linear regression). Bühlmann (2013) constructs p-values for coefficients in high-dimensional regression models, starting with ridge estimation and then employing a bias-correction term that uses the lasso. This initial work was followed by van de Geer et al. (2013), Javanmard and Montanari (2014), and Javanmard and Montanari (2013), who all present approaches for debiasing the lasso estimate based on estimates of the inverse covariance matrix of the predictors. (The latter work focuses on the special case of a predictor matrix $\mathbf{X}$ with i.i.d. Gaussian rows; the first two consider a general matrix $\mathbf{X}$.) These debiased lasso estimates are asymptotically normal, which allows one to compute p-values both marginally for an individual coefficient, and simultaneously for a group of coefficients. The PoSI (Post-Selection Inference) method was proposed in Berk et al. (2013).

Exercises

Ex. 6.1

(a) Show that in the orthonormal design setting $\mathbf{X}^T\mathbf{X} = \mathbf{I}_{p \times p}$, the covariance test (6.5) reduces to the simple form

$$T_k = \frac{1}{\sigma^2} \cdot \lambda_k(\lambda_k - \lambda_{k+1}). \tag{6.25}$$

for all steps k.

(b) Show that for general $\mathbf{X}$, the covariance test (6.5) reduces to (6.25) for the first step ($k = 1$)

Ex. 6.2 Show that R_k in Equation (6.4) can be written as a covariance statistic

$$R_k = \frac{1}{\sigma^2} \cdot \left(\langle \mathbf{y}, \mathbf{X}\hat{\beta}_k \rangle - \langle \mathbf{y}, \mathbf{X}\hat{\beta}_{k-1} \rangle \right), \tag{6.26}$$

where $\hat{\beta}_k$ is the coefficient vector after k steps of forward stepwise regression (with the coefficients of those variables not included set to 0).

Ex. 6.3 *Sequential control of FDR.* Suppose that we carry out tests of a set of hypotheses $H_0^1, H_0^2, \ldots, H_0^m$, using p-values $p_1, p_2, \ldots, p_m$. Let the ordered p-values be $p_{(1)} < p_{(2)} < \ldots < p_{(m)}$. If we apply a procedure that rejects R of the hypotheses and there are V false positives among these, then the *false discovery rate* of the procedure is defined to be $E(V/R)$. Given a target FDR of α, the Benjamini–Hochberg (BH) procedure (Benjamini and Hochberg 1995) rejects the R hypotheses with the smallest R p-values, where R is the largest j such that $p_{(j)} \leq \alpha \cdot j/m$. If the p-values are independent, this procedure has an FDR of at most α.

(a) Compute the univariate regression coefficients $\widehat{\beta}_j$ and standard errors $\widehat{\mathrm{se}}_j$ for each predictor in the `diabetes` data. Hence obtain approximate normal scores $z_j = \widehat{\beta}_j / \widehat{\mathrm{se}}_j$ and associated (two)-tailed p-values. Apply the BH procedure to find a list of significant predictors at an FDR of 5%.

(b) Now suppose that our hypotheses have to be considered in order. That is, we must reject a contiguous initial block of K of the hypotheses $H_0^1, H_0^2, \ldots, H_0^K$ (or we could reject none of them). The covariance or spacing test are examples of this. The BH procedure cannot be applied in this setting, as it does not respect the ordering. For example in Table 6.2, the BH procedure might tell us to reject the null hypothesis for `ltg`, but not reject that for `bmi`. This is not helpful, because we seek a model consisting of the first k predictors that enter, for some $k \geq 0$. There is a generalization of the BH procedure that can be applied here. Let the p-values from the covariance or spacing test be $p_1, p_2, \ldots p_m$ and let $r_k = -\sum_{j=1}^k \log(1 - p_j)/k$. Then the so-called *ForwardStop* rule rejects $p_1, p_2, \ldots p_{\hat{k}}$ where $\hat{k}$ is the largest k such that $r_k \leq \alpha$ (Grazier G'Sell et al. 2015). Apply the ForwardStop rule to the covariance or spacing test p-values with a target FDR of 5%.

Ex. 6.4 Here we derive a fact about the multivariate normal distribution, and then in (c) we apply it to derive the spacing test for LAR in the global null case. Suppose that the random vector $Z = (Z_1, \ldots, Z_p)$ follows the multivariate normal distribution $N(0, \boldsymbol{\Sigma})$ with $\Sigma_{jj} = 1$ for all j.

(a) Let

$$(j_1, s_1) = \underset{j \in \{1,2,\ldots,p\}, \, s \in \{-1,1\}}{\arg\max} (sZ_j)$$

and assume that these indices are uniquely attained. Define the random variables

$$M_j = \max_{1 \leq i \leq p, \, i \neq j, \, s \in \{-1,1\}} \left\{ \frac{sZ_i - s\Sigma_{ij}Z_j}{1 - ss_j\Sigma_{ij}} \right\}. \tag{6.27}$$

with $s_j = \arg\max_{s \in \{-1,1\}}(sZ_j)$. Show that M_j is independent of Z_j, for $j = 1, 2, \ldots p$.

(b) Let $\Phi(x)$ be the CDF of a standard Gaussian, and

$$U(z, m) = \frac{1 - \Phi(z)}{1 - \Phi(m)}. \tag{6.28}$$

Verify that $j_1 = j$ if and only if $Z_j \geq M_j$, and prove that $U(Z_{j_1}, M_{j_1})$ is uniformly distributed on $(0, 1)$.

(c) In the LAR procedure with standardized predictors, let $\mathbf{\Sigma} = \frac{1}{N}\mathbf{X}^T\mathbf{X}$, and $Z_j = \frac{1}{N}\mathbf{x}_j^T\mathbf{y}$. Show that $\lambda_1 = \max_{j,s}(sZ_j)$ and $\lambda_2 = M_{j_1}$ (difficult). Hence derive the spacing test (6.11).

Ex. 6.5 Show that as $N, p \to \infty$, the covariance test (6.5) and the spacing test (6.11) are asymptotically equivalent. [Hint: send $\lambda_2 \to \infty$ at a rate such that $\lambda_1/\lambda_2 \to 1$ and apply Mill's ratio.]

Ex. 6.6 Consider the residual sum of squares function $J(\beta) = \|\mathbf{y} - \mathbf{X}\beta\|^2$ and construct a Newton step for minimizing $J(\beta)$ of the form

$$\beta^{new} \leftarrow \beta + \left(\frac{\partial J}{\partial \beta}\right)^{-1}\frac{\partial J}{\partial \beta} \tag{6.29}$$

where β is the lasso estimate at some λ. Show that this has the form (6.16) with $(\mathbf{X}^T\mathbf{X})^{-1}$ replaced by the estimate $\hat{\mathbf{\Theta}}$ from (6.20).

Ex. 6.7 *General inference for the LAR algorithm and the lasso.* Let $\mathbf{y} \sim N(\boldsymbol{\mu}, \sigma^2\mathbf{I})$, and consider the distribution of $\mathbf{y}$ conditional on the selection event $\{\mathbf{A}\mathbf{y} \leq b\}$.

(a) Show that

$$\{\mathbf{A}\mathbf{y} \leq b\} = \{\mathcal{V}^-(\mathbf{y}) \leq \boldsymbol{\eta}^T\mathbf{y} \leq \mathcal{V}^+(\mathbf{y}), \ \mathcal{V}^0(\mathbf{y}) \geq 0\} \tag{6.30}$$

with the variables above defined as follows:

$$\alpha = \frac{\mathbf{A}\boldsymbol{\eta}}{\|\boldsymbol{\eta}\|^2}$$

$$\mathcal{V}^-(\mathbf{y}) = \max_{j:\alpha_j<0} \frac{b_j - (\mathbf{A}\mathbf{y})_j + \alpha_j\boldsymbol{\eta}^T\mathbf{y}}{\alpha_j}$$

$$\mathcal{V}^+(\mathbf{y}) = \min_{j:\alpha_j>0} \frac{b_j - (\mathbf{A}\mathbf{y})_j + \alpha_j\boldsymbol{\eta}^T\mathbf{y}}{\alpha_j}$$

$$\mathcal{V}^0(\mathbf{y}) = \min_{j:\alpha_j=0}(b_j - (\mathbf{A}\mathbf{y})_j) \tag{6.31}$$

[Hint: subtract $E(\mathbf{A}\mathbf{y}|\boldsymbol{\eta}^T\mathbf{y})$ from both sides of the inequality $\mathbf{A}\mathbf{y} \leq b$. Simplify and examine separately the cases $\alpha_j < 0$, $= 0$ and > 0.]

(b) Let

$$F_{\mu,\sigma^2}^{c,d}(x) = \frac{\Phi((x-\mu)/\sigma) - \Phi((c-\mu)/\sigma)}{\Phi((d-\mu)/\sigma) - \Phi((c-\mu)/\sigma)}. \tag{6.32}$$

This is the truncated normal distribution, with support on $[c, d]$. Show that

$$F_{\boldsymbol{\eta}^T\boldsymbol{\mu},\, \sigma^2\|\boldsymbol{\eta}\|^2}^{\mathcal{V}^-,\mathcal{V}^+}(\boldsymbol{\eta}^T\mathbf{y}) \mid \{\mathbf{A}\mathbf{y} \leq b\} \sim U(0, 1). \tag{6.33}$$

This result can be used to make inferences about parameter $\eta^T\mu$ at a given step of the LAR algorithm, or for a lasso solution computed at a fixed value of λ.

(c) Use result (6.33) to provide an alternate proof of the spacing test result (6.11).

Ex. 6.8 The kth knot in the LAR algorithm is the value λ_k at which the kth variable enters the model. At λ_k the coefficient of this variable is zero (about to grow from zero). Using the KKT optimality conditions, verify expression (6.12).

Ex. 6.9 With η_k defined in (6.12), show that $\tilde{\theta}_k$ in (6.14) is the absolute standardized partial regression coefficient of y on x_{j_k}, adjusted for $X_{\mathcal{A}_{k-1}}$.

Ex. 6.10 Consider a solution to the lasso problem

$$\underset{\beta}{\text{minimize}} \frac{1}{2}\|y - X\beta\|_2^2 + \lambda\|\beta\|_1,$$

and let $E \subset \{1,\ldots,p\}$ denote a candidate active set, and $s_E \in \{-1,1\}^{|E|}$ the signs of the active variables. The KKT conditions corresponding to any solution $\widehat{\beta}_E$ with the same E and S_E are given by

$$-X_E^T(y - X_E\widehat{\beta}_E) \quad + \quad \lambda s_E = 0 \tag{6.34}$$

$$-X_{-E}^T(y - X_E\widehat{\beta}_E) \quad + \quad \lambda s_{-E} = 0, \tag{6.35}$$

with $\text{sign}(\widehat{\beta}_E) = s_E$ and $\|s_{-E}\|_\infty < 1$. Eliminate $\widehat{\beta}_E$ in these equations, and show that the set of values of y with solution characterized by (E, s_E) can be defined by a set of linear inequalities

$$Ay \leq b.$$

Ex. 6.11 Consider the setup in Example 6.1, and assume $x_{j_1}^T y$ is positive. Using simple inequalities, derive an expression for $\mathcal{V}^-(y)$. Show that this is equal to λ_2, the second LAR knot.

Chapter 7

Matrix Decompositions, Approximations, and Completion

7.1 Introduction

This chapter is devoted to problems of the following type: given data in the form of an $m \times n$ matrix $\mathbf{Z} = \{z_{ij}\}$, find a matrix $\widehat{\mathbf{Z}}$ that approximates $\mathbf{Z}$ in a suitable sense. One purpose might be to gain an understanding of the matrix $\mathbf{Z}$ through an approximation $\widehat{\mathbf{Z}}$ that has simple structure. Another goal might be to impute or fill in any missing entries in $\mathbf{Z}$, a problem known as *matrix completion*.

Our general approach is to consider estimators based on optimization problems of the form

$$\widehat{\mathbf{Z}} = \arg\min_{\mathbf{M} \in \mathbb{R}^{m \times n}} \|\mathbf{Z} - \mathbf{M}\|_F^2 \text{ subject to } \Phi(\mathbf{M}) \leq c, \qquad (7.1)$$

where $\| \cdot \|_F^2$ is the (squared) Frobenius norm of a matrix (defined as the element-wise sum of squares), and $\Phi(\cdot)$ is a constraint function that encourages $\widehat{\mathbf{Z}}$ to be sparse in some general sense. The manner in which we impose sparsity leads to a variety of useful procedures, many of which are discussed in this chapter. One can regularize the overall approximating matrix $\widehat{\mathbf{Z}}$, or factor it and regularize the components of its factorization. Of course, there are variations: for instance, the observed matrix $\mathbf{Z}$ might have missing entries, so that the squared Frobenius norm $\| \cdot \|_F^2$ is modified accordingly. In other settings, we might impose multiple constraints on the approximating matrix $\widehat{\mathbf{Z}}$.

Table 7.1 provides a summary of the methods discussed in this chapter. Method (a) is based on a simple ℓ_1-norm constraint on all of the entries on the matrix $\widehat{\mathbf{Z}}$; this constraint leads to a soft-thresholded version of the original matrix—that is, the optimal solution to our general problem (7.1) takes the form $\widehat{z}_{ij} = \text{sign}(z_{ij})(|z_{ij}| - \gamma)_+$, where the scalar $\gamma > 0$ is chosen so that $\sum_{i=1}^m \sum_{j=1}^n |\widehat{z}_{ij}| = c$. The resulting estimate $\widehat{\mathbf{Z}}$ can be useful in the context of sparse covariance estimation. Method (b) bounds the rank of $\widehat{\mathbf{Z}}$, or in other words, the number of nonzero singular values in $\widehat{\mathbf{Z}}$. Although the matrix approximation problem (7.1) with such a rank constraint is nonconvex,

Table 7.1 *Different formulations for the matrix approximation problem.*

	Constraint	Resulting method
(a)	$\|\widehat{\mathbf{Z}}\|_{\ell_1} \leq c$	Sparse matrix approximation
(b)	$\mathrm{rank}(\widehat{\mathbf{Z}}) \leq k$	Singular value decomposition
(c)	$\|\widehat{\mathbf{Z}}\|_\star \leq c$	Convex matrix approximation
(d)	$\widehat{\mathbf{Z}} = \mathbf{UDV}^T,$	
	$\Phi_1(\mathbf{u}_j) \leq c_1,\ \Phi_2(\mathbf{v}_k) \leq c_2$	Penalized SVD
(e)	$\widehat{\mathbf{Z}} = \mathbf{LR}^T,$	
	$\Phi_1(\mathbf{L}) \leq c_1,\ \Phi_2(\mathbf{R}) \leq c_2$	Max-margin matrix factorization
(f)	$\widehat{\mathbf{Z}} = \mathbf{L} + \mathbf{S},$	
	$\Phi_1(\mathbf{L}) \leq c_1,\ \Phi_2(\mathbf{S}) \leq c_2$	Additive matrix decomposition

the optimal solution is easily found by computing the singular value decomposition (SVD) and truncating it to its top k components. In method (c), we relax the rank constraint to a *nuclear norm* constraint, namely an upper bound on the sum of the singular values of the matrix. The nuclear norm is a convex matrix function, so that the problem in (c) is convex and can be solved by computing the SVD, and soft-thresholding its singular values. This modification—from a rank constraint in (b) to the nuclear norm constraint in (c)—becomes important when the methods are applied to matrices with missing elements. In such settings, we can solve the corresponding problem (c) exactly, whereas methods based on (b) are more difficult to solve in general.

The approach in (d) imposes penalties on the left and right singular vectors of $\widehat{\mathbf{Z}}$. Examples of the penalty functions or regularizers Φ_1 and Φ_2 include the usual ℓ_2 or ℓ_1 norms, the latter choice yielding sparsity in the elements of the singular vectors. This property is useful for problems where interpretation of the singular vectors is important. Approach (e) imposes penalties directly on the components of the LR-matrix factorization; although ostensibly similar to approach (d), we will see it is closer to (c) when Φ_1 and Φ_2 are the Frobenius norm. Finally, approach (f) seeks an additive decomposition of the matrix, imposing penalties on both components in the sum.

Matrix decompositions also provide an approach for constructing sparse versions of popular multivariate statistical methods such as principal component analysis, canonical correlation analysis and linear discriminant analysis. In this case, the matrix $\mathbf{Z}$ is not the raw data, but is derived from the raw data. For example, principal components are based on the sample covariance matrix (or the column-centered data matrix), canonical correlation uses the cross-products matrix from two sets of measurements, while clustering starts with inter-point distances. We discuss these multivariate methods, and related approaches to these problems, in Chapter 8.

7.2 The Singular Value Decomposition

Given an $m \times n$ matrix $\mathbf{Z}$ with $m \geq n$, its *singular value decomposition* takes the form

$$\mathbf{Z} = \mathbf{U}\mathbf{D}\mathbf{V}^T. \tag{7.2}$$

This decomposition is standard in numerical linear algebra, and many algorithms exist for computing it efficiently (see, for example, the book by Golub and Loan (1996)). Here $\mathbf{U}$ is an $m \times n$ orthogonal matrix ($\mathbf{U}^T\mathbf{U} = \mathbf{I}_n$) whose columns $\mathbf{u}_j \in \mathbb{R}^m$ are called the *left singular vectors*. Similarly, the matrix $\mathbf{V}$ is an $n \times n$ orthogonal matrix ($\mathbf{V}^T\mathbf{V} = \mathbf{I}_n$) whose columns $\mathbf{v}_j \in \mathbb{R}^n$ are called the *right singular vectors*. The $n \times n$ matrix $\mathbf{D}$ is diagonal, with diagonal elements $d_1 \geq d_2 \geq \cdots \geq d_n \geq 0$ known as the *singular values*. If these diagonal entries $\{d_\ell\}_{\ell=1}^n$ are unique, then so are $\mathbf{U}$ and $\mathbf{V}$, up to column-wise sign flips. If the columns of $\mathbf{Z}$ (the variables) are centered, then the right singular vectors $\{\mathbf{v}_j\}_{j=1}^n$ define the *principal components* of $\mathbf{Z}$. Consequently, the unit vector $\mathbf{v}_1$ yields the linear combination $\mathbf{s}_1 = \mathbf{Z}\mathbf{v}_1$ with highest sample variance among all possible choices of unit vectors. Here $\mathbf{s}_1$ is called the *first principal component* of $\mathbf{Z}$, and $\mathbf{v}_1$ is the corresponding *direction* or *loading* vector. Similarly, $\mathbf{s}_2 = \mathbf{Z}\mathbf{v}_2$ is the second principal component, with maximal sample variance among all linear combinations uncorrelated with $\mathbf{s}_1$, and so on. See Exercise 7.1 and Section 8.2.1 for further details.

The singular value decomposition provides a solution to the rank-q matrix approximation problem. Suppose $r \leq \mathrm{rank}(\mathbf{Z})$, and let $\mathbf{D}_r$ be a diagonal matrix with all but the first r diagonal entries of the diagonal matrix $\mathbf{D}$ set to zero. Then the optimization problem

$$\underset{\mathrm{rank}(\mathbf{M})=r}{\mathrm{minimize}} \|\mathbf{Z} - \mathbf{M}\|_F \tag{7.3}$$

actually has a closed form solution $\widehat{\mathbf{Z}}_r = \mathbf{U}\mathbf{D}_r\mathbf{V}^T$, a decomposition known as the rank-r SVD (see Exercise 7.2). The estimate $\widehat{\mathbf{Z}}_r$ is sparse in the sense that all but r singular values are zero. A fuller discussion of the SVD—in the context of principal components analysis—is given in Section 8.2.1.

7.3 Missing Data and Matrix Completion

What if some of the entries of $\mathbf{Z}$ are missing? In general, the problem of filling in or imputing missing values in a matrix is known as *matrix completion* (Laurent 2001). Of course, the matrix completion problem is ill-specified unless we impose additional constraints on the unknown matrix $\mathbf{Z}$, and one common choice is a rank constraint. Low-rank forms of matrix completion arise in the problem of collaborative filtering and can be used to build recommender systems.

The SVD provides an effective method for matrix completion. Formally, suppose that we observe all entries of the matrix $\mathbf{Z}$ indexed by the subset

$\Omega \subset \{1, \ldots, m\} \times \{1, \ldots, n\}$. Given such observations, a natural approach is to seek the lowest rank approximating matrix $\widehat{\mathbf{Z}}$ that interpolates the observed entries of $\mathbf{Z}$—namely

$$\text{minimize rank}(\mathbf{M}) \quad \text{subject to } m_{ij} = z_{ij} \text{ for all } (i,j) \in \Omega. \qquad (7.4)$$

Unlike its fully observed counterpart, this rank-minimization problem is computationally intractable (NP-hard), and cannot be solved in general even for moderately large matrices.

In addition, forcing the estimate $\mathbf{M}$ to interpolate each of the observed entries z_{ij} will often be too harsh and can lead to overfitting; it is generally better to allow $\mathbf{M}$ to make some errors on the observed data as well. Accordingly, consider the optimization problem

$$\text{minimize rank}(\mathbf{M}) \quad \text{subject to } \sum_{(i,j) \in \Omega} (z_{ij} - m_{ij})^2 \leq \delta, \qquad (7.5)$$

or its equivalent form

$$\underset{\text{rank}(\mathbf{M}) \leq r}{\text{minimize}} \sum_{(i,j) \in \Omega} (z_{ij} - m_{ij})^2. \qquad (7.6)$$

In words, we seek the matrix $\widehat{\mathbf{Z}} = \widehat{\mathbf{Z}}_r$ of rank at most r that best approximates the observed entries of our matrix $\mathbf{Z}$, with the other entries of $\widehat{\mathbf{Z}}_r$ serving to fill in the missing values. The family of solutions generated by varying δ in optimization problem (7.5) is the same as that generated by varying r in problem (7.6).

Unfortunately, both optimization problems (7.5) and (7.6) are nonconvex, and so exact solutions are in general not available. However, there are useful heuristic algorithms that can be used to find local minima. For instance, suppose that we start with an initial guess for the missing values, and use them to complete $\mathbf{Z}$. We then compute the rank-r SVD approximation of the filled-in matrix as in (7.3), and use it to provide new estimates for the missing values. This process is repeated till convergence. The missing value imputation for a missing entry x_{ij} is simply the $(i,j)^{th}$ entry of the final rank-r approximation $\widehat{\mathbf{Z}}$. See Mazumder, Hastie and Tibshirani (2010) for further details. In Section 7.3.2, we discuss convex relaxations of these optimization problems based on the nuclear norm, for which exact solutions are available.

7.3.1 The Netflix Movie Challenge

The Netflix movie-rating challenge has become one of the canonical examples for matrix completion (Bennett and Lanning 2007). Netflix is a movie-rental company that launched a competition in 2006 to try to improve their system for recommending movies to their customers. The Netflix dataset has $n = 17,770$ movies (columns) and $m = 480,189$ customers (rows). Customers

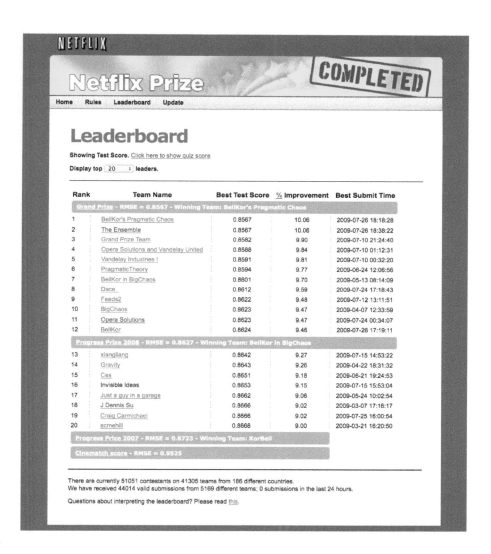

Figure 7.1 *The Netflix competition leaderboard at the close of the competition.*

have rated some of the movies on a scale from 1 to 5, where 1 is worst and 5 is best. The data matrix is very sparse with "only" 100 million (1%) of the ratings present in the training set. The goal is to predict the ratings for unrated movies, so as to better recommend movies to customers. In 2006, the "Cinematch" algorithm used by Netflix had a root-mean-square error of 0.9525 over a large test set. A competition was held starting in 2006, with the winner being the first algorithm that could improve this RMSE by at least 10%. The competition was finally won in 2009 by a large group of researchers called "Bellkor's Pragmatic Chaos," which was the combined effort of three individual groups. The winning algorithm used a combination of a large number of statistical techniques, but as with many of the competing algorithms, the SVD played a central role. Figure 7.1 shows the leaderboard at the close of the competition.

Table 7.2 *Excerpt of the Netflix movie rating data. The movies are rated from 1 (worst) to 5 (best). The symbol ◦ represents a missing value: a movie that was not rated by the corresponding customer.*

	Dirty Dancing	Meet the Parents	Top Gun	The Sixth Sense	Catch Me If You Can	The Royal Tenenbaums	Con Air	Big Fish	The Matrix	A Few Good Men
Customer 1	◦	◦	◦	◦	4	◦	◦	◦	◦	◦
Customer 2	◦	◦	3	◦	◦	◦	3	◦	◦	3
Customer 3	◦	2	◦	4	◦	◦	◦	◦	2	◦
Customer 4	3	◦	◦	◦	◦	◦	◦	◦	◦	◦
Customer 5	5	5	◦	◦	4	◦	◦	◦	◦	◦
Customer 6	◦	◦	◦	◦	◦	2	4	◦	◦	◦
Customer 7	◦	◦	5	◦	◦	◦	◦	3	◦	◦
Customer 8	◦	◦	◦	◦	◦	2	◦	◦	◦	3
Customer 9	3	◦	◦	◦	5	◦	◦	5	◦	◦
Customer 10	◦	◦	◦	◦	◦	◦	◦	◦	◦	◦

A low rank model provides a good heuristic for rating movies: in particular, suppose that we model the rating of user i on movie j by a model of the form

$$z_{ij} = \sum_{\ell=1}^{r} c_{i\ell} g_{j\ell} + w_{ij}, \tag{7.7}$$

or in matrix form $\mathbf{Z} = \mathbf{C}\mathbf{G}^T + \mathbf{W}$, where $\mathbf{C} \in \mathbb{R}^{m \times r}$ and $\mathbf{G} \in \mathbb{R}^{n \times r}$. In this model, there are r *genres* of movies, and corresponding to each is a *clique* of viewers who like them. Here viewer i has a membership weight of $c_{i\ell}$ for the ℓth clique, and the genre associated with this clique has a score $g_{j\ell}$ for movie j. The overall user rating is obtained by summing these products over

ℓ (cliques/genres), and then adding some noise. Table 7.2 shows the data for the ten customers and ten movies with the most ratings.

The competition identified a "probe set" of ratings, about 1.4 million of the entries, for testing purposes. These were not a random draw, rather movies that had appeared chronologically later than most. Figure 7.2 shows the root-mean-squared error over the training and test sets as the rank of the SVD was varied. Also shown are the results from an estimator based on nuclear norm regularization, discussed in the next section. Here we double centered the training data, by removing row and column means. This amounts to fitting the model

$$z_{ij} = \alpha_i + \beta_j + \sum_{\ell=1}^{r} c_{i\ell}g_{j\ell} + w_{ij}; \qquad (7.8)$$

However, the row and column means can be estimated separately, using a simple two-way ANOVA regression model (on unbalanced data).

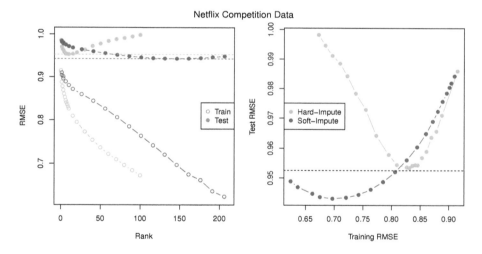

Figure 7.2 *Left: Root-mean-squared error for the Netflix training and test data for the iterated-SVD (*HARD-IMPUTE*) and the convex spectral-regularization algorithm (*SOFT-IMPUTE*). Each is plotted against the rank of the solution, an imperfect calibrator for the regularized solution. Right: Test error only, plotted against training error, for the two methods. The training error captures the amount of fitting that each method performs. The dotted line represents the baseline "Cinematch" score.*

While the iterated-SVD method is quite effective, it is not guaranteed to find the optimal solution for each rank. It also tends to overfit in this example, when compared to the regularized solution. In the next section, we present a convex relaxation of this setup that leads to an algorithm with guaranteed convergence properties.

7.3.2   Matrix Completion Using Nuclear Norm

A convenient convex relaxation of the nonconvex objective function (7.4) is given by

$$\text{minimize } \|\mathbf{M}\|_\star \text{ subject to } m_{ij} = z_{ij} \text{ for all } (i,j) \in \Omega, \qquad (7.9)$$

where $\|\mathbf{M}\|_\star$ is the nuclear norm, or the sum of the singular values of $\mathbf{M}$. It is also sometimes called the trace norm.[1] Figure 7.3 shows the level set of the nuclear norm of a symmetric 2×2 matrix, and depicts the convex problem (7.9).[2]

The nuclear norm is a convex relaxation of the rank of a matrix, and hence problem (7.9) is convex (Fazel 2002). Specifically, as shown in Exercise 7.3, it is a semi-definite program (SDP), a particular class of convex programs for which special purpose solvers can be applied. The underlying convexity is also theoretically useful, since one can characterize the properties of the observed matrix and sample size under which the method succeeds in exactly reconstructing the matrix, as discussed in Section 7.3.3.

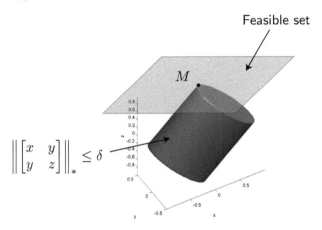

Figure 7.3 *The blue cylinder shows the level set of the nuclear norm unit-ball for a symmetric 2×2 matrix. The tangent plane is the feasible set $z = z_0$ for the matrix imputation problem where we observe z and wish to impute x and y. The point M is the solution that we seek, leading to the minimum value for δ. This figure is analogous to the lasso estimation picture in Figure 2.2 of Chapter 2.*

In practice, however, it is unrealistic to model the observed entries as being noiseless. Accordingly, a more practical method is based on the following

---

[1]This terminology can be confusing: for symmetric, positive semi-definite matrices, the trace is the sum of the eigenvalues. For general matrices, "trace norm" refers to $\text{trace}\sqrt{\mathbf{A}^T \mathbf{A}}$, which is the sum of the singular values.

[2]Thanks to Emmanuel Candes and Benjamin Recht for providing Figure 7.3.

relaxed version of the program (7.9):

$$\underset{\mathbf{M}}{\text{minimize}} \left\{ \frac{1}{2} \sum_{(i,j)\in\Omega} (z_{ij} - m_{ij})^2 + \lambda\|\mathbf{M}\|_\star \right\}, \tag{7.10}$$

called *spectral regularization*. As in our relaxation from problem (7.4) to (7.6), this modification allows for solutions $\widehat{\mathbf{Z}}$ that do not fit the observed entries exactly, thereby reducing potential overfitting in the case of noisy entries. The parameter λ is a tuning parameter that must be chosen from the data, typically by cross-validation. As in the previous section, we do not necessarily require the error $\sum_{(i,j)\in\Omega} (z_{ij} - m_{ij})^2$ to be zero, and this will only occur for a sufficiently small value of λ.

There is a simple algorithm for solving (7.10), similar to the iterated SVD for missing-data imputation in the previous section. First consider the case where there is no missing data, so that the set Ω of observed entries includes all $m \cdot n$ pairs $(i,j) \in \{1,\ldots,m\} \times \{1,\ldots,n\}$. Then to solve (7.10), we simply compute the SVD of $\mathbf{Z}$, soft-threshold the singular values by λ, and reconstruct the matrix. This observation leads to an obvious procedure for the setup with missing data. We start with an initial guess for the missing values, compute the (full rank) SVD, and then soft-threshold its singular values by an amount λ. We reconstruct the corresponding SVD approximation and obtain new estimates for the missing values. This process is repeated until convergence.

In order to describe this procedure more precisely, we require some more notation. Given an observed subset Ω of matrix entries, we can define the projection operator $\mathcal{P}_\Omega : \mathbb{R}^{m\times n} \mapsto \mathbb{R}^{m\times n}$ as follows:

$$[\mathcal{P}_\Omega(\mathbf{Z})]_{ij} = \begin{cases} z_{ij} & \text{if } (i,j) \in \Omega \\ 0 & \text{if } (i,j) \notin \Omega, \end{cases} \tag{7.11}$$

so that $\mathcal{P}_\Omega$ replaces the missing entries in $\mathbf{Z}$ with zeros, and leaves the observed entries alone. With this definition, we have the equivalence

$$\sum_{(i,j)\in\Omega} (z_{ij} - m_{ij})^2 = \|\mathcal{P}_\Omega(\mathbf{Z}) - \mathcal{P}_\Omega(\mathbf{M})\|_F^2. \tag{7.12}$$

Given the singular value decomposition[3] $\mathbf{W} = \mathbf{U}\mathbf{D}\mathbf{V}^T$ of a rank-r matrix $\mathbf{W}$, we define its soft-thresholded version as

$$S_\lambda(\mathbf{W}) \equiv \mathbf{U}\mathbf{D}_\lambda\mathbf{V}^T \quad \text{where} \quad \mathbf{D}_\lambda = \text{diag}\,[(d_1 - \lambda)_+, \ldots, (d_r - \lambda)_+] \tag{7.13}$$

(note that the soft-threshholding can reduce the rank even further). Using this operator, the procedure for solving (7.10) is given in Algorithm 7.1.

This algorithm was proposed and studied by Mazumder et al. (2010), where

---

[3]If a matrix has rank $r < \min(m,n)$, we assume its SVD is represented in the truncated form, discarding the singular values of zero, and the corresponding left and right vectors.

Algorithm 7.1 SOFT-IMPUTE FOR MATRIX COMPLETION.

---

1. Initialize $\mathbf{Z}^{\text{old}} = \mathbf{0}$ and create a decreasing grid $\lambda_1 > \ldots > \lambda_K$.

2. For each $k = 1, \ldots, K$, set $\lambda = \lambda_k$ and iterate until convergence:

   Compute $\widehat{\mathbf{Z}}_\lambda \leftarrow \mathcal{S}_\lambda \big(P_\Omega(\mathbf{Z}) + P_\Omega^\perp(\mathbf{Z}^{\text{old}}) \big)$.

   Update $\mathbf{Z}^{\text{old}} \leftarrow \widehat{\mathbf{Z}}_\lambda$

3. Output the sequence of solutions $\widehat{\mathbf{Z}}_{\lambda_1}, \ldots, \widehat{\mathbf{Z}}_{\lambda_K}$.

---

its convergence to the global solution is established. In Exercise 7.4, the reader is asked to verify that a fixed point of the algorithm satisfies the zero sub-gradient equations associated with the objective function (7.10). It can also be derived as a first-order Nesterov algorithm (see Exercise 7.5). Each itera-tion requires an SVD computation of a (potentially large) dense matrix, even though $\mathcal{P}_\Omega(\mathbf{Z})$ is sparse. For "Netflix-sized" problems, such large dense matri-ces can typically not even be stored in memory (68Gb with 8 bytes per entry). Note, however, that we can write

$$\mathcal{P}_\Omega(\mathbf{Z}) + \mathcal{P}_\Omega^\perp(\mathbf{Z}^{\text{old}}) = \underbrace{\mathcal{P}_\Omega(\mathbf{Z}) - \mathcal{P}_\Omega(\mathbf{Z}^{\text{old}})}_{\text{sparse}} + \underbrace{\mathbf{Z}^{\text{old}}}_{\text{low rank}} . \qquad (7.14)$$

The first component is sparse, with $|\Omega|$ nonmissing entries. The second com-ponent is a soft-thresholded SVD, so can be represented using the correspond-ing components. Moreover, for each component, we can exploit their special structure to efficiently perform left and right multiplications by a vector, and thereby apply iterative Lanczos methods to compute a (low rank) SVD effi-ciently. It can be shown that this iterative algorithm converges to the solution of the problem

$$\underset{\mathbf{M} \in \mathbb{R}^{m \times n}}{\text{minimize}} \left\{ \frac{1}{2} \|\mathcal{P}_\Omega(\mathbf{Z}) - \mathcal{P}_\Omega(\mathbf{M})\|_F^2 + \lambda \|\mathbf{M}\|_\star \right\}, \qquad (7.15)$$

which is another way of writing the objective function in (7.10).

Figure 7.2 shows the results of SOFT-IMPUTE applied to the Netflix ex-ample. We see that the regularization has paid off, since it outperforms the iterated SVD algorithm HARD-IMPUTE. It takes longer to overfit, and because of the regularization, is able to use a higher rank solution. Taking advantage of the warm starts in Algorithm 7.1, it took under 5 hours of computing to pro-duce the solution path in Figure 7.2, using the R package `softImpute` (Hastie and Mazumder 2013). See also Figure 7.5 in Section 7.3.3, which illustrates the performance of the SOFT-IMPUTE algorithm for noisy matrix completion over a range of different ranks and sample sizes. We discuss this figure at more length in that section.

In terms of convergence speed, Mazumder et al. (2010) show that the SOFT-IMPUTE algorithm is guaranteed to converge at least sub-linearly, mean-ing that $\mathcal{O}(1/\delta)$ iterations are sufficient to compute a solution that is δ-close to the global optimum. In the absence of additional structure (such as strong convexity), this is the fastest rate that can be expected from a first-order gradient method (Nemirovski and Yudin 1983). Interestingly, in certain settings, it can be shown that simple first-order methods converge at a much faster geometric rate, meaning that $\mathcal{O}(\log(1/\delta))$ iterations are sufficient to compute a δ-optimum. For instance, Agarwal, Negahban and Wainwright (2012a) analyze an algorithm closely related to the SOFT-IMPUTE algorithm; they show that under the same conditions that guarantee good statistical performance of the nuclear norm estimator, this first-order algorithm is guaranteed to converge at the geometric rate.

7.3.3 Theoretical Results for Matrix Completion

There are a variety of theoretical results for matrix completion using nuclear-norm regularization. Beginning with the simpler "no-noise" case, suppose that we sample N entries of a $p \times p$ matrix uniformly at random. How large does N need to be, as a function of the matrix dimension p and rank r, for the nuclear norm relaxation (7.9) to recover the matrix exactly? Of course, this is always possible if $N \geq p^2$, so that our interest is in guarantees based on $N \ll p^2$ samples.

A first easy observation is that if there are no observed entries in some row (or column) of the matrix, then it is impossible to recover the matrix exactly, even if it is rank one. In Exercise 7.8, we show how this argument implies that any method—not only nuclear norm relaxation—needs at least $N > p \log p$ samples, even for a rank one matrix. This phenomenon is an instance of the famous "coupon collector" problem (Erdos and Renyi 1961). As for the effect of the rank, note that we need roughly $\mathcal{O}(rp)$ parameters to specify an arbitrary $p \times p$ matrix with rank r, since it has $\mathcal{O}(r)$ singular vectors, each with p components. As we will see, under certain restrictions on the "coherence" of the matrices, nuclear norm relaxation succeeds in exact recovery based on a sample size just a logarithmic factor larger. Coherence measures the extent to which the singular vectors of a matrix are aligned with the standard basis.

In order to appreciate the need for coherence constraints, consider the rank-one matrix $\mathbf{Z} = \mathbf{e}_1 \mathbf{e}_1^T$, with a single one in its upper left corner, as shown on the left side of Equation (7.16) below:

$$\mathbf{Z} = \begin{pmatrix} 1 & 0 & 0 & 0 \\ 0 & 0 & 0 & 0 \\ 0 & 0 & 0 & 0 \\ 0 & 0 & 0 & 0 \end{pmatrix} \quad \text{and} \quad \mathbf{Z}' = \begin{pmatrix} v_1 & v_2 & v_3 & v_4 \\ 0 & 0 & 0 & 0 \\ 0 & 0 & 0 & 0 \\ 0 & 0 & 0 & 0 \end{pmatrix}. \quad (7.16)$$

If we are allowed to observe only $N \ll p^2$ entries of this matrix, with the entries chosen uniformly at random, then with high probability, we will *not* observe the single nonzero entry, and hence have no hope of distinguishing it from the all-zeroes matrix. Similar concerns apply to a matrix of the form

$\mathbf{Z}' = \mathbf{e}_1\mathbf{v}^T$, where $\mathbf{v} \in \mathbb{R}^p$ is an arbitrary p vector, as shown on the right side of Equation (7.16). Thus, any theoretical guarantees on nuclear norm regularization must somehow account for these pathological cases. Both the matrices $\mathbf{Z}$ and $\mathbf{Z}'$ have maximal coherence with the standard basis of $\mathbb{R}^4$, meaning that some subset of their left and/or right singular vectors are perfectly aligned with some standard basis vector $\mathbf{e}_j$.

One way to exclude troublesome matrices is by drawing matrices from some random ensemble; for instance, we might construct a random matrix of the form $\mathbf{Z} = \sum_{j=1}^r \mathbf{a}_j \mathbf{b}_j^T$, where the random vectors $\mathbf{a}_j \sim N(0, I_p)$ and $\mathbf{b}_j \sim N(0, I_p)$ are all independently drawn. Such random matrices are extremely unlikely to have singular vectors that are highly coherent with standard basis vectors. For this ensemble, Gross (2011) shows that with high probability over the randomness in the ensemble and sampling, the nuclear norm relaxation succeeds in exact recovery if the number of samples satisfies

$$N \geq Crp\log p, \tag{7.17}$$

where $C > 0$ is a fixed universal constant. See also Candès and Recht (2009) for earlier but somewhat weaker guarantees. More generally, it is possible to give exact recovery guarantees in which the pre-factor C depends on the singular vector incoherence, as measured by the maximal alignment between the singular vectors and the standard basis. We refer the reader to the papers by Candès and Recht (2009), Gross (2011), and Recht (2011) for further details on results of this type, as well as to Keshavan, Oh and Montanari (2009) for related results on a slightly different estimator.

We carried out a small simulation study to better understand what result (7.17) is saying. We generated matrices $\mathbf{U}, \mathbf{V}$ each of size $p \times r$ and with i.i.d standard normal entries and defined $\mathbf{Z} = \mathbf{U}\mathbf{V}^T$. Then we set to missing a fixed proportion of entries, and applied SOFT-IMPUTE with λ chosen small enough so that $\|\mathcal{P}_\Omega(\mathbf{Z}-\widehat{\mathbf{Z}})\|_F^2/\|\mathcal{P}_\Omega(\mathbf{Z})\|_F^2 < 10^{-5}$; in other words, the observed entries are (effectively) reproduced. Then we checked to see if

$$\|\mathcal{P}_\Omega^\perp(\mathbf{Z} - \widehat{\mathbf{Z}})\|_2^2/\|\mathcal{P}_\Omega^\perp(\mathbf{Z})\|_2^2 < 10^{-5}, \tag{7.18}$$

that is, the missing data was interpolated. The process was repeated 100 times for various values of the rank r and the proportion set to missing.

The proportion of times that the missing data was successfully interpolated is shown in Figure 7.4. We see that when the rank is a small fraction of the matrix dimension, one can reproduce the missing entries with fairly high probability. But this gets significantly more difficult when the true rank is higher.

Of course, the "exact" setting is often not realistic, and it might be more reasonable to assume some subset of the entries are observed with additional noise, as in observation model (7.7)—that is, $\mathbf{Z} = \mathbf{L}^* + \mathbf{W}$, where $\mathbf{L}^*$ has rank r. In this setting, exact matrix completion is not generally possible, and we would be interested in how well we can approximate the low-rank matrix $\mathbf{L}^*$ using the

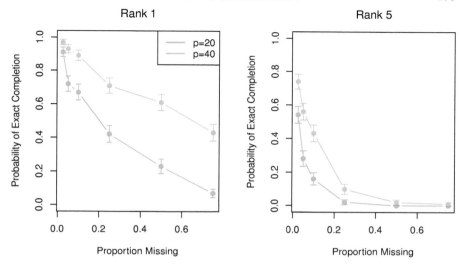

Figure 7.4 *Convex matrix completion in the no-noise setting. Shown are probabilities of exact completion (mean ± one standard error) as a function of the proportion missing, for $n \times n$ matrices with $n \in \{20, 40\}$. The true rank of the complete matrix is one in the left panel and five in the right panel.*

estimator (7.10). Singular vector incoherence conditions are less appropriate for noisy observations, because they are not robust to small perturbations. To understand this issue, suppose that we start with a matrix $\mathbf{B}$ that has rank $r - 1$, Frobenius norm one, and is *maximally incoherent*, meaning that all its singular vectors are orthogonal to the standard basis vectors. Recalling the troublesome matrix $\mathbf{Z}$ from Equation (7.16), now consider the perturbed matrix $\mathbf{L}^* = \mathbf{B} + \delta\mathbf{Z}$ for some $\delta > 0$. The matrix $\mathbf{L}^*$ always has rank r, and no matter how small we choose the parameter δ, it is always *maximally coherent*, since it has the standard basis vector $\mathbf{e}_1 \in \mathbb{R}^p$ as one of its singular vectors.

An alternative criterion that is not sensitive to such small perturbations is based on the "spikiness" ratio of a matrix (Negahban and Wainwright 2012). In particular, for any nonzero matrix $\mathbf{L} \in \mathbb{R}^{p \times p}$, we define $\alpha_{\mathrm{sp}}(\mathbf{L}) = \frac{p\|\mathbf{L}\|_\infty}{\|\mathbf{L}\|_{\mathrm{F}}}$, where $\|\mathbf{L}\|_\infty$ is the element-wise maximum absolute value of the matrix entries. This ratio is a measure of the uniformity (or lack thereof) in the spread of the matrix entries; it ranges between 1 and p. For instance, any matrix $\mathbf{L}$ with all equal entries has $\alpha_{\mathrm{sp}}(\mathbf{L}) = 1$, the minimal value, whereas the spikiest possible matrix such as $\mathbf{Z}$ from Equation (7.16) achieves the maximal spikiness ratio $\alpha_{\mathrm{sp}}(\mathbf{Z}) = p$. In contrast to singular vector incoherence, the spikiness ratio involves the singular values (as well as the vectors). Thus, the matrix $\mathbf{L}^* = \mathbf{B} + \delta\mathbf{Z}$. will have a low spikiness ratio whenever the perturbation $\delta > 0$ is sufficiently small.

For the nuclear-norm regularized estimator (7.10) with a bound on the spikiness ratio, Negahban and Wainwright (2012) show that the estimate $\widehat{\mathbf{L}}$ satisfies a bound of the form

$$\frac{\|\widehat{\mathbf{L}} - \mathbf{L}^*\|_{\mathrm{F}}^2}{\|\mathbf{L}^*\|_{\mathrm{F}}^2} \leq C \max\left\{\sigma^2,\, \alpha_{\mathrm{sp}}^2(\mathbf{L}^*)\right\} \frac{rp\log p}{N} \tag{7.19}$$

with high probability over the sampling pattern, and random noise (assumed i.i.d., zero-mean with all moments finite, and variance σ^2). See also Keshavan, Montanari and Oh (2010) and Koltchinskii, Lounici and Tsybakov (2011), who prove related guarantees for slightly different estimators.

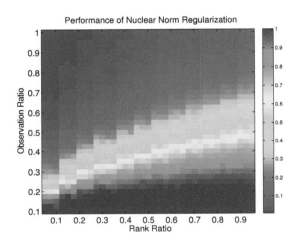

Figure 7.5 *Performance of the nuclear-norm regularized estimator* (7.10), *solved via the* SOFT-IMPUTE *algorithm, for noisy matrix completion under the model* (7.7) *with matrix* $\mathbf{L}^* = \mathbf{C}\mathbf{G}^T$ *of rank* r. *Plots of the relative Frobenius norm error* $\|\widehat{\mathbf{L}} - \mathbf{L}^*\|_{\mathrm{F}}^2 / \|\mathbf{L}^*\|_{\mathrm{F}}^2$ *for* $p = 50$ *as a function of the rank ratio* $\delta = \frac{r\log p}{p}$ *and observation ratio* $\nu = \frac{N}{p^2}$, *corresponding to the fraction of observed entries in a* $p \times p$ *matrix. Observations were of the linear form* (7.7) *with* $w_{ij} \sim N(0, \sigma^2)$ *where* $\sigma = 1/4$, *and we used the* SOFT-IMPUTE *algorithm to solve the program* (7.10) *with* $\lambda/N = 2\sigma\sqrt{\frac{p}{N}}$, *the latter choice suggested by theory. The theory also predicts that the Frobenius error should be low as long as* $\nu \succsim \delta$, *a prediction confirmed in this plot.*

In order to better understand the guarantee (7.19), we carried out a simulation. Let us define the ratio $\nu = \frac{N}{p^2} \in (0, 1)$, corresponding to the fraction of observed entries in a $p \times p$ matrix, and the rank ratio $\delta = \frac{r\log p}{p}$, corresponding to the relative rank of the matrix (up to a logarithmic factor). For a constant noise variance and spikiness ratio, the bound predicts that the estimator (7.10) should have low relative mean-squared error whenever $\nu > \delta$. Figure 7.5 confirms this prediction, and shows that the theory is actually somewhat conservative; see the figure caption for further details.

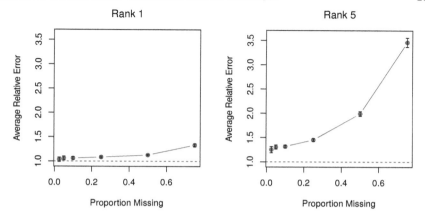

Figure 7.6 *Matrix completion via* SOFT-IMPUTE *in the noisy setting. The plots show the imputation error from matrix completion as a function of the proportion missing, for 40×40 matrices. Shown are the mean absolute error(± one standard error) over 100 simulations, all relative to the noise standard deviation. In each case we chose the penalty parameter to minimize the imputation error, and the results would be somewhat worse if that parameter were chosen by cross-validation. The true rank of the complete matrix is one in the left panel and five in the right panel. The average absolute size of each matrix entry was 0.80 and 1.77 in the left and right panels, respectively.*

Figure 7.6 is another illustration of the imputation error from matrix completion in the noisy setting. Here we use SOFT-IMPUTE on 40×40 matrices, with entries generated from a standard Gaussian matrix with rank $r = 1$ or 5, plus noise of standard deviation $\sigma = 0.5$. We see that for rank one, we can impute the missing values with average error close to σ even when the proportion missing is as high as 50%. However when the true rank increases to five, the procedure starts to break down at about 30% missing.

7.3.4   Maximum Margin Factorization and Related Methods

Here we discuss a class of techniques that are close in spirit to the method of the previous section. These are known as *maximum margin matrix factorization* methods (MMMF), and use a factor model for approximating the matrix $\mathbf{Z}$ (Rennie and Srebro 2005).[4] Consider a matrix factorization of the form $\mathbf{M} = \mathbf{AB}^T$, where $\mathbf{A}$ and $\mathbf{B}$ are $m \times r$ and $n \times r$, respectively. One way to

---

[4]The "maximum margin" refers to the particular margin-based loss used by these authors; although we use squared-error loss, our focus is on the penalty, so we use the same acronym.

estimate such a factorization is by solving the optimization problem

$$\underset{\substack{\mathbf{A}\in\mathbb{R}^{m\times r}\\\mathbf{B}\in\mathbb{R}^{n\times r}}}{\text{minimize}}\left\{\|\mathcal{P}_{\Omega}(\mathbf{Z})-\mathcal{P}_{\Omega}(\mathbf{AB}^{T})\|_{\mathrm{F}}^{2}+\lambda\left(\|\mathbf{A}\|_{\mathrm{F}}^{2}+\|\mathbf{B}\|_{\mathrm{F}}^{2}\right)\right\}. \tag{7.20}$$

Interestingly, this problem turns out to be equivalent to the nuclear norm regularized problem (7.10) for sufficiently large r, in a way that we now make precise. First, for any matrix $\mathbf{M}$, it can be shown (Rennie and Srebro 2005, Mazumder et al. 2010) that

$$\|\mathbf{M}\|_{\star}=\underset{\substack{\mathbf{A}\in\mathbb{R}^{m\times r},\ \mathbf{B}\in\mathbb{R}^{n\times r}\\\mathbf{M}=\mathbf{AB}^{T}}}{\min}\frac{1}{2}\left(\|\mathbf{A}\|_{\mathrm{F}}^{2}+\|\mathbf{B}\|_{\mathrm{F}}^{2}\right) \tag{7.21}$$

As shown in Exercise 7.6, the solution to the problem (7.21) need not be unique. However, the equivalence (7.21) implies that the family of solutions $\widehat{\mathbf{M}}=\widehat{\mathbf{A}}\widehat{\mathbf{B}}^{T}$ of the biconvex problems (7.20) for $r\geq\min(m,n)$ are the same as those for the family of convex problems (7.10). To be more specific, we have the following result:

Theorem 1. Let $\mathbf{Z}$ be an $m\times n$ matrix with observed entries indexed by Ω.

(a) The solutions to the MMMF criterion (7.20) with $r=\min\{m,n\}$ and the nuclear norm regularized criterion (7.10) coincide for all $\lambda\geq 0$.

(b) For some fixed $\lambda^{*}>0$, suppose that the objective (7.10) has an optimal solution with rank r^{*}. Then for any optimal solution $(\widehat{\mathbf{A}},\widehat{\mathbf{B}})$ to the problem (7.20) with $r\geq r^{*}$ and $\lambda=\lambda^{*}$, the matrix $\widehat{\mathbf{M}}=\widehat{\mathbf{A}}\widehat{\mathbf{B}}^{T}$ is an optimal solution for the problem (7.10). Consequently, the solution space of the objective (7.10) is contained in that of (7.20).

The MMMF criterion (7.20) defines a two-dimensional family of models indexed by the pair (r,λ), while the SOFT-IMPUTE criterion (7.10) defines a one-dimensional family. In light of Theorem 1, this family is a special path in the two-dimensional grid of solutions $\left(\widehat{\mathbf{A}}_{(r,\lambda)},\widehat{\mathbf{B}}_{(r,\lambda)}\right)$. Figure 7.7 depicts the situation. Any MMMF model at parameter combinations above the red points are redundant, since their fit is the same at the red point. However, in practice the red points are not known to MMMF, nor is the actual rank of the solution. Further orthogonalization of $\widehat{\mathbf{A}}$ and $\widehat{\mathbf{B}}$ would be required to reveal the rank, which would only be approximate (depending on the convergence criterion of the MMMF algorithm). In summary, the formulation (7.10) is preferable for two reasons: it is convex and it does both rank reduction and regularization at the same time. Using (7.20) we need to choose the rank of the approximation and the regularization parameter λ.

In a related approach, Keshavan et al. (2010) propose the criterion

$$\|\mathcal{P}_{\Omega}(\mathbf{Z})-\mathcal{P}_{\Omega}(\mathbf{USV}^{T})\|_{\mathrm{F}}^{2}+\lambda\|\mathbf{S}\|_{\mathrm{F}}^{2}, \tag{7.22}$$

to be minimized over the triplet $(\mathbf{U},\mathbf{V},\mathbf{S})$, where $\mathbf{U}^{T}\mathbf{U}=\mathbf{V}^{T}\mathbf{V}=\mathbf{I}_{r}$ and

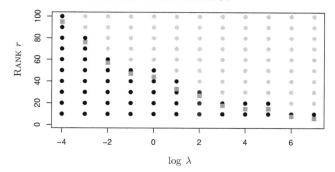

Figure 7.7 *Comparison of the parameter space for* MMMF *(gray and black points), and* SOFT-IMPUTE *(red points) for a simple example.* MMMF *solutions for ranks above the red points are identical to the* SOFT-IMPUTE *solutions at the red points and hence the gray points are redundant. On the other hand, fixing a rank for* MMMF *(for a given* λ*) that is less than that of the* SOFT-IMPUTE *solution leads to a non-convex problem.*

S is an $r \times r$ matrix. For a fixed rank r, they minimize the criterion (7.22) by gradient descent. This criterion is similar to the MMMF criterion (7.20), except that the matrices $\mathbf{U}, \mathbf{V}$ are constrained to be orthonormal, so that the "signal" and corresponding regularization are shifted to the (full) matrix **S**. Like MMMF, the problem is nonconvex so that gradient descent is not guaranteed to converge to the global optimum; moreover, it must be solved separately for different values of the rank r.

Keshavan et al. (2010) provide some asymptotic theory for the estimator (7.22) when applied to noisy matrix completion, using a scaling in which the aspect ratio m/n converges to some constant $\alpha \in (0, 1)$. Here is a rough description of one such result. Consider an $m \times n$ matrix **Z** that can be written as a sum of the form $\mathbf{Z} = \mathbf{U}\boldsymbol{\Sigma}\mathbf{V} + \mathbf{W}$, where $\boldsymbol{\Sigma} \in \mathbb{R}^{r \times r}$ is a diagonal matrix. Here the term **W** is a random matrix with i.i.d. entries, each with zero mean and variance $\sigma^2 \sqrt{mn}$. Each entry of the matrix **Z** is assumed to be observed independently with probability ρ. Let $\widehat{\mathbf{Z}}$ be the estimate obtained by minimizing the criterion (7.22) using the optimal value for λ. For this criterion, Keshavan et al. (2010) show that the relative error $\|\widehat{\mathbf{Z}} - \mathbf{Z}\|_{\mathrm{F}}^2 / \|\mathbf{Z}\|_{\mathrm{F}}^2$ converges in probability to a quantity $1 - c(\rho)$ as as $m/n \to \alpha \in (0, 1)$. The constant $c(\rho)$ is zero if $\sigma^2/\rho \geq \max_{jj} \Sigma_{jj}$ and nonzero otherwise. This shows that the estimator undergoes a phase transition: if the noise and probability of missing entries are low relative to the signal strength, then the missing entries can be recovered successfully. Otherwise they are essentially useless in reconstructing the missing entries. Full details may be found in Keshavan et al. (2009) and Keshavan et al. (2010).

7.4   Reduced-Rank Regression

In this section we briefly revisit a topic touched on in Section 4.3, namely *multivariate regression*. We have vector-valued responses $y_i \in \mathbb{R}^K$ and covariates $x_i \in \mathbb{R}^p$, and we wish to build a series of K linear regression models. With N observations on (y_i, x_i), we can write these regression models in matrix form as

$$\mathbf{Y} = \mathbf{X\Theta} + \mathbf{E}, \tag{7.23}$$

with $\mathbf{Y} \in \mathbb{R}^{N \times K}$, $\mathbf{X} \in \mathbb{R}^{N \times p}$, $\mathbf{\Theta} \in \mathbb{R}^{p \times K}$ a matrix of coefficients, and $\mathbf{E} \in \mathbb{R}^{N \times K}$ a matrix of errors.

The simplest approach would be to fit K separate models, perhaps via the lasso or elastic net. However, the idea is that the responses may have a lot in common, and these similarities can be used to *borrow strength* when fitting the K regression models. In Section 4.3, we used the group lasso to select variables simultaneously for each response; i.e., we used the group lasso to set whole rows of $\mathbf{\Theta}$ to zero. In this section we instead assume $\mathbf{\Theta}$ has low rank. The same ideas underlie *multitask* machine learning. Hence we entertain models of the form

$$\mathbf{Y} = \mathbf{XAB}^T + \mathbf{E}, \tag{7.24}$$

with $\mathbf{A} \in \mathbb{R}^{p \times r}$ and $\mathbf{B} \in \mathbb{R}^{K \times r}$. One can think of having $r < K$ derived features $\mathbf{Z} = \mathbf{X}\widehat{\mathbf{A}}$ which are then distributed among the responses via K separate regressions $\widehat{\mathbf{Y}} = \mathbf{Z}\widehat{\mathbf{B}}^T$. Although fitting (7.24) by least-squares is a nonconvex optimization problem, with $N > p$ closed-form solutions are available through a form of canonical-correlation analysis (Hastie et al. 2009).

Example 7.1. As an example, we consider the problem of video denoising. Figure 7.8 shows four representative images of a video taken by a helicopter flying over the desert. Each column j of the matrix $\mathbf{Y}$ represents an image frame (in a vectorized form) at a time k, and the full matrix $\mathbf{Y}$ represents a video consisting of K image frames. The p columns of $\mathbf{X}$ represent a dictionary of image basis functions (e.g.,unions of orthonormal bases; see Chapter 10). Imposing a low rank model on $\mathbf{\Theta}$ is reasonable when the video sequence changes relatively slowly over time (as they do in this sequence), so that most of its variation can be described by linear combinations of a small number of representative images.

Figure 7.9 shows the SVD computed using $K = 100$ frames from the video in Figure 7.8; although the matrix $\mathbf{Y}$ is not exactly low-rank, its singular values decay rapidly, suggesting that it can be well-approximated by a low-rank matrix. $\diamond$

As before, the nuclear norm is a useful convex penalty for enforcing low-rank structure on an estimate. In this case we would solve the optimization problem

$$\underset{\mathbf{\Theta} \in \mathbb{R}^{p \times K}}{\text{minimize}} \left\{ \|\mathbf{Y} - \mathbf{X\Theta}\|_F^2 + \lambda \|\mathbf{\Theta}\|_\star \right\}, \tag{7.25}$$

and for sufficiently large values of λ the solution $\widehat{\mathbf{\Theta}}$ would have rank less than $\min(N, K)$.

Figure 7.8 *Four* 352×640 *image frames from a video sequence of a helicopter flying over the desert. Each image was converted to a vector in* $\mathbb{R}^N$ *with* $N = 352 \times 640 = 225280$ *elements, and represented by one column of the matrix* **Y**.

7.5   A General Matrix Regression Framework

In this section we present a general "trace" regression framework, that includes matrix completion and reduced-rank regression as special cases. This general framework allows for a unified theoretical treatment.

Let's start with matrix completion. Let **M** represent a model underlying a partially observed $m \times n$ matrix **Z** that we wish to complete. Then we consider observations $(\mathbf{X}_i, y_i), i = 1, 2, \ldots, |\Omega|$ from the model

$$y_i = \text{trace}(\mathbf{X}_i^T \mathbf{M}) + \varepsilon_i. \tag{7.26}$$

Here $\mathbf{X}_i$ are $m \times n$ matrices and y_i and ε_i are scalars. The observation model (7.26) can be viewed as a regression with inputs the matrices $\mathbf{X}_i$ and outputs the y_i. The trace inner product on matrices plays the role of an ordinary inner product on vectors, but otherwise everything is conceptually the same as in a usual regression model.[5]

To relate this to matrix completion, let $[a(i), b(i)]$ be the row-column indices of the matrix entry observed in observation i. We then define

---

[5]Recall that if **A** and **B** are both $m \times n$ matrices, $\text{trace}(\mathbf{A}^T \mathbf{B}) = \sum_{i=1}^{m} \sum_{j=1}^{n} a_{ij} b_{ij}$.

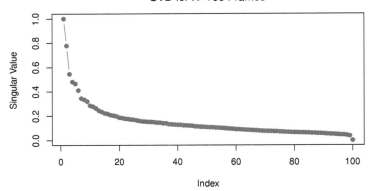

Figure 7.9 *Singular values of the matrix* $\mathbf{Y} \in \mathbb{R}^{p \times K}$ *using* $K = 100$ *frames from the video sequence. Note the rapid decay, showing that a low-rank approximation is possible.*

$\mathbf{X}_i = e_{a(i)}^n e_{b(i)}^m{}^T$, where $e_\ell^m \in \mathbb{R}^m$ denotes the unit m-vector with a single one in the ℓth coordinate, so that $\mathbf{X}_i$ is zero everywhere except at position $[a(i), b(i)]$. With this choice, we have $\text{trace}(\mathbf{X}_i^T \mathbf{M}) = m_{a(i)\,b(i)}$, so that the observation model (7.26) provides us with certain entries of $\mathbf{M}$—those in the training set Ω—each contaminated with noise ε_i. Our goal is to predict the unobserved entries in $\mathbf{Z}$ via $\widehat{\mathbf{M}}$, which can be thought of as $\mathbb{E}(y^* | \mathbf{X}^*)$ for feature values $\mathbf{X}^*$ that are distinct from those in the training set.

The trace observation model (7.26) is also relevant in a more general setting, since with different choices of the covariate matrices $\{\mathbf{X}_i\}$, it can also be used to model other types of matrix estimation problems involving low-rank constraints.

The multiresponse regression model of the previous section is another example. The response and covariate vectors are linked via the equation $y_i = \mathbf{\Theta}^T x_i + \epsilon_i$, where $\mathbf{\Theta} \in \mathbb{R}^{p \times K}$ is a matrix of regression coefficients, and $\epsilon_i \in \mathbb{R}^K$ is a noise vector. Since each response y_i is a K-vector of observations, it can be rewritten as a collection of K separate observations in the trace form: if we set $\mathbf{X}_{ij} = x_i e_j^{K\,T}$ where $e_j^K \in \mathbb{R}^K$ is the unit vector with a single one in the jth position, then the jth component of y_i can be expressed in the form $y_{ij} = \text{trace}(\mathbf{X}_{ij}^T \mathbf{\Theta}) + \epsilon_{ij}$.

In the context of multivariate regression, the matrix lasso takes the form

$$\underset{\mathbf{\Theta}}{\text{minimize}} \left\{ \frac{1}{2N} \sum_{i=1}^N \sum_{j=1}^K \left(y_{ij} - \text{trace}(\mathbf{X}_{ij}^T \mathbf{\Theta}) \right)^2 + \lambda \|\mathbf{\Theta}\|_\star \right\}. \qquad (7.27)$$

Exercise 7.9 explores another example. See the papers by Yuan, Ekici, Lu and Monteiro (2007), Negahban and Wainwright (2011a), and Rohde and Tsybakov (2011) for further details and benefits of this unified approach. See also Bunea, She and Wegkamp (2011) for analysis of an alternative procedure for reduced-rank multivariate regression.

7.6   Penalized Matrix Decomposition

Maximum-margin matrix factorization methods lead naturally to other forms of regularization such as the ℓ_1-penalized version

$$\underset{\substack{\mathbf{U}\in\mathbb{R}^{m\times r},\ \mathbf{V}\in\mathbb{R}^{n\times r}\\ \mathbf{D}\in\mathbb{R}^{r\times r}}}{\text{minimize}} \left\{ \|\mathbf{Z}-\mathbf{U}\mathbf{D}\mathbf{V}^T\|_{\mathrm{F}}^2 + \lambda_1\|\mathbf{U}\|_1 + \lambda_2\|\mathbf{V}\|_1 \right\}, \qquad (7.28)$$

where $\mathbf{D}$ is diagonal and nonnegative. Here we assume that all values of $\mathbf{Z}$ are observed, and apply an ℓ_1 penalty to the left and right singular vectors of the decomposition. The idea is to obtain sparse versions of the singular vectors for interpretability.

Before discussing how to optimize the criterion (7.28), let's see how it can be used. Returning to the Netflix example, we created a smaller matrix consisting of the 1000 users and the 100 movies, each with the most ratings. We imputed the missing values using an iterated rank 10 SVD (Section 7.3). Then we set the rank of $\mathbf{U}$ and $\mathbf{V}$ to two, and minimized a version of the criterion (7.28) for values of λ_1 and λ_2 that yielded a very sparse solution. The resulting solution $\widehat{\mathbf{V}}$ had 12 nonzero entries, all with the same sign, corresponding to the movies in Table 7.3. The first group looks like a mixture

Table 7.3 *Movies with nonzero loadings, from a two-dimensional penalized matrix decomposition.*

First Component	Second Component
The Wedding Planner	Lord of the Rings: The Fellowship of the Ring
Gone in 60 Seconds	The Last Samurai
The Fast and the Furious	Lord of the Rings: The Two Towers
Pearl Harbor	Gladiator
Maid in Manhattan	Lord of the Rings: The Return of the King
Two Weeks Notice	
How to Lose a Guy in 10 Days	

of romantic comedies and action movies, while the second group consists of historical action/fantasy movies.

How do we solve the optimization problem (7.28)? Let us first consider the one-dimensional case, written in the constrained rather than Lagrangian form:

$$\underset{\substack{\mathbf{u}\in\mathbb{R}^m,\ \mathbf{v}\in\mathbb{R}^n\\ d\geq 0}}{\text{minimize}} \|\mathbf{Z}-d\mathbf{u}\mathbf{v}^T\|_{\mathrm{F}}^2 \text{ subject to } \|\mathbf{u}\|_1 \leq c_1 \text{ and } \|\mathbf{v}\|_1 \leq c_2. \qquad (7.29)$$

It turns out that the estimator (7.29) is not very useful, as it tends to produce solutions that are too sparse, as illustrated in Figure 7.10 (right panel). In order to fix this problem, we augment our formulation with additional ℓ_2-norm constraints, thereby obtaining the optimization problem

$$\underset{\substack{\mathbf{u}\in\mathbb{R}^m,\,\mathbf{v}\in\mathbb{R}^n \\ d\geq 0}}{\text{minimize}} \ \|\mathbf{Z} - d\mathbf{u}\mathbf{v}^T\|_{\mathrm{F}}^2 \ \text{subject to} \ \|\mathbf{u}\|_1 \leq c_1, \ \|\mathbf{v}\|_1 \leq c_2, \ \|\mathbf{u}\|_2 \leq 1, \ \|\mathbf{v}\|_2 \leq 1.$$

$$(7.30)$$

It may seem surprising that *adding* constraints can make the solution *sparse*, but Figure 7.10 provides some insight.

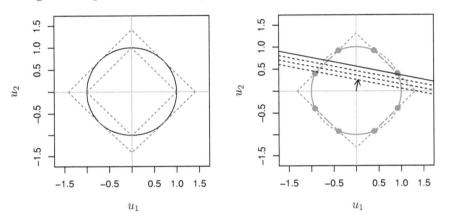

Figure 7.10 *A graphical representation of the ℓ_1 and ℓ_2 constraints on $\mathbf{u} \in \mathbb{R}^2$ in the PMD(ℓ_1, ℓ_1) criterion. The constraints are as follows: $\|\mathbf{u}\|_2^2 \leq 1$ and $\|\mathbf{u}\|_1 \leq c$. The gray lines indicate the coordinate axes u_1 and u_2. Left panel: The ℓ_2 constraint is the solid circle. For both the ℓ_1 and ℓ_2 constraints to be active, the constraint radius c must be between 1 and $\sqrt{2}$. The constraints $\|\mathbf{u}\|_1 = 1$ and $\|\mathbf{u}\|_1 = \sqrt{2}$ are shown using dashed lines. Right panel: The ℓ_2 and ℓ_1 constraints on $\mathbf{u}$ are shown for some c between 1 and $\sqrt{2}$. Red dots indicate the points where both the ℓ_1 and the ℓ_2 constraints are active. The red contour shows the boundary of the constraint region. The black lines are the linear contours of the criterion (7.30) as a function of $\mathbf{u}$, which increase as we move to the upper right in this example. The solid red arcs indicate the solutions that occur when $\lambda_1 = 0$ in Algorithm 7.2 (ℓ_2 active, ℓ_1 not). The figure shows that in two dimensions, the points where both the ℓ_1 and ℓ_2 constraints are active have neither u_1 nor u_2 equal to zero. We also see that without the ℓ_2 constraints, we would always end up at a corner; this would lead to trivial solutions.*

If we fix the second component $\mathbf{v}$, the criterion (7.30) is linear in $\mathbf{u}$. Suppose that the linear contour of the criterion is angled as in Figure 7.10 and not exactly parallel to a side of the polyhedral constraint region. Then to solve the problem, we move the linear contour toward the top right as far as possible while still remaining inside the constraint region. The solution occurs at one of the open circles, or on the solid red contour. Notice that without the ℓ_2 constraint, the solution will occur at a corner of the polyhedron where only one coefficient is nonzero. As shown in the left panel of Figure 7.10, the problem is well defined as long as $1 \leq c_1 \leq \sqrt{m}$ and $1 \leq c_2 \leq \sqrt{n}$.

Since the criterion (7.30) is biconvex, we can minimize it in an alternating fashion. It is easy to verify that the solution in each direction is a soft-thresholding operation. For example, the update for $\mathbf{v} \in \mathbb{R}^n$ takes the form

$$\mathbf{u} \leftarrow \frac{\mathcal{S}_{\lambda_1}(\mathbf{Zv})}{\|\mathcal{S}_{\lambda_1}(\mathbf{Zv})\|_2}. \tag{7.31}$$

Here we apply our soft-thresholding operator $\mathcal{S}$ element-wise on its vector

---

Algorithm 7.2 ALTERNATING SOFT-THRESHOLDING FOR RANK-ONE PENALIZED MATRIX DECOMPOSITION.

1. Set $\mathbf{v}$ to the top left singular vector from the SVD of $\mathbf{Z}$.

2. Perform the update $\mathbf{u} \leftarrow \dfrac{\mathcal{S}_{\lambda_1}(\mathbf{Zv})}{\|\mathcal{S}_{\lambda_1}(\mathbf{Zv})\|_2}$, with λ_1 being the smallest value such that $\|\mathbf{u}\|_1 \leq c_1$;

3. Perform the update $\mathbf{v} \leftarrow \dfrac{\mathcal{S}_{\lambda_2}(\mathbf{Z}^T\mathbf{u})}{\|\mathcal{S}_{\lambda_2}(\mathbf{Z}^T\mathbf{u})\|_2}$, with λ_2 being the smallest value such that $\|\mathbf{v}\|_1 \leq c_2$;

4. Iterate steps 2 and 3 until convergence.

5. Return $\mathbf{u}$, $\mathbf{v}$ and $d = \mathbf{u}^T\mathbf{Zv}$.

---

argument. The threshold λ_1 in Equation (7.31) must be chosen adaptively to satisfy the constraints: it is set to zero if this results in $\|\mathbf{u}\|_1 \leq c_1$, and otherwise λ_1 is chosen to be a positive constant such that $\|\mathbf{u}\|_1 = c_1$. (See Exercise 7.7). The overall procedure is summarized in Algorithm 7.2. We note that if $c_1 > \sqrt{m}$ and $c_2 > \sqrt{n}$ so that the ℓ_1 constraints have no effect, then Algorithm 7.2 reduces to the power method for computing the largest singular vectors of the matrix $\mathbf{Z}$. See Section 5.9 for further discussion of the (ordinary) power method. Some recent work has established theoretical guarantees for iterative soft thresholding updates related to Algorithm 7.2; see the bibliographic section for further details.

The criterion (7.30) is quite useful and may be used with other penalties (in addition to the ℓ_1-norm) for either $\mathbf{u}$ or $\mathbf{v}$, such as the fused lasso penalty

$$\Phi(\mathbf{u}) = \sum_{j=2}^{m} |u_j - u_{j-1}|, \tag{7.32}$$

where $\mathbf{u} = (u_1, u_2, \ldots u_m)$. This choice is useful for enforcing smoothness along a one-dimensional ordering $j = 1, 2, \ldots m$, such as chromosomal position in a genomics application. Depending on this choice of penalty, the corresponding minimization in Algorithm 7.2 must change accordingly. In addition, one can modify Algorithm 7.2 to handle missing matrix entries, for example by omitting missing values when computing the inner products $\mathbf{Z}\mathbf{v}$ and $\mathbf{Z}^T\mathbf{u}$.

To obtain a multifactor penalized matrix decomposition, we apply the rank-one Algorithm (7.2) successively to the matrix $\mathbf{Z}$, as given in Algorithm 7.3. If ℓ_1 penalties on $\mathbf{u}_k$ and $\mathbf{v}_k$ are *not* imposed—equivalently, if we

---

Algorithm 7.3 MULTIFACTOR PENALIZED MATRIX DECOMPOSITION

1. Let $\mathbf{R} \leftarrow \mathbf{Z}$.

2. For $k = 1, \ldots K$:

   (a) Find $\mathbf{u}_k$, $\mathbf{v}_k$, and d_k by applying the single-factor Algorithm 7.2 to data $\mathbf{R}$.

   (b) Update $\mathbf{R} \leftarrow \mathbf{R} - d_k\mathbf{u}_k\mathbf{v}_k^T$.

---

set $\lambda_1 = \lambda_2 = 0$ in Algorithm 7.2—then it can be shown that the K-factor PMD algorithm leads to the rank-K SVD of $\mathbf{Z}$. In particular, the successive solutions are orthogonal. With penalties present, the solutions are no longer in the column and row spaces of $\mathbf{Z}$, and so the orthogonality does not hold.

It is important to note the difference between sparse matrix decomposition and matrix completion, discussed earlier. For successful matrix completion, we required that the singular vectors of $\mathbf{Z}$ have low coherence; that is, they need to be *dense*. In sparse matrix decomposition, we seek *sparse* singular vectors, for interpretability. Matrix completion is not the primary goal in this case.

Unlike the minimization of convex functions, alternating minimization of biconvex functions is not guaranteed to find a global optimum. In special cases, such as the power method for computing the largest singular vector, one can show that the algorithm converges to a desired solution, as long as the starting vector is not orthogonal to this solution. But in general, these procedures are only guaranteed to move downhill to a partial local minimum of the function; see Section 5.9 for discussion of this issue. Based on our experience, however, they behave quite well in practice, and some recent theoretical work provides rigorous justification of this behavior. See the bibliographic section for discussion.

Lee, Shen, Huang and Marron (2010) suggest the use of the penalized matrix decomposition for biclustering of two-way data. In Chapter 8 we describe applications of the penalized matrix decomposition to derive penalized multivariate methods such as sparse versions of principal components, canonical correlation, and clustering.

7.7 Additive Matrix Decomposition

In the problem of additive matrix decomposition, we seek to decompose a matrix into the sum of two or more matrices. The components in this additive composition should have complementary structures; for instance, one of the most widely studied cases involves decomposing into the sum of a low-rank matrix with a sparse matrix (see also Section 9.5). Additive matrix decompositions arise in a wide variety of applications, among them factor analysis, robust forms of PCA and matrix completion, and multivariate regression problems, as discussed below.

Most of these applications can be described in terms of the noisy linear observation model $\mathbf{Z} = \mathbf{L}^* + \mathbf{S}^* + \mathbf{W}$, where the pair $(\mathbf{L}^*, \mathbf{S}^*)$ specify the additive matrix decomposition into low rank and sparse components, and $\mathbf{W}$ is a noise matrix. In certain cases, we consider a slight generalization of this model, in which we observe a noisy version of $\mathfrak{X}(\mathbf{L}^* + \mathbf{S}^*)$, where $\mathfrak{X}$ is some type of linear operator on the matrix sum (e.g., the projection operator $\mathcal{P}_\Omega$ in the case of matrix completion, or multiplication via the model matrix $\mathbf{X}$ in the case of matrix regression.

Given such observations, we consider estimators of the pair $(\mathbf{L}^*, \mathbf{S}^*)$ based on the criterion

$$\underset{\substack{\mathbf{L} \in \mathbb{R}^{m \times n} \\ \mathbf{S} \in \mathbb{R}^{m \times n}}}{\text{minimize}} \left\{ \frac{1}{2} \|\mathbf{Z} - (\mathbf{L} + \mathbf{S})\|_F^2 + \lambda_1 \Phi_1(\mathbf{L}) + \lambda_2 \Phi_2(\mathbf{S}) \right\}, \qquad (7.33)$$

where Φ_1 and Φ_2 are penalty functions each designed to enforce a different type of generalized sparsity. For instance, in the case of low rank and sparse matrices, we study the choices $\Phi_1(\mathbf{L}) = \|\mathbf{L}\|_*$ and $\Phi_2(\mathbf{S}) = \|\mathbf{S}\|_1$.

We now turn to some applications of additive matrix decompositions.

Factor Analysis with Sparse Noise: Factor analysis is a widely used form of linear dimensionality reduction that generalizes principal component analysis. Factor analysis is easy to understand as a generative model: we generate random vectors $y_i \in \mathbb{R}^p$ using the "noisy subspace" model

$$y_i = \mu + \mathbf{\Gamma} u_i + w_i, \quad \text{for } i = 1, 2, \dots, N. \qquad (7.34)$$

Here $\mu \in \mathbb{R}^p$ is a mean vector, $\mathbf{\Gamma} \in \mathbb{R}^{p \times r}$ is a loading matrix, and the random vectors $u_i \sim N(0, \mathbf{I}_{r \times r})$ and $w_i \sim N(0, \mathbf{S}^*)$ are independent. Each vector y_i drawn from model (7.34) is obtained by generating a random element in the r-dimensional subspace spanned by the columns of $\mathbf{\Gamma}$. Given N samples from this model, the goal is to estimate the column of the loading matrix $\mathbf{\Gamma}$, or equivalently, the rank r matrix $\mathbf{L}^* = \mathbf{\Gamma}\mathbf{\Gamma}^T \in \mathbb{R}^{p \times p}$ that spans the column space of $\mathbf{\Gamma}$.

A simple calculation shows that the covariance matrix of y_i has the form $\mathbf{\Sigma} = \mathbf{\Gamma}\mathbf{\Gamma}^T + \mathbf{S}^*$. Consequently, in the special case when $\mathbf{S}^* = \sigma^2 \mathbf{I}_{p \times p}$, then the column span of $\mathbf{\Gamma}$ is equivalent to the span of the top r eigenvectors of $\mathbf{\Sigma}$, and so we can recover it via standard principal components analysis. In particular, one way to do so is by computing the SVD of the data matrix $\mathbf{Y} \in \mathbb{R}^{N \times p}$, as discussed in Section 7.2. The right singular vectors of $\mathbf{Y}$ specify the eigenvectors of the sample covariance matrix, which is a consistent estimate of $\mathbf{\Sigma}$.

What if the covariance matrix $\mathbf{S}^*$ is not a multiple of the identity? A typical assumption in factor analysis is that $\mathbf{S}^*$ is diagonal, but with the noise variance depending on the component of the data. More generally, it might have nonzero entries off the diagonal as well, but perhaps a relatively small number, so that it could be represented as a sparse matrix. In such settings, we no longer have any guarantees that the top r eigenvectors of $\mathbf{\Sigma}$ are close to the column span of $\mathbf{\Gamma}$. When this is not the case, PCA will be *inconsistent*—meaning that it will fail to recover the true column span even if we have an infinite sample size.

Nonetheless, when $\mathbf{S}^*$ is a sparse matrix, the problem of estimating $\mathbf{L}^* = \mathbf{\Gamma}\mathbf{\Gamma}^T$ can be understood as an instance of our general observation model with $p = N$. In particular, given our observations $\{y_i\}_{i=1}^N$, we can let our observation matrix $\mathbf{Z} \in \mathbb{R}^{p \times p}$ be the sample covariance matrix $\frac{1}{N} \sum_{i=1}^N y_i y_i^T$. With this algebra, we can then write $\mathbf{Z} = \mathbf{L}^* + \mathbf{S}^* + \mathbf{W}$, where $\mathbf{L}^* = \mathbf{\Gamma}\mathbf{\Gamma}^T$ is of rank r, and the random matrix $\mathbf{W}$ is a re-centered form of Wishart noise—in particular, the zero-mean matrix $\mathbf{W} := \frac{1}{N} \sum_{i=1}^N y_i y_i^T - \{\mathbf{L}^* + \mathbf{S}^*\}$.

Robust PCA: As discussed in Section 7.2, standard principal component analysis is based on performing an SVD of a (column centered) data matrix $\mathbf{Z} \in \mathbb{R}^{N \times p}$, where row i represents the ith sample of a p-dimensional data vector. As shown there, the rank-r SVD can be obtained by minimizing the squared Frobenius norm $\|\mathbf{Z} - \mathbf{L}\|_F^2$ subject to a rank constraint on $\mathbf{L}$. What if some entries of the data matrix $\mathbf{Z}$ are corrupted? Or even worse, what if some subset of the rows (data vectors) are corrupted? Since PCA is based on a quadratic objective function, its solution (the rank r SVD) can be very sensitive to these types of perturbations.

Additive matrix decompositions provide one way in which to introduce some robustness to PCA. In particular, instead of approximating $\mathbf{Z}$ with a low-rank matrix, we might approximate it with the sum $\mathbf{L} + \mathbf{S}$ of a low-rank matrix with a sparse component to model the corrupted variables. In the case of element-wise corruption, the component $\mathbf{S}$ would be modeled as element-wise sparse, having relatively few nonzero entries, whereas in the more challenging setting of having entirely corrupted rows, it would be modeled as a row-sparse matrix. Given some target rank r and sparsity k, the direct approach would be to try and solve the optimization problem

$$\underset{\substack{\text{rank}(\mathbf{L}) \leq r \\ \text{card}(\mathbf{S}) \leq k}}{\text{minimize}} \frac{1}{2}\|\mathbf{Z} - (\mathbf{L} + \mathbf{S})\|_F^2. \tag{7.35}$$

Here card denotes a cardinality constraint, either the total number of nonzero entries (in the case of element-wise corruption), or the total number of nonzero rows (in the case of row-wise corruption). Of course, the criterion (7.35) is doubly nonconvex, due to both the rank and cardinality constraints, but a natural convex relaxation is provided by our general estimator (7.33) with $\Phi_1(\mathbf{L}) = \|\mathbf{L}\|_*$ and $\Phi_2(\mathbf{S}) = \sum_{i,j} |s_{ij}|$ for element-wise sparsity.

Figure 7.11 shows an example of robust PCA with the above penalties, taken from an unpublished paper by Mazumder and Hastie and using images from Li, Huang, Gu and Tian (2004). The columns of the data matrix $\mathbf{Z}$ are frames from a video surveillance camera, and are noisy and have missing pixel values (next section). The last two columns show the reconstructed frames; the low-rank part represents the static background, while the sparse component changes in each frame, and in this case represent people moving.

| True Image | Training Image | Low-Rank ($\widehat{\mathbf{L}}$) | Sparse ($\widehat{\mathbf{S}}$) |

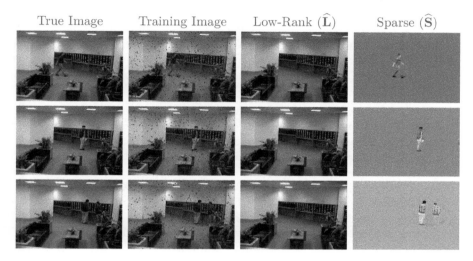

Figure 7.11 *Video surveillance. Shown are the true image, noisy training image with missing-values, the estimated low-rank part, and the sparse part aligned side by side. The true images were sampled from the sequence and include ones with varying illumination and some benchmark test sequences. Despite the missingness and added noise the procedure succeeds in separating the moving components (people) from the fixed background.*

Robust Matrix Completion: Robustness is also a concern for matrix-completion methods (Section 7.3), which are used in collaborative filtering and recommender systems. Ratings may be corrupted for various reasons: for instance, users might try to "game" the system (e.g., a movie star would like to have his/her movies more highly recommended by Netflix). Alternatively, a subset of users might simply be playing pranks with their ratings; for instance, in 2002, *The New York Times* reported how the Amazon system had been manipulated by adversarial users, so that it would recommend a sex manual to users who expressed interest in Christian literature (Olsen 2002).

As we did for robust PCA, we can build in robustness to matrix completion by introducing a sparse component $\mathbf{S}$ to our representation. The nature of sparsity depends on how we model the adversarial behavior: if we believed that only a small fraction of *entries* were corrupted, then it would be appropriate to impose element-wise sparsity via the ℓ_1-norm. On the other hand, if we wished to model users (rows) as being adversarial, then it would be appropriate to impose a row-wise sparsity penalty, such as the group lasso norm $\|\mathbf{S}\|_{1,2} = \sum_{i=1}^{m} \|\mathbf{S}_i\|_2$, where $\mathbf{S}_i \in \mathbb{R}^n$ denotes the ith row of the matrix. This choice would lead to the following modification of our earlier estimator (7.10):

$$\underset{\mathbf{L},\mathbf{S}\in\mathbb{R}^{m\times n}}{\text{minimize}} \left\{ \frac{1}{2} \sum_{(i,j)\in\Omega} (z_{ij} - (\mathbf{L}_{ij} + \mathbf{S}_{ij}))^2 + \lambda_1\|\mathbf{L}\|_\star + \lambda_2 \sum_{i=1}^{m} \|\mathbf{S}_i\|_2 \right\}. \quad (7.36)$$

Exercise 7.10 shows an equivalence between this criterion and a nuclear-norm regularized robust Huber loss. Hence one can develop an algorithm along the lines of SOFT-IMPUTE in Section 7.1, replacing the squared-error loss with the Huber loss. Figure 7.11 shows the results of this approach on some video-surveillance data.

Multivariate Regression: Recall the multivariate linear regression model $y_i = \mathbf{\Theta}^T x_i + \epsilon_i$, where $\mathbf{\Theta} \in \mathbb{R}^{p\times K}$ is a matrix of regression coefficients used to predict the multivariate response vector $y \in \mathbb{R}^K$. As discussed in Section 7.5, in one application of matrix regression, each column of the response matrix $\mathbf{Y}$ represents a vectorized image, so that the full matrix represents a video sequence consisting of K frames. The model matrix $\mathbf{X}$ represents p image basis functions, one per column; for example, an orthonormal basis of two-dimensional wavelets, at different scales and locations (see Section 10.2.3). Figure 7.8 showed that for certain types of video sequences, the matrix $\mathbf{Y}$ exhibits rapid decay in its singular values, and so can be well-approximated by a low-rank matrix.

In a more realistic setting, a video sequence consists of both a background, and various types of foreground elements. The background component is often slowly varying, so that the low-rank model is appropriate, whereas foreground elements vary more rapidly, and may disappear and reappear. (The "helicopter" sequence in Figure 7.8 can be viewed as pure background.) Consequently, a more realistic model for the video sequence is based on decomposition $\mathbf{\Theta} = \mathbf{L} + \mathbf{S}$, where $\mathbf{L}$ is low-rank, and $\mathbf{S}$ is a relatively sparse matrix. Active entries of $\mathbf{S}$ correspond to the basis functions (rows) and time positions (columns) for representing the foreground elements that play a role in the video.

Of course, these types of decompositions also arise in other applications of multivariate regression. In the general setting, we try to recover the decomposition using the estimator

$$\underset{\mathbf{L},\mathbf{S}}{\text{minimize}} \left\{ \frac{1}{2N} \sum_{i=1}^{N} \sum_{j=1}^{K} \left(y_{ij} - \text{trace}(\mathbf{X}_{ij}^T (\mathbf{L} + \mathbf{S})) \right)^2 + \lambda_1\|\mathbf{L}\|_\star + \lambda_2\|\mathbf{S}\|_1 \right\}$$
$$(7.37)$$

where $\mathbf{X}_{ij} = x_i e_j^{K^T}$ for $i = 1,\ldots,N$ and $j = 1,\ldots,K$. Note that this is a natural generalization of our earlier estimator (7.27) for nuclear-norm regularized multivariate regression.

Bibliographic Notes

Early work by Fazel (2002) studied the use of the nuclear norm as a surrogate for a rank constraint. Srebro, Alon and Jaakkola (2005) studied the nuclear norm as well as related relaxations of rank constraints in the context of matrix completion and collaborative filtering. Bach (2008) derived some asymptotic theory for consistency of nuclear norm regularization. Recht, Fazel and Parrilo (2010) derived nonasymptotic bounds on the performance of nuclear norm relaxation in the compressed sensing observation model. See also the papers Negahban and Wainwright (2011a), Rohde and Tsybakov (2011) for nonasymptotic analysis of the nuclear norm relaxation for more general observation models.

Maximum margin matrix factorization is discussed in Srebro and Jaakkola (2003), Srebro, Alon and Jaakkola (2005), and Srebro, Rennie and Jaakkola (2005). Spectral regularization and the SOFT-IMPUTE algorithm were developed by Mazumder et al. (2010). The penalized matrix decomposition is described in Witten, Tibshirani and Hastie (2009). Matrix completion using the nuclear norm has been studied by various authors, with initial results on prediction-error bounds by Srebro, Alon and Jaakkola (2005). The first theoretical results on exact recovery with noiseless observations for exactly low-rank matrices were established by Candès and Recht (2009), with subsequent refinements by various authors. Gross (2011) developed a general dual-witness scheme for proving exactness of nuclear norm relaxations given noiseless observations in arbitrary bases, generalizing the case of entry-wise sampling; see also Recht (2011) for related arguments. Keshavan et al. (2009) provide exact recovery guarantees for a slightly different two-stage procedure, involving trimming certain rows and columns of the matrix and then applying the SVD. The more realistic noisy observation model has also been studied by various authors (e.g., Candès and Plan (2010), Negahban and Wainwright (2012), Keshavan et al. (2010)).

The problem of additive matrix decomposition was first considered by Chandrasekaran, Sanghavi, Parrilo and Willsky (2011) in the noiseless setting, who derived worst-case incoherence conditions sufficient for exact recovery of an arbitrary low-rank/sparse pair. Subsequent work by Candès, Li, Ma and Wright (2011) studied the case of random sparse perturbations to the low-rank matrix, with applications to robust PCA. Xu, Caramanis and Sanghavi (2012) proposed an alternative approach to robust PCA, based on modeling the corruptions in terms of a row-sparse matrix. Chandrasekaran, Parrilo and Willsky (2012) developed the use of sparse/low-rank decompositions for the problem of latent Gaussian graphical model selection. In the more general noisy setting, Hsu, Kakade and Zhang (2011) and Agarwal, Negahban and Wainwright (2012b) provide bounds on relatives of the estimator (7.33).

A recent line of work has provide some theory for alternating minimization algorithms in application to particular nonconvex problems, including matrix completion (Netrapalli, Jain and Sanghavi 2013), phase retrieval (Netrapalli et al. 2013), mixtures of regression (Yi, Caramanis and Sanghavi 2014), and dictionary learning (Agarwal, Anandkumar, Jain, Netrapalli and Tandon 2014). These papers show that given suitable initializations, alternating minimization schemes do converge (with high probability) to estimates with similar statistical accuracy to a global minimum. Similarly, there are also theoretical guarantees for variants of the power method with soft thresholding for recovering sparse eigenvectors (Ma 2013, Yuan and Zhang 2013).

Exercises

Ex. 7.1 Recall the singular value decomposition (7.2) of a matrix.

(a) Show that the SVD of the column-centered matrix $\mathbf{Z}$ gives the principal components of $\mathbf{Z}$.

(b) Show that the condition that successive PCs are uncorrelated is equivalent to the condition that the vectors $\{\mathbf{v}_j\}$ are orthogonal. What is the relationship between the vectors $\{\mathbf{s}_j\}$ in Section 7.2 and the components of the SVD?

Ex. 7.2 In this exercise, we work through the proof of assertion (7.3), namely that

$$\widehat{\mathbf{Z}}_r = \underset{\text{rank}(\mathbf{M})=r}{\arg\min} \ \|\mathbf{Z} - \mathbf{M}\|_F^2,$$

where $\widehat{\mathbf{Z}}_r = \mathbf{U}\mathbf{D}_r\mathbf{V}^T$ is the SVD truncated to its top r components. (In detail, the SVD is given by $\mathbf{Z} = \mathbf{U}\mathbf{D}\mathbf{V}^T$, and $\mathbf{D}_r$ is the same as $\mathbf{D}$ except all but the first r diagonal elements are set to zero.) Here we assume that $m \leq n$ and rank$(\mathbf{Z}) = m$.

We begin by noting that any rank r matrix $\mathbf{M}$ can be factored as $\mathbf{M} = \mathbf{Q}\mathbf{A}$, where $\mathbf{Q} \in \mathbb{R}^{m \times r}$ is an orthogonal matrix, and $\mathbf{A} \in \mathbb{R}^{r \times n}$.

(a) Show that given $\mathbf{Q}$, the optimal value for $\mathbf{A}$ is given by $\mathbf{Q}^T\mathbf{Z}$.

(b) Using part (a), show that minimizing $\|\mathbf{Z} - \mathbf{M}\|_F^2$ is equivalent to solving

$$\underset{\mathbf{Q} \in \mathbb{R}^{m \times r}}{\text{maximize}} \, \text{trace}(\mathbf{Q}^T \mathbf{\Sigma} \mathbf{Q}) \text{ subject to } \mathbf{Q}^T\mathbf{Q} = \mathbf{I}_r, \tag{7.38}$$

where $\mathbf{\Sigma} = \mathbf{Z}\mathbf{Z}^T$.

(c) Show that this is equivalent to the problem

$$\underset{\mathbf{Q} \in \mathbb{R}^{m \times r}}{\text{maximize}} \, \text{trace}(\mathbf{Q}^T \mathbf{D}^2 \mathbf{Q}) \text{ subject to } \mathbf{Q}^T\mathbf{Q} = \mathbf{I}_r. \tag{7.39}$$

(d) Given an orthonormal matrix $\mathbf{Q} \in \mathbb{R}^{m \times r}$, define $\mathbf{H} = \mathbf{Q}\mathbf{Q}^T$ with diagonal elements h_{ii} for $i = 1, \ldots, m$. Show that $h_{ii} \in [0, 1]$ and that $\sum_{i=1}^m h_{ii} = r$. Conclude that problem (7.39) is equivalent to

$$\underset{\substack{h_{ii} \in [0,1] \\ \sum_{i=1}^m h_{ii} = r}}{\text{maximize}} \sum_{i=1}^m h_{ii} \, d_i^2. \tag{7.40}$$

(e) Assuming that $d_1^2 \geq d_2^2 \geq \ldots d_m^2 \geq 0$, show that the solution to problem (7.40) is obtained by setting $h_{11} = h_{22} = \ldots = h_{rr} = 1$, and setting the remaining coefficients zero. If the $\{d_i^2\}$ are strictly ordered, show that this solution is unique.

(f) Conclude that an optimal choice for $\mathbf{Q}$ in problem (7.38) is $\mathbf{U}_1$, the matrix formed from the first r columns of $\mathbf{U}$. This completes the proof.

Ex. 7.3

(a) ℓ_1 *norm as an LP:* For any vector $\beta \in \mathbb{R}^p$, show that

$$\|\beta\|_1 = \max_{u \in \mathbb{R}^p} \sum_{j=1}^{p} u_j \beta_j \text{ subject to } \|u\|_\infty \leq 1. \tag{7.41}$$

This relation expresses the fact that the ℓ_∞ norm is dual to the ℓ_1 norm.

(b) *Nuclear norm as an SDP:* For any matrix $\mathbf{B} \in \mathbb{R}^{m \times n}$, show that

$$\|\mathbf{B}\|_\star = \max_{\mathbf{U} \in \mathbb{R}^{m \times n}} \text{trace}(\mathbf{U}^T \mathbf{B}) \text{ subject to } \|\mathbf{U}\|_{\text{op}} \leq 1,$$

where $\|\mathbf{U}\|_{\text{op}}$ is the maximum singular value of the matrix $\mathbf{U}$, a quantity known as the *spectral norm* or ℓ_2 *operator norm*. This relation expresses the fact that the spectral norm is dual to the nuclear norm. (*Hint:* use the SVD of $\mathbf{Z}$ and cyclical properties of trace operator in order to reduce this to an instance of part (a).)

(c) Given a matrix $\mathbf{U} \in \mathbb{R}^{m \times n}$, show that the inequality $\|\mathbf{U}\|_{\text{op}} \leq 1$ is equivalent to the constraint

$$\begin{pmatrix} \mathbf{I}_m & \mathbf{U} \\ \mathbf{U}^T & \mathbf{I}_n \end{pmatrix} \succeq 0. \tag{7.42}$$

Since this constraint is a linear matrix inequality, it shows nuclear norm minimization can be formulated as an SDP. (*Hint:* The Schur-complement formula might be useful.)

Ex. 7.4 *Subgradients of the nuclear norm:* Subgradients, as previously defined in Section 5.2, extend the notion of a gradient to nondifferentiable functions.

(a) Given a matrix $\mathbf{A} \in \mathbb{R}^{m \times n}$ with rank $r \leq \min(m, n)$, write its singular value decomposition as $\mathbf{A} = \mathbf{U}\mathbf{D}\mathbf{V}^T$. With this notation, show that the subgradient of the nuclear norm is

$$\partial\|\mathbf{A}\|_\star = \left\{ \mathbf{U}\mathbf{V}^T + \mathbf{W} \mid \mathbf{U}^T\mathbf{W} = \mathbf{W}\mathbf{V} = 0, \|\mathbf{W}\|_{\text{op}} \leq 1 \right\}. \tag{7.43}$$

(b) Use part (a) to show that a fixed point of the SOFT-IMPUTE procedure (7.1) satisfies the subgradient equation of the criterion (7.10).

Ex. 7.5 From Chapter 5, recall our description (5.21) of Nesterov's generalized gradient procedure. Show that the SOFT-IMPUTE procedure (7.1) corresponds to this algorithm applied to the criterion (7.10).

Ex. 7.6 Construct a solution to the maximum-margin problem (7.21), in the case $\text{rank}(\mathbf{M}) = r \leq \min(m, n)$, of the form $\widehat{\mathbf{M}} = \widehat{\mathbf{A}}_{m \times r} \widehat{\mathbf{B}}_{r \times n}^T$. Show that this solution is not unique. Suppose we restrict $\mathbf{A}$ and $\mathbf{B}$ to have $r' > r$ columns. Show how solutions of this enlarged problem might not reveal the rank of $\mathbf{M}$.

Ex. 7.7 Consider the convex optimization problem

$$\underset{\mathbf{u} \in \mathbb{R}^p}{\text{maximize}}\, \mathbf{u}^T \mathbf{Z} \mathbf{v} \text{ subject to } \|\mathbf{u}\|_2 \leq 1 \text{ and } \|\mathbf{u}\|_1 \leq c. \qquad (7.44)$$

Show that the solution is given by

$$\mathbf{u} = \frac{\mathcal{S}_\lambda(\mathbf{Z}\mathbf{v})}{\|\mathcal{S}_\lambda(\mathbf{Z}\mathbf{v})\|_2}, \qquad (7.45)$$

where $\lambda \geq 0$ is the smallest positive value such that $\|\mathbf{u}\|_1 \leq c$.

Ex. 7.8 In this exercise, we demonstrate that, in the context of exact completion of a $n \times n$ matrix $\mathbf{M}$ from noiseless entries, it is necessary to observe at least $N > n \log n$ entries, even for a rank one matrix. We begin by noting that if we fail to observe any entries from some row (or column) of $\mathbf{M}$, then it is impossible to recover $\mathbf{M}$ exactly (even if we restrict to incoherent matrices with rank one). We let $\mathcal{F}$ be the event that there exists some row with no observed entries, under the sampling model in which we choose N entries from the matrix uniformly at random *with replacement*.

(a) For each row $j = 1, \ldots, p$, let Z_j be a binary indicator variable for the event that *no entries* of j are observed, and define $Z = \sum_{j=1}^n Z_j$. Show that

$$\mathbb{P}[\mathcal{F}] \;=\; \mathbb{P}[Z > 0] \geq \frac{(\mathbb{E}[Z])^2}{\mathbb{E}[Z^2]}.$$

(*Hint:* The Cauchy–Schwarz inequality could be useful.)

(b) Show that $\mathbb{E}[Z] = n(1 - 1/n)^N$.

(c) Show that $\mathbb{E}[Z_i Z_j] \leq \mathbb{E}[Z_i]\mathbb{E}[Z_j]$ for $i \neq j$.

(d) Use parts (b) and (c) so show that $\mathbb{E}[Z^2] \leq n(1 - 1/n)^N + n^2(1 - 1/n)^{2N}$.

(e) Use the previous parts to show that $\mathbb{P}[\mathcal{F}]$ stays bounded away from zero unless $N > n \log n$.

Ex. 7.9 Quadratic polynomial regression in high dimensions is dangerous, because the number of parameters is proportional to the square of the dimension. Show how to represent this problem as a matrix regression (Section 7.5), and hence suggest how the parameter explosion can be controlled.

Ex. 7.10 . In Exercise 2.11 of Chapter 2, we show that a regression model that allows for a sparse perturbation of each prediction is equivalent to a robust regression using Huber's ρ function. Here we establish an analogous result for robust PCA.

Recall the sparse plus low-rank version of PCA:

$$\underset{\mathbf{L},\mathbf{S}}{\text{minimize}} \, \frac{1}{2}\|\mathbf{Z} - (\mathbf{L} + \mathbf{S})\|_{\text{F}}^2 + \lambda_1\|\mathbf{L}\|_\star + \lambda_2\|\mathbf{S}\|_1. \tag{7.46}$$

Now consider a robustified version of PCA

$$\underset{\mathbf{L}}{\text{minimize}} \, \frac{1}{2} \sum_{i=1}^{N} \sum_{j=1}^{p} \rho\left(z_{ij} - \ell_{ij}; \lambda_2\right) + \lambda_1\|\mathbf{L}\|_\star, \tag{7.47}$$

where

$$\rho(t; \lambda) = \begin{cases} \lambda|t| - \lambda^2/2 & \text{if } |t| > \lambda \\ t^2/2 & \text{if } |t| \le \lambda. \end{cases} \tag{7.48}$$

is Huber's loss function. Show that problem (7.47) has the same solution for $\mathbf{L}$ as does problem (7.46).

Chapter 8

Sparse Multivariate Methods

8.1 Introduction

In this chapter, we discuss some popular methods for multivariate analysis and explore how they can be "sparsified": that is, how the set of features can be reduced to a smaller set to yield more interpretable solutions. Many standard multivariate methods are derived from the singular value decomposition of an appropriate data matrix. Hence, one systematic approach to sparse multivariate analysis is through a sparse decomposition of the same data matrix. The penalized matrix decomposition of Section 7.6 is well-suited to this task, as it delivers sparse versions of the left and/or right singular vectors.

For example, suppose that we have a data matrix $\mathbf{X}$ of dimension $N \times p$, and assume that the columns each have mean zero. Then the principal components of $\mathbf{X}$ are derived from its singular value decomposition (SVD) $\mathbf{X} = \mathbf{UDV}^T$: the columns of $\mathbf{V}$ are the principal component direction vectors (in order), and the columns of $\mathbf{U}$ are the standardized principal components. Hence we can derive *sparse* principal components by applying instead the penalized matrix decomposition to $\mathbf{X}$, with sparsity enforced on the right vectors. In a similar way, many multivariate methods can be derived by appropriate application of the penalized matrix decomposition. These methods are summarized in Table 8.1.

Table 8.1 *The penalized matrix decomposition of Section 7.6 applied to appropriate input matrices leads to sparse versions of classical multivariate methods.*

Input Matrix	Result
Data matrix	sparse SVD and principal components
Variance-covariance	sparse principal components
Cross-products	sparse canonical variates
Dissimilarity	sparse clustering
Between-class covariance	sparse linear discriminants

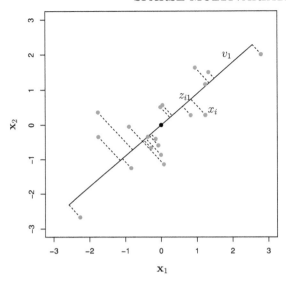

Figure 8.1 *A two-dimensional illustration of principal components analysis, show-ing the first principal component* $v_1 \in \mathbb{R}^2$ *of a collection of data points* $x_i = (x_{i1}, x_{i2})$, *shown in green circles. Letting* $\bar{x} = (\bar{x}_1, \bar{x}_2)^T$ *denote the sample mean, the line* $\bar{x} + \lambda^T v_1$ *maximizes the variance of the projected points along the line, and min-imizes the total squared distance from each point to its orthogonal projection onto the line. Here* $z_{i1} = u_{i1} d_1$ *is the scalar valued representation of observation* x_i *in the first principal component* $\mathbf{z}_1$.

8.2   Sparse Principal Components Analysis

We begin our exploration with the problem of sparse principal component analysis, which is a natural extension of PCA well-suited to high-dimensional data. To set the stage, we first review principal components.

8.2.1   Some Background

Given a data matrix $\mathbf{X}$ of dimension $N \times p$, consisting of N vectors $\{x_1, \ldots, x_N\}$ in $\mathbb{R}^p$, principal component analysis provides a sequence of linear approximations, indexed by a rank $r \le \min\{p, N\}$.

There are two different but equivalent ways of viewing and deriving prin-cipal components. The first approach is based on the directions of *maximal variance*. Any unit-norm vector $\alpha \in \mathbb{R}^p$ leads to a one-dimensional projection of the data, namely the N-vector $\mathbf{X}\alpha$.[1] Assuming that the columns of $\mathbf{X}$ have been centered, the sample variance of the projected data vector is given by

---

[1] In this chapter we deal with multivariate methods applied to a data matrix $\mathbf{X} \in \mathbb{R}^{N \times p}$; we hence adhere to our convention of representing N-vectors and all matrices in boldface, and p-vectors in plain text.

$\widehat{\text{Var}}(\mathbf{X}\alpha) = \frac{1}{N}\sum_{i=1}^{N}(x_i^T\alpha)^2$. Principal components analysis finds the direction that maximizes the sample variance

$$v_1 = \arg\max_{\|\alpha\|_2=1}\left\{\widehat{\text{Var}}(\mathbf{X}\alpha)\right\} = \arg\max_{\|\alpha\|_2=1}\left\{\alpha^T\frac{\mathbf{X}^T\mathbf{X}}{N}\alpha\right\}. \qquad (8.1)$$

Hence the first principal component direction corresponds to the largest eigenvalue of the sample covariance $\mathbf{X}^T\mathbf{X}/N$, which provides the link to the notion of maximal variance at the population level. See Exercise 8.1 for further details. Figure 8.1 illustrates the geometry of this optimization problem. The resulting projection $\mathbf{z}_1 = \mathbf{X}v_1$ is called the first principal component of the data $\mathbf{X}$, and the elements of v_1 are called the *principal component loadings*. The vector v_1 is easily seen to be the right singular vector corresponding to the largest singular value d_1 of $\mathbf{X}$. Similarly $\mathbf{z}_1 = \mathbf{u}_1 d_1$, where $\mathbf{u}_1$ is the corresponding left singular vector.

Subsequent principal-component directions (eigen-vectors) $v_2, v_3, \ldots, v_p$ correspond to maxima of $\widehat{\text{Var}}(\mathbf{X}v_j)$ subject to $\|v_j\|_2 = 1$ and v_j orthogonal to $v_1, \ldots v_{j-1}$. This property also implies that the $\mathbf{z}_j$ are mutually uncorrelated (see Exercise 8.2). In fact, after r steps of this procedure, we obtain a rank r matrix that solves the optimization problem

$$\mathbf{V}_r = \arg\max_{\mathbf{A}:\,\mathbf{A}^T\mathbf{A}=\mathbf{I}_r}\text{trace}(\mathbf{A}^T\mathbf{X}^T\mathbf{X}\mathbf{A}) \qquad (8.2)$$

See Exercise 7.2 for further details on this property. Thus, even though they are defined sequentially, the collection of loading vectors in $\mathbf{V}_r$ also maximize the total variance among all such collections.

A second derivation of principal components is based on minimizing the *reconstruction error* associated with a particular generative model for the data. Suppose that the rows of the data matrix can be modeled as $x_i \approx f(\lambda_i)$, where the function

$$f(\lambda) = \mu + \mathbf{A}_r\lambda \qquad (8.3)$$

parametrizes an affine set of dimension r. Here $\mu \in \mathbb{R}^p$ is a location vector, $\mathbf{A}_r \in \mathbb{R}^{p\times r}$ is a matrix with orthonormal columns corresponding to directions, and $\lambda \in \mathbb{R}^r$ is a parameter vector that varies over samples. It is natural to choose the parameters $\{\mu, \mathbf{A}_r, \{\lambda_i\}_{i=1}^N\}$ to minimize the average reconstruction error

$$\frac{1}{N}\sum_{i=1}^{N}\|x_i - \mu - \mathbf{A}_r\lambda_i\|_2^2. \qquad (8.4)$$

This interpretation of PCA is illustrated in Figure 8.1. As we explore in Exercise 8.3, when the data has been precentered (so that we may take $\mu = 0$), the criterion (8.4) can be reduced to

$$\frac{1}{N}\sum_{i=1}^{N}\|x_i - \mathbf{A}_r\mathbf{A}_r^T x_i\|_2^2, \qquad (8.5)$$

and the value of $\mathbf{A}_r$ that minimizes the reconstruction error (8.5) can again be obtained from the singular value decomposition of the data matrix. In operational terms, we compute the SVD $\mathbf{X} = \mathbf{U}\mathbf{D}\mathbf{V}^T$, and then form $\hat{\mathbf{A}}_r = \mathbf{V}_r$ by taking the r columns of $\mathbf{V}$ corresponding to the top r singular values. The estimates for λ_i are given by the rows of $\mathbf{Z}_r = \mathbf{U}_r\mathbf{D}_r$. So maximizing total variance within the affine surface corresponds to minimizing total distance from the surface. Again we observe that the successive solutions are *nested*; this property is special, and is not necessarily inherited by the generalizations that we discuss in this chapter.

8.2.2 Sparse Principal Components

We often interpret principal components by examining the loading vectors $\{v_j\}_{j=1}^r$ so as to determine which of the variables play a significant role. In this section, we discuss some methods for deriving principal components with sparse loadings. Such sparse principal components are especially useful when the number of variables p is large relative to the sample size. With a large number of variables, it is often desirable to select a smaller subset of relevant variables, as revealed by the loadings. At the theoretical level, in the $p \gg N$ regime, ordinary PCA is known to break down very badly, in that the eigenvectors of the sample covariance need not be close to the population eigenvectors (Johnstone 2001). Imposing sparsity on the principal components makes the problem well-posed in the "large p, small N" regime. In this section, we discuss a number of methods for obtaining sparse principal components, all based on lasso-type (ℓ_1) penalties. As with ordinary PCA, we start with an $N \times p$ data matrix $\mathbf{X}$ with centered columns. The proposed methods focus on either the maximum variance property of principal components, or minimum reconstruction error. For ease of exposition, we begin by discussing the rank-one case for each method, deferring the case of higher ranks until Section 8.2.3.

8.2.2.1 Sparsity from Maximum Variance

We begin by discussing how the maximum variance characterization of PCA can be modified to incorporate sparsity. The most natural modification would be to impose an ℓ_0-restriction on the criterion, leading to the problem

$$\underset{\|v\|_2=1}{\text{maximize}} \left\{ v^T \mathbf{X}^T \mathbf{X} v \right\} \text{ subject to } \|v\|_0 \le t, \qquad (8.6)$$

where $\|v\|_0 = \sum_{j=1}^p \mathbb{I}[v_j \ne 0]$ simply counts the number of nonzeros in the vector v. However, this problem is doubly nonconvex, since it involves maximizing (as opposed to minimizing) a convex function with a combinatorial constraint. The SCoTLASS procedure of Jolliffe, Trendafilov and Uddin (2003) is a natural relaxation of this objective, based on replacing the ℓ_0-norm by the ℓ_1-norm, leading to

$$\underset{\|v\|_2=1}{\text{maximize}} \left\{ v^T \mathbf{X}^T \mathbf{X} v \right\} \text{ subject to } \|v\|_1 \le t. \qquad (8.7)$$

The ℓ_1-constraint encourages some of the loadings to be zero and hence v to be sparse. Although the ℓ_1-norm is convex, the overall problem remains nonconvex, and moreover is not well-suited to simple iterative algorithms.

There are multiple ways to address these challenges. One approach draws on the SVD version of principal components; we re-express the problem, leaving it nonconvex but leading to a computationally efficient algorithm for finding local optima. Recall the penalized matrix criterion (7.28) on page 187; applying it with no constraint on $\mathbf{u}$—that is, with $c_1 = \infty$—leads to the optimization problem

$$\underset{\|\mathbf{u}\|_2=\|v\|_2=1}{\text{maximize}} \left\{\mathbf{u}^T\mathbf{X}v\right\} \text{ subject to } \|v\|_1 \le t. \tag{8.8}$$

Any optimal solution $\widehat{v}$ to this problem is also optimal for the original SCoT-LASS program (8.7). The advantage of this reformulation is that the objective function (8.8) is biconvex in the pair $(\mathbf{u}, v)$, so that we can apply alternating minimization to solve it—in particular, recall Algorithm 7.2 in Chapter 7 for the penalized matrix decomposition. Doing so leads to Algorithm 8.1, which consists of the following steps:

Algorithm 8.1 ALTERNATING ALGORITHM FOR RANK ONE SPARSE PCA.

1. Initialize $v \in \mathbb{R}^p$ with $\|v\|_2 = 1$.
2. Repeat until changes in $\mathbf{u}$ and v are sufficiently small:
   (a) Update $\mathbf{u} \in \mathbb{R}^N$ via $\mathbf{u} \leftarrow \frac{\mathbf{X}v}{\|\mathbf{X}v\|_2}$.
   (b) Update $v \in \mathbb{R}^p$ via

$$v \leftarrow v(\lambda, \mathbf{u}) = \frac{\mathcal{S}_\lambda(\mathbf{X}^T\mathbf{u})}{\|\mathcal{S}_\lambda(\mathbf{X}^T\mathbf{u})\|_2}, \tag{8.9}$$

where $\lambda = 0$ if $\|\mathbf{X}^T\mathbf{u}\|_1 \le t$, and otherwise $\lambda > 0$ is chosen such that $\|v(\lambda, \mathbf{u})\|_1 = t$.

Here $\mathcal{S}_\lambda(x) = \text{sign}(x)\,(|x| - \lambda)_+$ is the familiar soft-thresholding operator at level λ. In Exercise 8.6, we show that any fixed point of this algorithm represents a local optimum of the criterion (8.7), and moreover, that the updates can be interpreted as a *minorization-maximization*, or simply a minorization algorithm for the objective function (8.7).

An alternative approach, taken by d'Aspremont, El Ghaoui, Jordan and Lanckriet (2007), is to further relax the SCoTLASS objective to a convex program, in particular by lifting it to a linear optimization problem over the space of positive semidefinite matrices. Such optimization problems are known as semidefinite programs. In order to understand this method, let us begin with an exact reformulation of the nonconvex objective function (8.7). By the properties of the matrix trace, we can rewrite the quadratic form $v^T\mathbf{X}^T\mathbf{X}v$ in terms of a trace operation—specifically

$$v^T \mathbf{X}^T \mathbf{X} v = \text{trace}(\mathbf{X}^T \mathbf{X} vv^T). \tag{8.10}$$

In terms of the rank one matrix $\mathbf{M} = vv^T$, the constraint $\|v\|_2^2 = 1$ is equivalent to the linear constraint $\text{trace}(\mathbf{M}) = 1$, and the constraint $\|v\|_1 \leq t$ can be expressed as $\text{trace}(|\mathbf{M}|\mathbf{E}) \leq t^2$, where $\mathbf{E} \in \mathbb{R}^{p \times p}$ is a matrix of all ones, and $|\mathbf{M}|$ is the matrix obtained by taking absolute values entry-wise. Putting together the pieces, we conclude that the nonconvex SCoTLASS objective has the equivalent reformulation

$$\underset{\mathbf{M} \succeq 0}{\text{maximize trace}}(\mathbf{X}^T \mathbf{X} \mathbf{M})$$
$$\text{subject to trace}(\mathbf{M}) = 1, \text{trace}(|\mathbf{M}|\mathbf{E}) \leq t^2, \text{ and rank}(\mathbf{M}) = 1. \tag{8.11}$$

By construction, any optimal solution to this problem is a positive semidefinite matrix of rank one, say $\mathbf{M} = vv^T$, and the vector v is an optimal solution to the original problem (8.7). However, the optimization problem (8.11) is still nonconvex, due to the presence of the constraint $\text{rank}(\mathbf{M}) = 1$. By dropping this constraint, we obtain the semidefinite program proposed by d'Aspremont et al. (2007), namely

$$\underset{\mathbf{M} \succeq 0}{\text{maximize trace}}(\mathbf{X}^T \mathbf{X} \mathbf{M})$$
$$\text{subject to trace}(\mathbf{M}) = 1, \text{ trace}(|\mathbf{M}|\mathbf{E}) \leq t^2. \tag{8.12}$$

Since this problem is convex, it has no local optima, and a global optimum can be obtained by various standard methods. These include interior point methods (Boyd and Vandenberghe 2004); see also d'Aspremont et al. (2007) for a special-purpose and more efficient method for solving it.

In general, solving the SDP (8.12) is computationally more intensive than finding a local optimum of the biconvex criterion (8.8). However, since it is a convex relaxation of an exact reformulation, it has an attractive theoretical guarantee: if we solve the SDP and do obtain a rank-one solution, then we have in fact obtained the global optimum of the nonconvex SCoTLASS criterion. For various types of spiked covariance models, it can be shown that the SDP (8.12) will have a rank-one solution with high probability, as long as the sample size N is sufficiently large relative to the sparsity and dimension (but still allowing for $N \ll p$); see Section 8.2.6 for further discussion. Thus, for all of these problems, we are guaranteed to have found the *global* optimum of the SCoTLASS criterion.

8.2.2.2   *Methods Based on Reconstruction*

We now turn to methods for sparse PCA that are based on its reconstruction interpretation. In the case of a single sparse principal component, Zou, Hastie and Tibshirani (2006) proposed the optimization problem

$$\underset{\substack{\theta, v \in \mathbb{R}^p \\ \|\theta\|_2 = 1}}{\text{minimize}} \left\{ \frac{1}{N} \sum_{i=1}^{N} \|x_i - \theta v^T x_i\|_2^2 + \lambda_1 \|v\|_1 + \lambda_2 \|v\|_2^2 \right\}, \tag{8.13}$$

where λ_1, λ_2 are nonnegative regularization parameters. Let's examine this formulation in more detail.

- If we set $\lambda_1 = \lambda_2 = 0$, then it is easy to show that the program (8.13) achieves its optimum at a pair $\hat{\theta} = \hat{v} = v_1$, corresponding to a maximum eigenvector of $\mathbf{X}^T \mathbf{X}$, so that we recover the usual PCA solution.

- When $p \gg N$ the solution is not necessarily unique unless $\lambda_2 > 0$. If we set $\lambda_1 = 0$, then for any $\lambda_2 > 0$, the optimal solution $\hat{v}$ is proportional to the largest principal component direction.

- In the general setting with both λ_1 and λ_2 strictly positive, the ℓ_1-penalty weighted by λ_1 encourages sparsity of the loadings.

Like the objective (8.8), criterion (8.13) is not jointly convex in v and θ, but is biconvex. Minimization over v with θ fixed is equivalent to an elastic-net problem (see Section 4.2) and can be computed efficiently. On the other hand, minimization over θ with v fixed has the simple solution

$$\theta = \frac{\mathbf{X}^T \mathbf{z}}{\|\mathbf{X}^T \mathbf{z}\|_2}, \tag{8.14}$$

where $z_i = v^T x_i$ for $i = 1, \ldots, N$ (see Exercise 8.8). Overall, this procedure is reasonably efficient, but not as simple as Algorithm 8.1, which involves just soft-thresholding.

It turns out that the original SCoTLASS criterion (8.7) and the regression-based objective function (8.13) are intimately related. Focusing on the rank-one case (8.13), consider the constrained as opposed to the Lagrangian form of the optimization problem—namely

$$\underset{\|v\|_2 = \|\theta\|_2 = 1}{\text{minimize}} \|\mathbf{X} - \mathbf{X} v \theta^T\|_F^2 \quad \text{subject to} \quad \|v\|_1 \leq t. \tag{8.15}$$

If we add the extra ℓ_1-constraint $\|\theta\|_1 \leq t$, then as shown in Exercise 8.7, the resulting optimization problem is equivalent to the SCoTLASS criterion (8.7). Consequently, it can be solved conveniently by Algorithm 8.1. Note that adding this ℓ_1-constraint is quite natural, as it just symmetrizes the constraints in problem (8.15).

8.2.3 Higher-Rank Solutions

In Section 8.2.1, we presented a sequential approach for standard principal components analysis, based on successively solving the rank-one problem, restricting each candidate to be orthogonal to all previous solutions. This sequential approach also solves the multirank problem (8.2).

How about in the sparse setting? This sequential approach is also used in the SCoTLASS procedure, where each candidate solution for rank k is restricted to be orthogonal to all previous solutions for ranks $< k$. However, here the sequential approach will typically not solve a multirank criterion.

For the sparse PCA approach (8.8), we can apply the multifactor penalized matrix decomposition (7.3) of Chapter 7. Given the rank-one solution $(\mathbf{u}_1, v_1, d_1)$, we simply compute the residual $\mathbf{X}' = \mathbf{X} - d_1\mathbf{u}_1 v_1^T$ and apply the rank-one Algorithm (8.8) to $\mathbf{X}'$ to obtain the next solution.[2] Doing so ensures neither orthogonality of the principal components $\{(\mathbf{u}_1 d_1), (\mathbf{u}_2 d_2), \ldots (\mathbf{u}_k d_k)\}$ nor the sparse loading vectors $\{v_1, v_2 \ldots, v_k\}$. But the solutions do tend to be somewhat orthogonal in practice.

However, there is a subtle issue here: it is not clear that orthogonality of the vectors $\{v_1, v_2, \ldots v_k\}$ is desirable in the setting of sparse PCA, as orthogonality may be at odds with sparsity. Otherwise stated, enforcing orthogonality might result in less sparse solutions. A similar issue arises with sparse coding, as discussed in Section 8.2.5.

Interestingly, one can modify the approach (8.8) to constrain the vectors $\mathbf{u}_j$ to be orthogonal, with no such constraints on the vectors v_j. This modification can improve the interpretability of the set of solutions while still allowing the v_j to be sparse. In detail, consider the problem

$$\underset{\mathbf{u}_k,\, v_k}{\text{maximize}} \left\{\mathbf{u}_k^T \mathbf{X} v_k\right\} \text{ subject to } \|v_k\|_2 \leq 1, \ \|v_k\|_1 \leq c,$$
$$\text{and } \|\mathbf{u}_k\|_2 \leq 1 \text{ with } \mathbf{u}_k^T \mathbf{u}_j = 0 \text{ for all } j = 1, \ldots, k-1. \tag{8.16}$$

The solution for $\mathbf{u}_k$ with v_k fixed is

$$\mathbf{u}_k = \frac{\mathbf{P}_{k-1}^{\perp} \mathbf{X} v_k}{\|\mathbf{P}_{k-1}^{\perp} \mathbf{X} v_k\|_2} \tag{8.17}$$

where $\mathbf{P}_{k-1}^{\perp} = \mathbf{I} - \sum_{i=1}^{k-1} \mathbf{u}_i \mathbf{u}_i^T$, the projection onto the orthogonal complement of the space spanned by $\mathbf{u}_1, \mathbf{u}_2, \ldots \mathbf{u}_{k-1}$. This multifactor version of Algorithm 8.1 uses operation (8.17) in place of the rank-one projection $\mathbf{u} = \frac{\mathbf{X} v}{\|\mathbf{X} v\|_2}$.

The approach (8.13) of Zou et al. (2006) can be generalized to $r > 1$ components by minimizing the cost function

$$\frac{1}{N} \sum_{i=1}^{N} \|x_i - \mathbf{\Theta}\mathbf{V}^T x_i\|_2^2 + \sum_{k=1}^{r} \lambda_{1k}\|v_k\|_1 + \lambda_2 \sum_{k=1}^{r} \|v_k\|_2^2, \tag{8.18}$$

subject to $\mathbf{\Theta}^T\mathbf{\Theta} = \mathbf{I}_{r \times r}$. Here $\mathbf{V}$ is a $p \times r$ matrix with columns $\{v_1, \ldots, v_r\}$, and $\mathbf{\Theta}$ is also a matrix of dimension $p \times r$. Although this objective function (8.18) is not jointly convex in $\mathbf{V}$ and $\mathbf{\Theta}$, it is biconvex. Minimization over $\mathbf{V}$ with $\mathbf{\Theta}$ fixed is equivalent to solving r separate elastic net problems and

---

[2]Without sparsity constraints, this procedure would deliver exactly the usual sequence of principal components.

can be done efficiently. On the other hand, minimization over Θ with $\mathbf{V}$ fixed is a version of the so-called Procrustes problem, and can be solved by a simple SVD calculation (Exercise 8.10). These steps are alternated until convergence to a local optimum.

8.2.3.1 *Illustrative Application of Sparse PCA*

Here we demonstrate sparse principal components on a dataset of digitized handwritten digits. We have a training set of $N = 664$ gray-scale images of handwritten sevens. Each image contains 16×16 pixels, leading to a data matrix $\mathbf{X}$ of dimension 664×256. Panel (a) of Figure 8.2 shows some ex-

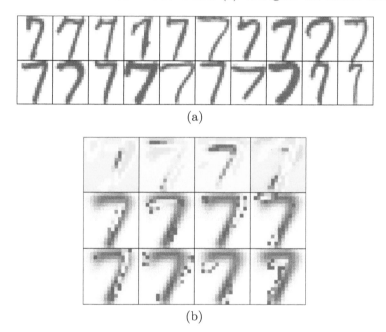

(a)

(b)

Figure 8.2 *(a) A sample of handwritten sevens from the zip code database. (b) Top row: first four principal components for "sevens" data (color shades represent negative loadings as yellow and positive loadings as blue); Bottom two rows: first eight sparse principal components, constrained to be positive. These are superimposed on the average seven to enhance interpretability.*

amples of these images, where panel (b) shows the results of sparse principal components, and contrasts them with standard PCA. The top row in panel (b) shows the first four standard principal components, which explain about 50% of the variance. To enhance interpretability, we compute sparse principal components with the loadings constrained to be nonnegative. In order to do so, we simply replace the soft-threshold operator $\mathcal{S}_\lambda(x)$ in Algorithm 8.1 by the nonnegative soft-threshold operator $\mathcal{S}_\lambda^+(x) = (x - \lambda)_+$.

The first eight sparse principal components are shown in the middle and bottom rows, and also explain about 50% of the variation. While more components are needed to explain the same amount of variation, the individual components are simpler and potentially more interpretable. For example, the 2nd and 6th sparse components appear to be capturing the "notch" style used by some writers, for example in the top left image of Figure 8.2(a).

8.2.4 Sparse PCA via Fantope Projection

Vu, Cho, Lei and Rohe (2013) propose another related approach to sparse PCA. Letting $\mathbf{S} = \mathbf{X}^T\mathbf{X}/N$, their proposal is to solve the semidefinite program

$$\underset{\mathbf{Z} \in \mathcal{F}^p}{\text{maximize}} \left\{ \text{trace}(\mathbf{SZ}) - \lambda \|\mathbf{Z}\|_1 \right\} \tag{8.19}$$

where the convex set $\mathcal{F}^p = \{\mathbf{Z} : \mathbf{0} \preceq \mathbf{Z} \preceq \mathbf{I},\ \text{trace}(\mathbf{Z}) = p\}$ is known as a *Fantope*. When $p = 1$ the spectral norm bound in $\mathcal{F}^p$ is redundant and (8.19) reduces to the direct approach of d'Aspremont et al. (2007). For $p > 1$, although the penalty in (8.19) only implies entry-wise sparsity of the solution, it can be shown (Lei and Vu 2015) that the solution is able to consistently select the nonzero entries of the leading eigenvectors under appropriate conditions.

8.2.5 Sparse Autoencoders and Deep Learning

In the neural network literature, an autoencoder generalizes the idea of principal components. Figure 8.3 provides a simple illustration of the idea, which is based on reconstruction, much like in the criterion (8.13). The autoen-

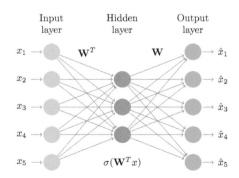
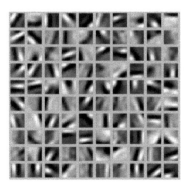

Figure 8.3 *Left: Network representation of an autoencoder used for unsupervised learning of nonlinear principal components. The middle layer of hidden units creates a bottleneck, and learns nonlinear representations of the inputs. The output layer is the transpose of the input layer, and so the network tries to reproduce the input data using this restrictive representation. Right: Images representing the estimated columns of $\mathbf{W}$ in an image modeling task.*

coder is based on a $p \times m$ matrix of weights $\mathbf{W}$ with $m < p$; it is used to create m linear combinations of the input vector x. Each such linear combination is passed through a nonlinear function σ, with the sigmoid function $\sigma(t) = 1/(1 + e^{-t})$ being one typical choice, as represented in Figure 8.3 via the vector function $h(x) = \sigma(\mathbf{W}^T x)$. The output layer is then modeled as $\mathbf{W}h(x) = \mathbf{W}\sigma(\mathbf{W}^T x)$.[3] Given input vectors x_i for $i = 1, \ldots, N$, the weight matrix $\mathbf{W}$ is then estimated by solving the (nonconvex) optimization problem

$$\underset{\mathbf{W} \in \mathbb{R}^{m \times p}}{\text{minimize}} \left\{ \frac{1}{2} \sum_{i=1}^{N} \|x_i - \mathbf{W}h(x_i)\|^2 \right\}. \tag{8.20}$$

If we restrict σ to be the identity function, then $h(x) = \mathbf{W}^T x$ and the solution to (8.20) is equivalent to principal components; i.e., $\mathbf{W}\mathbf{W}^T = \mathbf{V}_m \mathbf{V}_m^T$, where $\mathbf{V}_m$ is the $p \times m$ matrix consisting of the first m principal component loadings (see Exercise 8.12). Here the bottleneck in the network imposes a rank constraint on $\mathbf{W}$, forcing it to learn structure.

In modeling high-dimensional signals such as images, the vectors x_i might represent the pixels of a (sub) image. The columns of $\mathbf{W}$ represent a learned dictionary of image shapes, and $h(x_i)$ tries to represent x_i in this basis. Now the bottleneck might be seen as an unnecessary restriction, since many slightly different shapes are likely in an image. The idea is to replace this restriction by imposing sparseness on the coefficients $h(x)$, leading to so-called *sparse coding* (Olshausen and Field 1996). To build intuition, we first consider the linear case, but now with $m > p$. In the optimization problem

$$\underset{\mathbf{W} \in \mathbb{R}^{p \times m}, \{s_i\}_1^N \in \mathbb{R}^m}{\text{minimize}} \left\{ \frac{1}{2} \sum_{i=1}^{N} \left\{ \|x_i - \mathbf{W}s_i\|_2^2 + \lambda \|s_i\|_1 \right\} \right\} \tag{8.21}$$
$$\text{subject to } \|\mathbf{W}\|_F^2 \leq 1,$$

the individual s_i are forced to be sparse through the ℓ_1-penalties. The columns of $\mathbf{W}$ are not constrained to be uncorrelated, and their total size is kept in bound by the Frobenius norm. Exercise 8.13 examines the sparse linear coder (8.21) in more detail, and develops a natural alternating algorithm for solving it. The right panel of Figure 8.3 illustrates a typical solution for $\mathbf{W}$ in an image modeling problem, where each x_i is a vectorized version of an image. Each subimage represents a column of $\mathbf{W}$ (the codebook), and every image is modeled as a sparse superposition of elements of $\mathbf{W}$. Modern sparse encoders used in deep learning generalize this formulation in several ways (Le et al. 2012):

- They use multiple hidden layers, leading to a hierarchy of dictionaries;
- Nonlinearities that can be computed more rapidly than the sigmoid are used—for example $\sigma(t) = t_+$.

---

[3] In practice, bias terms are also included in each linear combination; we omit them here for simplicity.

- More general sparseness penalties are imposed directly on the coefficients $h(x_i)$ in the problem (8.20).

These models are typically fit by (stochastic) gradient descent, and often on very large databases of images (for example), using distributed computing with large clusters of processors.

One important use of the sparse autoencoder is for *pretraining*. When fitting a supervised neural network to labelled data, it is often advantageous to first fit an autoencoder to the data without the labels and then use the resulting weights as starting values for fitting the supervised neural network (Erhan et al. 2010). Because the neural-network objective function is nonconvex, these starting weights can significantly improve the quality of the final solution. Furthermore, if there is additional data available without labels, the autoencoder can make use of these data in the pretraining phase.

8.2.6   Some Theory for Sparse PCA

Here we give a brief overview of how standard principal component analysis breaks down in the high-dimensional setting ($p \gg N$), and why some structural assumption—such as sparsity in the principal components—is essential. One way of studying the behavior of (sparse) PCA is in terms of a *spiked covariance model*, meaning a p-dimensional covariance matrix of the form

$$\mathbf{\Sigma} = \sum_{j=1}^{M} \omega_j \theta_j \theta_j^T + \sigma^2 \mathbf{I}_{p \times p}, \tag{8.22}$$

where the vectors $\{\theta_j\}_{j=1}^M$ are orthonormal, and associated with positive weights $\omega_1 \geq \omega_2 \geq \cdots \geq \omega_M > 0$. By construction, the vectors $\{\theta_j\}_{j=1}^M$ are the top M eigenvectors of the population covariance, with associated eigenvalues $\{\sigma^2 + \omega_j\}_{j=1}^M$.

Given N i.i.d. samples $\{x_i\}_{i=1}^N$ from a zero-mean distribution with covariance $\mathbf{\Sigma}$, standard PCA is based on estimating the span of $\{\theta_j\}_{j=1}^M$ using the top M eigenvectors of the sample covariance matrix $\widehat{\mathbf{\Sigma}} = \frac{1}{N} \sum_{i=1}^N x_i x_i^T$. In the classical setting, in which the dimension p remains fixed while the sample size $N \to +\infty$, the sample covariance converges to the population covariance, so that the principal components are consistent estimators. More relevant for high-dimensional data analysis is a scaling in which *both* p and N tend to infinity, with $p/N \to c \in (0, \infty)$ with M and the eigenvalues remaining fixed.[4] Under this scaling, the sample eigenvectors or principal components do not converge to the population eigenvectors $\{\theta_j^{(p)}\}_{j=1}^M$. In fact, if the signal-to-noise ratios ω_j/σ^2 are sufficiently small, the sample eigenvectors are asymptotically orthogonal to the population eigenvectors! This poor behavior is caused by

---

[4]To be clear, for each $j = 1, \ldots, M$, we have a sequence $\{\theta_j^{(p)}\}$ of population eigenvectors, but we keep the signal-to-noise ratio ω_j/σ^2 fixed, independently of (p, N).

the $p - M$ dimensions of noise in the spiked covariance model (8.22), which can swamp the signal when $N \ll p$; see Johnstone and Lu (2009) for a precise statement of this phenomenon.

Given the breakdown of high-dimensional PCA without any structure on the eigenvectors, we need to make additional assumptions. A number of authors have explored how sparsity can still allow for consistent estimation of principal components even when $p \gg N$. Johnstone and Lu (2009) propose a two-stage procedure, based on thresholding the diagonal of the sample covariance matrix in order to isolate the highest variance coordinates, and then performing PCA in the reduced-dimensional space. They prove consistency of this method even when p/N stays bounded away from zero, but allow only polynomial growth of p as a function of sample size. Amini and Wainwright (2009) analyze the variable selection properties of both diagonal thresholding and the semidefinite programming relaxation (8.12) of the SCoTLASS problem (8.7). For a spiked covariance model (8.22) with a single leading eigenvector that is k-sparse, they show that the diagonal thresholding method (Johnstone and Lu 2009) succeeds in recovering sparsity pattern of the leading eigenvector if and only if the sample size $N \asymp k^2 \log p$. The SDP relaxation also performs correct variable selection with this scaling of the sample size, and in certain settings, can succeed with fewer samples. Amini and Wainwright (2009) show that no method—even one based on exhaustively enumerating all the subsets—can succeed with fewer than $N \asymp k \log p$ samples.

Other authors have studied the estimation of the eigenspaces themselves in ℓ_2 or related norms. Paul and Johnstone (2008) propose the augmented SPCA algorithm, a refinement of the two-stage method of Johnstone and Lu (2009); this algorithm is also analyzed by Birnbaum, Johnstone, Nadler and Paul (2013), who show that it achieves the minimax rates for models of weakly sparse eigenvectors in ℓ_q-balls. In independent work, Vu and Lei (2012) prove minimax lower bounds for the sparse PCA problem, and show that they can be achieved by computing the maximum eigenvalue of the sample covariance subject to an ℓ_q-constraint. Ma (2010, 2013) and Yuan and Zhang (2013) have studied algorithms for sparse PCA based on a combination of the power method (a classical iterative technique for computing eigenvectors) with intermediate soft-thresholding steps. When $M = 1$, the procedure of Ma (2013) is essentially the same as Algorithm 8.1, the only difference being the use of a fixed level λ in the soft-thresholding step, rather than the variable choice used in the latter to solve the bound version of the problem.

8.3 Sparse Canonical Correlation Analysis

Canonical correlation analysis extends the idea of principal components analysis to two data matrices. Suppose that we have data matrices $\mathbf{X}$, $\mathbf{Y}$ of dimensions $N \times p$ and $N \times q$, respectively, with centered columns. Given two vectors $\beta \in \mathbb{R}^p$ and $\theta \in \mathbb{R}^q$, they define one-dimensional projections of the two datasets, namely the variates (N-vectors) $\mathbf{X}\beta$ and $\mathbf{Y}\theta$, respectively. Canonical correlation analysis (CCA) chooses β and θ to maximize the correlation between these two variates.

In detail, the sample covariance between $\mathbf{X}\beta$ and $\mathbf{Y}\theta$ is given by

$$\widehat{\mathrm{Cov}}(\mathbf{X}\beta, \mathbf{Y}\theta) = \frac{1}{N}\sum_{i=1}^{N}(x_i^T\beta)(y_i^T\theta) = \frac{1}{N}\beta^T\mathbf{X}^T\mathbf{Y}\theta. \tag{8.23}$$

where x_i and y_i are the ith rows of $\mathbf{X}$ and $\mathbf{Y}$, respectively. CCA solves the problem

$$\begin{aligned} &\underset{\beta \in \mathbb{R}^p, \, \theta \in \mathbb{R}^q}{\text{maximize}} \left\{ \widehat{\mathrm{Cov}}(\mathbf{X}\beta, \mathbf{Y}\theta) \right\} \\ &\text{subject to } \widehat{\mathrm{Var}}(\mathbf{X}\beta) = 1 \text{ and } \widehat{\mathrm{Var}}(\mathbf{Y}\theta) = 1. \end{aligned} \tag{8.24}$$

The solution set (β_1, θ_1) are called the first canonical vectors, and the corresponding linear combinations $\mathbf{z}_1 = \mathbf{X}\beta_1$ and $\mathbf{s}_1 = \mathbf{Y}\theta_1$ the first canonical variates. Subsequent pairs of variates can be found by restricting attention to vectors such that the resulting variates are uncorrelated with the earlier ones. All solutions are given by a generalized SVD of the matrix $\mathbf{X}^T\mathbf{Y}$ (see Exercise 8.14).

Canonical correlation analysis fails when the sample size N is strictly less than $\max(p, q)$: in this case, the problem is degenerate, and one can find meaningless solutions with correlations equal to one. One approach to avoiding singularity of the sample covariance matrices $\frac{1}{N}\mathbf{X}^T\mathbf{X}$ and $\frac{1}{N}\mathbf{Y}^T\mathbf{Y}$ is by imposing additional restrictions. For instance, the method of *ridge regularization* is based on adding some positive multiple λ of the identity to each sample covariance matrix; see Exercise 8.17 for further discussion. An alternative method is based on taking only the diagonal entries of the sample covariance matrices, an approach that we adopt below.

Sparse canonical vectors can be derived by imposing ℓ_1-constraints on β and θ in the criterion (8.24), leading to the modified objective

$$\begin{aligned} &\underset{\beta, \, \theta}{\text{maximize}} \left\{ \widehat{\mathrm{Cov}}(\mathbf{X}\beta, \mathbf{Y}\theta) \right\} \\ &\text{subject to } \mathrm{Var}(\mathbf{X}\beta) = 1, \, \|\beta\|_1 \leq c_1, \mathrm{Var}(\mathbf{Y}\theta) = 1, \, \|\theta\|_1 \leq c_2. \end{aligned} \tag{8.25}$$

Note that one can use either the bound form for the ℓ_1-constraints (as above), or add corresponding Lagrangian terms. For numerical solution of this problem, we note that the standard CCA problem (8.24) can be solved by alternating least squared regressions (see Exercises 8.14–8.17). Not surprisingly then, the sparse version (8.25) can be solved by alternating elastic-net procedures, as we explore in Exercise 8.19.

The sparse formulation (8.25) is useful when $N > \max(p, q)$, but can fail in high-dimensional situations just as before. Again, ridging the individual covariance matrices will resolve the issue, and can be absorbed in the alternating elastic-net regressions. When the dimensions are very high—as in genomic problems—the cross-covariance between $\mathbf{X}$ and $\mathbf{Y}$ is of primary interest, and the internal covariance among the columns of $\mathbf{X}$ and among the columns of $\mathbf{Y}$ are nuisance parameters which can add to the estimation variance. In this case, it is convenient to standardize the variables, and then assume the internal covariance matrices are the identity. We are thus led to the problem

$$
\underset{\beta,\theta}{\text{maximize}} \ \widehat{\text{Cov}}(\mathbf{X}\beta, \mathbf{Y}\theta)
$$
$$
\text{subject to} \|\beta\|^2 \le 1, \ \|\theta\|^2 \le 1, \ \|\beta\|_1 \le c_1, \ \|\theta\|_1 \le c_2. \tag{8.26}
$$

This objective has the same form as the penalized matrix decomposition (7.6) previously discussed in Chapter 7, but using as input the data matrix $\mathbf{X}^T\mathbf{Y}$. We can thus apply Algorithm 7.2 directly, using alternating soft-thresholding to compute the solutions.

Higher-order sparse canonical variates are obtained from the higher-order PMD components: as in Algorithm 7.3, after computing a solution, we take residuals and then apply the procedure to what remains.

8.3.1   Example: Netflix Movie Rating Data

Let us illustrate the behavior of sparse CCA by applying it to the Netflix movie-ratings data. As originally described in detail in Section 7.3.1, the full dataset consists of 17,770 movies and 480,189 customers. Customers have rated some (around 1%) of the movies on a scale from 1 to 5. For this example, we selected the $p = N = 500$ movies and customers with the most ratings, and imputed the missing values with the movie means.

Among the 500 films, we identified those that were action movies (59 in all) and those that were romantic movies (73 in all). The remaining movies were discarded. We then applied the sparse CCA procedure to the data. The idea was to correlate each customer's ratings on the action movies with their ratings on the romantic movies. We divided the 500 customers into two equal-sized training and test groups at random, and applied sparse CCA to the training set. We constrained the weight vectors to be nonnegative for interpretability. The movies receiving positive weights in the first sparse pair of components are shown in Table 8.3.1. Perhaps a movie buff could tell us why the ratings on these particular movies should correlate; for example, the action movies may be relatively "tame" compared to films like Terminator. In Figure 8.4, we plot the average ratings for the seven action movies on the test set, plotted against the average rating for the 16 romantic movies for each customer. The correlation is quite high—about 0.7. Hence for a given customer, we can do a reasonable job of predicting his/her average rating for the seven action movies from his/her average rating on the 16 romantic movies, and vice versa.

Table 8.2 *Small Netflix dataset: Action and romantic movies with nonzero weights in the first sparse canonical covariates.*

Action Movies

Speed	S.W.A.T.	Men in Black II
The Fast and the Furious	Behind Enemy Lines	Charlies Angels
Con Air		

Romantic Movies

What Women Want	Ghost	The Family Man
The Bodyguard	Miss Congeniality	Pretty Woman
Sister Act	Dirty Dancing	Runaway Bride
Just Married	Maid in Manhattan	Two Weeks Notice
Legally Blonde 2: Red	13 Going on 30	Father of the Bride
Legally Blonde		

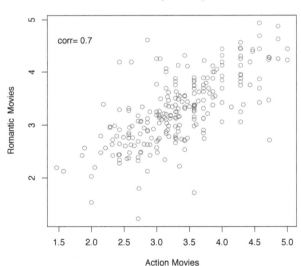

Figure 8.4 *Sparse canonical correlation analysis applied to a subset of the Netflix movie rating data. The plot shows the average rating for the seven action movies on the test data versus the average for the 16 romantic movies having nonzero weights in the first sparse CCA components.*

8.4   Sparse Linear Discriminant Analysis

Linear discriminant analysis (LDA) is an important technique for classification. There is a variety of different proposals for sparse linear discriminant analysis, in part because there are at least three different ways to approach classical discriminant analysis. These are the normal theory model, Fisher's between-to-within variance criterion, and optimal scoring. In addition, in the high-dimensional regime $p \gg N$, some form of regularization is needed for the within-class covariance estimate, and the form of this estimate leads to different methods for sparse LDA.

8.4.1   Normal Theory and Bayes' Rule

Consider a response variable G falling into one of K classes $\{1, 2, \ldots, K\}$, and a predictor vector $X \in \mathbb{R}^p$. Suppose that $f_k(x)$ is the class-conditional density of X in class $G = k$, and let π_k be the prior probability of class k, with $\sum_{k=1}^{K} \pi_k = 1$. A simple application of Bayes' rule gives us

$$\Pr(G = k \mid X = x) = \frac{\pi_k f_k(x)}{\sum_{\ell=1}^{K} \pi_\ell f_\ell(x)}. \tag{8.27}$$

Suppose moreover that each class density is modeled as a multivariate Gaussian $N(\mu_k, \boldsymbol{\Sigma}_w)$, with density

$$f_k(x) = \frac{1}{(2\pi)^{p/2} |\boldsymbol{\Sigma}_w|^{1/2}} e^{-\frac{1}{2}(x - \mu_k)^T \boldsymbol{\Sigma}_w^{-1}(x - \mu_k)}, \tag{8.28}$$

based on a common covariance matrix $\boldsymbol{\Sigma}_w$. In comparing two classes k and ℓ, it is sufficient to look at the log-ratio of their posterior probabilities (8.27), and we find that

$$\begin{aligned}
\log \frac{\Pr(G = k \mid X = x)}{\Pr(G = \ell \mid X = x)} &= \log \frac{f_k(x)}{f_\ell(x)} + \log \frac{\pi_k}{\pi_\ell} \\
&= \log \frac{\pi_k}{\pi_\ell} - \frac{1}{2}(\mu_k + \mu_\ell)^T \boldsymbol{\Sigma}_w^{-1}(\mu_k - \mu_\ell) \\
&\quad + x^T \boldsymbol{\Sigma}_w^{-1}(\mu_k - \mu_\ell),
\end{aligned} \tag{8.29}$$

an equation linear in x. Consequently, the decision boundary between classes k and ℓ—i.e., all vectors x for which $\Pr(G = k \mid X = x) = \Pr(G = \ell \mid X = x)$—defines a hyperplane in $\mathbb{R}^p$. This statement holds for any pair of classes, so all the decision boundaries are linear. If we divide $\mathbb{R}^p$ into regions that are classified as class 1, class 2, and so on, these regions will be separated by hyperplanes.

Equation (8.29) shows us that these LDA models are also linear logistic regression models; the only difference is the way the parameters are estimated. In logistic regression, we use the conditional binomial/multinomial likelihoods, whereas estimation in LDA is based on the joint likelihood of X and G (Hastie et al. 2009, Chapter 4). From Equation (8.29), we see that the *linear discriminant functions*

$$\delta_k(x) = x^T \boldsymbol{\Sigma}_w^{-1} \mu_k - \frac{1}{2} \mu_k^T \boldsymbol{\Sigma}_w^{-1} \mu_k + \log \pi_k \qquad (8.30)$$

provide an equivalent description of the decision rule, leading to the classification function $G(x) = \arg\max_{k \in \{1,...,K\}} \delta_k(x)$.

In practice, the parameters of the Gaussian class-conditional distributions are not known. However, given N samples $\{(x_1, g_1), \ldots, (x_N, g_N)\}$ of feature-label pairs, we can estimate the parameters as follows. Let C_k denote the subset of indices i for which $g_i = k$, and let $N_k = |C_k|$ denote the total number of class-k samples. We then form the estimates $\hat{\pi}_k = N_k/N$, and

$$\hat{\mu}_k = \frac{1}{N_k} \sum_{i \in C_k} x_i, \text{ and} \qquad (8.31a)$$

$$\hat{\boldsymbol{\Sigma}}_w = \frac{1}{N-K} \sum_{k=1}^{K} \sum_{i \in C_k} (x_i - \hat{\mu}_k)(x_i - \hat{\mu}_k)^T. \qquad (8.31b)$$

Note that $\hat{\boldsymbol{\Sigma}}_w$ is an unbiased estimate of the pooled within-class covariance.

In the high-dimensional setting with $p > N$, the sample within-class covariance matrix $\hat{\boldsymbol{\Sigma}}_w$ is singular, and so we must regularize it in order to proceed. As before, there are many ways to do so; later in this section, we describe an approach based on quadratic regularization (Hastie, Buja and Tibshirani 1995).

In very high dimensions, it is often effective to assume that predictors are uncorrelated, which translates into a diagonal form for $\boldsymbol{\Sigma}_w$. Doing so yields the so-called *naive Bayes* classifier, or alternatively *diagonal linear discriminant analysis* (see Exercise 8.20). Letting $\hat{\sigma}_j^2 = s_j^2$ be the pooled within-class variance for feature j, the estimated classification rule simplifies to

$$\hat{G}(x) = \arg\min_{\ell=1,...,K} \left\{ \sum_{j=1}^{p} \frac{(x_j - \hat{\mu}_{j\ell})^2}{\hat{\sigma}_j^2} - \log \hat{\pi}_k \right\}, \qquad (8.32)$$

known as the *nearest centroid rule*.

8.4.2   Nearest Shrunken Centroids

Notice that the classification rule (8.32) will typically involve all features; when p is large, while one might expect that only a subset of these features is informative. This subset can be revealed by reparametrizing the model, and imposing a sparsity penalty. More specifically, suppose that we decompose the mean vector for class k into the sum $\mu_k = \bar{x} + \alpha_k$, where $\bar{x} = \frac{1}{N} \sum_{i=1}^{N} x_i$ is the overall mean vector, and $\alpha_k \in \mathbb{R}^p$, $k = 1, \ldots, K$ denotes the contrast for class k, together satisfying the constraint $\sum_{k=1}^{K} \alpha_k = 0$. We then consider optimizing the ℓ_1-regularized criterion

$$\underset{\alpha_k \in \mathbb{R}^p,\, k=1,\ldots,K}{\text{minimize}} \left\{ \frac{1}{2N} \sum_{k=1}^{K} \sum_{i \in C_k} \sum_{j=1}^{p} \frac{(x_{ij} - \bar{x}_j - \alpha_{jk})^2}{s_j^2} + \lambda \sum_{k=1}^{K} \sum_{j=1}^{p} \frac{\sqrt{N_k}}{s_j} |\alpha_{jk}| \right\}$$

$$\text{subject to } \sum_{k=1}^{K} \alpha_{jk} = 0 \text{ for } j = 1, \ldots, p.$$

$$(8.33)$$

The solutions for α_{jk} amount to simple soft-thresholding of particular class-wise contrasts. In detail, we define the contrasts

$$d_{jk} = \frac{\widetilde{x}_{jk} - \bar{x}_j}{m_k s_j}, \qquad (8.34)$$

where $\widetilde{x}_{jk} = \frac{1}{N_k} \sum_{i \in C_k} x_{ij}$, the quantity $\bar{x}_j$ denotes the jth component of the global mean $\bar{x}$, and[5] $m_k^2 = \frac{1}{N_k} - \frac{1}{N}$. We then apply the soft-thresholding operator

$$d'_{jk} = \mathcal{S}_\lambda(d_{jk}) = \text{sign}(d_{jk})(|d_{jk}| - \lambda)_+, \qquad (8.35a)$$

and reverse the transformation to obtain the shrunken centroid estimates

$$\widehat{\mu}'_{jk} = \bar{x}_j + m_k s_j d'_{kj}. \qquad (8.35b)$$

Finally, we use these shrunken centroids for the estimates for μ_{jk} in the nearest centroid rule (8.32).

Suppose for a given feature j, the contrasts d'_{jk} are set to zero by the soft-thresholding for each of the k classes. Then that feature does not participate in the nearest centroid rule (8.32), and is ignored. In this way, the nearest shrunken centroid procedure does automatic feature selection. Alternatively, a feature might have $d'_{jk} = 0$ for some classes but not others, and hence would only play a role for those classes.

The nearest shrunken centroid classifier is very useful for high-dimensional classification problems, like those that occur in genomic and proteomic data. The publicly available software (Hastie, Tibshirani, Narasimhan and Chu 2003) includes some additional bells and whistles: a small constant s_0 is added to each s_j, to stabilize the contrasts when s_j is close to zero; class-specific shrinkage rates, to name a few.

Figure 8.5 shows the results of this procedure applied to some Lymphoma cancer data (Tibshirani, Hastie, Narasimhan and Chu 2003). These data consist of expression measurements on 4026 genes from samples of 59 lymphoma patients. The samples are classified into diffuse large B-cell lymphoma (DL-BCL), follicular lymphoma (FL), and chronic lymphocytic lymphoma (CLL).

---

[5] The quantity m_k is a standardization constant, based on the variance of the numerator, which makes d_{jk} a t-statistic.

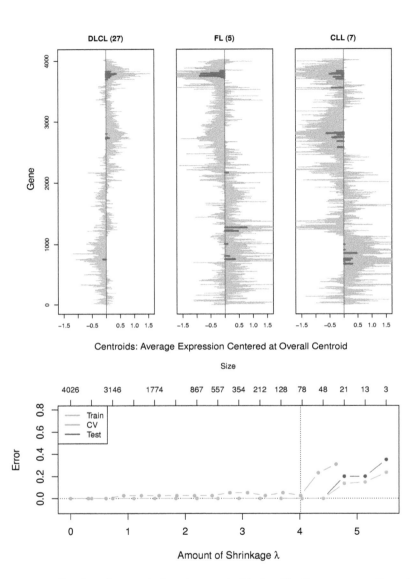

Figure 8.5 *Results of nearest-shrunken-centroid classification on some Lymphoma data, with three classes. The top plot shows the class-specific mean expression for each gene (gray lines), and their shrunken versions (blue). Most of the genes are shrunk to the overall mean (0 here). The lower plot shows training, cross-validated, and test misclassification error as a function of the shrinkage threshold λ. The chosen model includes 79 genes, and makes 0 test errors.*

The data are divided into a training set of 39 (27, 5, 7) samples, and a test set of 20. The genes have been organized by hierarchical clustering. All but 79 of the genes have been shrunk to zero. Notice that the deviations of the smaller classes are larger, since the biggest class DLBCL mostly determines the overall mean. In Section 8.4.3.1, we compare the nearest shrunken centroid classifier to a sparse version of Fisher's linear discriminant analysis, discussed next.

8.4.3 Fisher's Linear Discriminant Analysis

A different approach to sparse discriminant analysis arises from Fisher's discriminant framework. Here the idea is to produce low-dimensional projections of the data that preserve the class separation. Although these projections are primarily intended for visualization, one can also perform Gaussian classification in the subspace produced.

Let $\mathbf{X}$ be an $N \times p$ matrix of observations, and assume that its columns, corresponding to features, have been standardized to have mean zero. Given such an observation matrix, we seek a low-dimensional projection such that the *between-class* variance is large relative to the *within-class* variance. As before, let $\widehat{\boldsymbol{\Sigma}}_w$ be the pooled within-class covariance matrix and $\widehat{\mu}_k$ the class-specific centroids. The between-class covariance matrix $\widehat{\boldsymbol{\Sigma}}_b$ is the covariance matrix of these centroids, given by

$$\widehat{\boldsymbol{\Sigma}}_b = \sum_{k=1}^{K} \widehat{\pi}_k \widehat{\mu}_k \widehat{\mu}_k^T, \tag{8.36}$$

treating them as multivariate observations with mass $\widehat{\pi}_k$. Note that

$$\widehat{\boldsymbol{\Sigma}}_t = \frac{1}{N} \mathbf{X}^T \mathbf{X} = \widehat{\boldsymbol{\Sigma}}_b + \widehat{\boldsymbol{\Sigma}}_w. \tag{8.37}$$

For now we assume that $\widehat{\boldsymbol{\Sigma}}_w$ is of full rank (which implies that $p \leq N$); we treat the non-full rank case below. For a linear combination $\mathbf{z} = \mathbf{X}\beta$, Fisher's between-to-within variance criterion is captured by the ratio

$$R(\beta) = \frac{\beta^T \widehat{\boldsymbol{\Sigma}}_b \beta}{\beta^T \widehat{\boldsymbol{\Sigma}}_w \beta}, \tag{8.38}$$

which is to be maximized. Fisher's LDA proceeds by sequentially solving the following problem:

$$\underset{\beta \in \mathbb{R}^p}{\text{maximize}} \left\{ \beta^T \widehat{\boldsymbol{\Sigma}}_b \beta \right\} \text{ such that } \beta^T \widehat{\boldsymbol{\Sigma}}_w \beta \leq 1, \text{ and } \beta^T \widehat{\boldsymbol{\Sigma}}_w \widehat{\beta}_\ell = 0 \text{ for all } \ell < k.$$
$$\tag{8.39}$$

for $k = 1, 2, \ldots, \min(K - 1, p)$. Although the problem (8.39) is generally written with the inequality constraint replaced with an equality constraint, the two programs are equivalent if $\widehat{\boldsymbol{\Sigma}}_w$ has full rank. The solution $\widehat{\beta}_k$ is called the kth *discriminant vector*, and $\mathbf{z}_k = \mathbf{X}\widehat{\beta}_k$ the corresponding discriminant variable. Note that LDA essentially does principal components on the class centroids, but using a normalization metric that respects the within-class variances (Hastie et al. 2009, Chapter 4). In practice, we do not need to solve the problem sequentially, because as with PCA we can get all the solutions with a single eigen-decomposition: the first k discriminant vectors are the k leading eigenvectors of $\widehat{\mathbf{\Sigma}}_w^{-1}\widehat{\mathbf{\Sigma}}_b$.

Witten and Tibshirani (2011) proposed a way to "sparsify" the objective (8.39), in particular by solving

$$\underset{\beta}{\text{maximize}} \left\{ \beta^T \widehat{\mathbf{\Sigma}}_b \beta - \lambda \sum_{j=1}^{p} \hat{\sigma}_j |\beta_j| \right\} \text{ subject to } \beta^T \widetilde{\mathbf{\Sigma}}_w \beta \leq 1, \qquad (8.40)$$

where $\hat{\sigma}_j^2$ is the jth diagonal element of $\widehat{\mathbf{\Sigma}}_w$, and $\widetilde{\mathbf{\Sigma}}_w$ is a positive definite estimate for $\mathbf{\Sigma}_w$. This produces a first sparse discriminant vector $\widehat{\beta}_1$ with level of sparsity determined by the choice of λ. Further components can be successively found by first removing the current solution from $\widehat{\mathbf{\Sigma}}_b$ before solving problem (8.40); see the reference for details.

The choice for the regularized within-covariance matrix $\widetilde{\mathbf{\Sigma}}_w$ depends on the setting. In some problems, we might choose $\widetilde{\mathbf{\Sigma}}_w$ to encourage spatial smoothness, for example when the data are images. Then we can take $\widetilde{\mathbf{\Sigma}}_w = \widehat{\mathbf{\Sigma}}_w + \mathbf{\Omega}$ where $\mathbf{\Omega}$ penalizes differences in spatially nearby values. This idea is studied in the *flexible and penalized discriminant analysis* approach of Hastie, Tibshirani and Buja (1994) and Hastie et al. (1995). In the sparse setting, this is conveniently implemented using the optimal-scoring approach of Section 8.4.4. In other cases we only require that $\widetilde{\mathbf{\Sigma}}_w$ makes the sample estimate $\widehat{\mathbf{\Sigma}}_w$ positive definite, and for that purpose we can use a ridged version $\widetilde{\mathbf{\Sigma}}_w = \widehat{\mathbf{\Sigma}}_w + \epsilon \, \text{diag}(\widehat{\mathbf{\Sigma}}_w)$ for some $\epsilon > 0$.

One simple case of particular interest is where $\widetilde{\mathbf{\Sigma}}_w$ is taken to be a diagonal matrix, for example $\text{diag}(\widehat{\mathbf{\Sigma}}_w)$. Then problem (8.40) can be cast as a penalized matrix decomposition applied to the between covariance matrix $\widehat{\mathbf{\Sigma}}_b$, and Algorithm 7.2 can be applied. In this case, with $K = 2$ classes, this method gives a solution that is similar to nearest shrunken centroids: details are in Witten and Tibshirani (2011, Section 7.2). With more than two classes, the two approaches are different. Nearest shrunken centroids produce sparse contrasts between each class and the overall mean, while the sparse LDA approach (8.40) produces sparse discriminant vectors for more general class contrasts. This distinction is explored in the next example.

8.4.3.1   *Example: Simulated Data with Five Classes*

We created two artificial scenarios to contrast the nearest shrunken centroids approach with sparse discriminant analysis (8.40). Figure 8.6 shows the results of nearest shrunken centroids classifier applied to the two different simulated

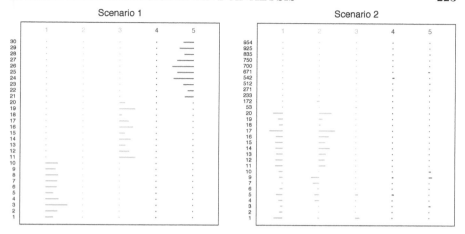

Figure 8.6 *Results of nearest shrunken centroid classifier applied to two different simulated datasets, as described in the text. Those features (rows in each plot) with nonzero estimated contrasts are shown. The length of each horizontal line segment is proportional to the size of the contrast, with positive values to the right and negative values to the left.*

datasets. In both cases there are $N = 100$ observations, with 20 observations falling into each of $K = 5$ classes involving $p = 1000$ features.

1. In the first scenario, the first 10 features are two units higher in class 1, features 11–20 are two units higher in class 3, and features 21–30 are two units lower in class 5. Thus, higher values in each of the first three block of 10 features characterize classes 1, 3, and 5.

2. In the second scenario, features 1–10 are one unit higher in classes 3–5 versus 1,2 and features 11–20 are one unit higher in class 2 and one unit lower in class 1. Hence higher values for the first 10 features discriminate classes 3–5 versus 1 and 2, while higher values for the second 10 features discriminate between classes 1 versus 2.

We applied the nearest shrunken centroid classifier, using cross-validation to choose the shrinkage parameter and show the features with nonzero estimated contrasts. The length of each horizontal line segment is proportional to the size of the contrast, with positive values to the right and negative values to the left. In the left panel, we see that nearest shrunken centroids has clearly revealed the structure of the data, while in the right panel, the structure does not come through as clearly. Figure 8.7 shows the results of rank-2 sparse linear discriminant analysis with diagonal within-class covariance in each of the two scenarios. In the first scenario (top row), the discriminant projection cleanly separates classes 1, 3, and 5 from the rest, but the discriminant loadings (top right) are forced to combine three pieces of information into two vectors, and hence give a cloudy picture. Of course one could use more than

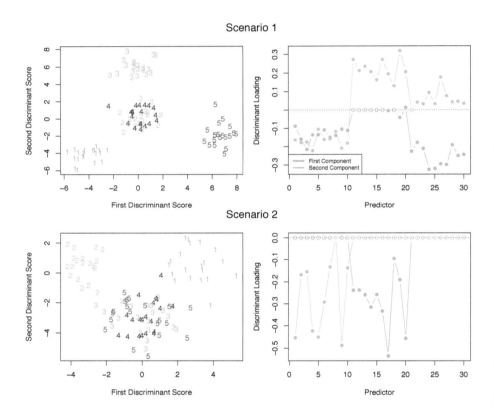

Figure 8.7 *Rank two sparse linear discriminant analysis with diagonal within-class covariance, applied to the same two scenarios (top and bottom panels) as in Figure 8.6. The projections onto the first two sparse discriminant vectors is shown in the left panels, while the right panels show the discriminant weights or loadings.*

$K = 2$ sparse components and this would help in this example, but this approach is less attractive if high-order discriminants are required. The second scenario is well suited to sparse LDA, as it cleanly separates classes 1 and 2 from the rest (bottom left), and reveals the features responsible for this separation (bottom right).

8.4.4 Optimal Scoring

A third approach to the derivation of linear discriminant analysis is called *optimal scoring*. It is based on a recasting of the problem in terms of a multivariate linear regression, where the codes for the output classes are chosen "optimally," as we detail next. Suppose that the membership of the samples are coded using a binary-valued $N \times K$ indicator matrix $\mathbf{Y}$, with entries

$$y_{ik} = \begin{cases} 1 & \text{if observation } i \text{ belongs to class } k \\ 0 & \text{otherwise.} \end{cases}$$

Using this notation, optimal scoring involves solving the sequence of problems for $k = 1, \ldots, K$, each of the form

$$(\widehat{\beta}_k, \widehat{\theta}_k) = \underset{\beta_k \in \mathbb{R}^p, \theta_k \in \mathbb{R}^K}{\arg\min} \left\{ \frac{1}{N} \|\mathbf{Y}\theta_k - \mathbf{X}\beta_k\|_2^2 \right\} \tag{8.41}$$

such that $\theta_k^T \mathbf{Y}^T \mathbf{Y} \theta_k = 1$ and $\theta_k^T \mathbf{Y}^T \mathbf{Y} \theta_j = 0$ for all $j = 1, 2, \ldots, k - 1$.

The optimal solution $\widehat{\beta}_k$ of this problem turns out to be proportional to the solution of the Fisher linear discriminant criterion (8.39) (Breiman and Ihaka 1984, Hastie et al. 1995). This equivalence is not too surprising. With just $K = 2$ classes, it is well known that the linear regression of a binary response $\tilde{y} = \mathbf{Y}\theta$ (with arbitrary coding θ) on $\mathbf{X}$ gives the same coefficient vector as linear discriminant analysis (up to a proportionality factor). For example, see Exercise 4.2 of Hastie et al. (2009). With more than two classes, a regression of $\tilde{y}_\ell$ on $\mathbf{X}$ will differ according to how we assign numerical scores $\theta_{\ell k}$ to the classes. We obtain the particular solution from linear regression that is equivalent to linear discriminant analysis by optimizing over the choice of scores, as in the problem (8.41).

As with the other methods for sparse discriminant analysis, adding an ℓ_1-penalty to the criterion (8.41) yields the modified optimization problem

$$\underset{\beta_k \in \mathbb{R}^p, \theta_k \in \mathbb{R}^K}{\text{minimize}} \left\{ \frac{1}{N} \|\mathbf{Y}\theta_k - \mathbf{X}\beta_k\|_2^2 + \beta_k^T \mathbf{\Omega} \beta_k + \lambda \|\beta_k\|_1 \right\} \tag{8.42}$$

such that $\theta_k^T \mathbf{Y}^T \mathbf{Y} \theta_k = 1$ and $\theta_k^T \mathbf{Y}^T \mathbf{Y} \theta_j = 0$ for all $j = 1, 2, \ldots, k - 1$

(Leng 2008, Clemmensen, Hastie, Witten and Ersboll 2011). In addition to the ℓ_1-penalty with nonnegative regularization weight λ, we have also added a quadratic penalty defined by a positive semidefinite matrix $\mathbf{\Omega}$, equivalent to the elastic net penalty in the special case $\boldsymbol{\Omega} = \gamma\mathbf{I}$. The resulting discriminant vectors will be sparse if the regularization weight λ on the ℓ_1-penalty is sufficiently large. At the other extreme, if $\lambda = 0$, then minimizing the criterion (8.42) is equivalent to the *penalized discriminant analysis* proposal of Hastie et al. (1995). Although the criterion is nonconvex (due to the quadratic constraints), a local optimum can be obtained via alternating minimization, using the elastic net to solve for β. In fact, if any convex penalties are applied to the discriminant vectors in the optimal scoring criterion (8.41), then it is easy to apply alternating minimization to solve the resulting problem. Moreover, there is a close connection between this approach and sparse Fisher LDA (8.40). In particular, they are essentially equivalent if we take $\widetilde{\boldsymbol{\Sigma}} = \widehat{\boldsymbol{\Sigma}}_w + \boldsymbol{\Omega}$. The qualification "essentially" is needed due to nonconvexity: we can only say that a stationary point for the one problem is also a stationary point for the other (see Exercise 8.22 for details).

Whether one approaches the sparse discriminant problem through Fisher LDA (8.40) or optimal scoring (8.42) depends on the nature of the problem. When $p \gg N$ and the features are not structured—a category containing many genomic problems—it is most attractive to set $\widetilde{\boldsymbol{\Sigma}}_w$ equal to $\text{diag}(\widehat{\boldsymbol{\Sigma}}_w)$. Since this matrix is positive definite, we can take $\boldsymbol{\Omega} = \mathbf{0}$, and the resulting problem is easily solved via the soft-thresholding algorithm for penalized matrix decomposition. When the problem has a spatial or temporal structure, the matrix $\boldsymbol{\Omega}$ can be chosen to encourage spatial or temporal smoothness of the solution. In this case the optimal scoring approach is attractive, since the quadratic term can be absorbed into the quadratic loss. Otherwise, the matrix $\boldsymbol{\Omega}$ can be chosen to be a diagonal matrix, as in the next example, and again optimal scoring is convenient. Both methods are implemented in packages in R: `penalizedLDA` (Witten 2011) for the criterion (8.40), and `sparseLDA` (Clemmensen 2012) for the criterion (8.42).

8.4.4.1   Example: Face Silhouettes

We illustrate sparse discriminant analysis based on the objective function (8.42) with a morphometric example taken from Clemmensen et al. (2011). The dataset consisting of 20 adult male and 19 adult female face-silhouettes. Following the work of Thodberg and Olafsdottir (2003), we apply a minimum description length (MDL) approach to annotate the silhouettes, and then perform Procrustes' alignment on the resulting 65 MDL landmarks of (x, y)-coordinates. These 65 pairs are vectorized resulting in $p = 130$ spatial features. We set $\boldsymbol{\Omega} = \mathbf{I}$ in the criterion (8.42); in this case the spatial features are already smooth, and the ridge penalty $\mathbf{I}$ is sufficient to deal with the strong spatial autocorrelations. For training, we used 22 of the silhouettes (11 female and 11 male), which left 17 silhouettes for testing (8 female and 9 male). The left and middle panels of Figure 8.8 illustrate the two classes of silhouettes.

Leave-one-out cross validation was performed on the training data, estimating an optimal value of λ that yielded 10 nonzero features. Since there are

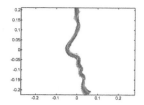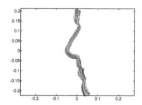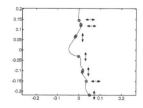

Figure 8.8 *The silhouettes and the 65 (x, y)-coordinates for females (left) and males (middle). Right: The mean shape of the silhouettes, and the 10 (x, y)-coordinates in the SDA model. The superimposed dots indicate the landmarks retained in the sparse discriminant vector. The arrows illustrate the directions of the differences between male and female classes.*

two classes, there is only one sparse direction. The nonzero weights are shown in the right panel of Figure 8.8. The few landmarks included in the model are placed near high curvature points in the silhouettes, suggesting that the important gender differences are located in these regions. The training and test classification rates were both 82%.

8.5   Sparse Clustering

In this section, we discuss methods for clustering observations that employ sparsity to filter out the uninformative features. We first give a brief background on clustering; more details can be found, for example, in Hastie et al. (2009, Chapter 14).

8.5.1   Some Background on Clustering

Suppose we wish to group or cluster a collection of N observations on p features, where $p \gg N$. Our goal is to find groups of observations that are similar with respect to the p features. A standard method for doing this is called "hierarchical clustering." More precisely, we are referring to agglomerative (or bottom-up) hierarchical clustering. This method starts with the individual observations, and then merges or agglomerates the pair that are closest according to some metric, with the Euclidean distance over the p features being one common choice. This process is continued, with the closest pair grouped together at each stage. Along the way, we consider merging not only individual pairs of observations, but also merging clusters of observations that were created at previous steps, with individual observations or other clusters. For this, we need to define a linkage measure—the distance between two clusters. Some common choices include *average linkage*, which define the distance between two clusters as the average distance between any two observations, one in each cluster; *complete linkage*, which uses the maximum pairwise distance; and *single linkage*, which uses the minimum pairwise distance.

The top panel of Figure 8.9 shows an example of hierarchical clustering applied to some artificial data with 120 observations and 2000 features. The figure shows the result of hierarchical clustering using Euclidean distance and complete linkage. The clustering tree or *dendrogram* summarizes the sequence of merges, leading to a single cluster at the top. The colors of the leaves of the tree were not used in the clustering, and are explained below.

Now suppose that the observations vary only over a subset of the features. Then we would like to isolate that subset both for interpretation and to improve the clustering. In the top panel of Figure 8.9, the data were actually generated so that the average levels of the first 200 features varied over six predefined classes, with the remaining 1800 features being standard Gaussian noise. These classes are not used in the clustering, but after carrying out the clustering, we have colored the leaves of the dendrogram according to the true class. We see that hierarchical clustering is confused by the uninformative features, and does a poor job of clustering the observations into classes. In this instance we would like to isolate the informative subset of features both for interpretability and to improve the clustering. One way of doing that is described next.

8.5.2   Sparse Hierarchical Clustering

We now describe an approach that introduces sparsity and feature selection to this problem. Given a data matrix $\mathbf{X} \in \mathbb{R}^{N \times p}$, standard clustering based on Euclidean distance uses the dissimilarity measure $D_{i,i'} = \sum_{j=1}^{p} d_{i,i',j}$ with $d_{i,i',j} = (x_{ij} - x_{i'j})^2$. The idea here is to find a set of feature weights $w_j \geq 0$ and use these to define a weighted dissimilarity measure $\tilde{D}_{i,i'} = \sum_{j=1}^{p} w_j d_{i,i',j}$; we want each weight to reflect the importance of that feature. Finally, this modified dissimilarity matrix is used as input into hierarchical clustering.

Denote by $\mathbf{\Delta}$ the $N^2 \times p$ matrix with column j containing the N^2 pairwise dissimilarities for feature j. Then $\mathbf{\Delta}1$ is the vectorized version of $D_{i,i'}$, and likewise $\mathbf{\Delta}w$ the vectorized version of $\tilde{D}_{i,i'}$. We now seek the vector w, subject to sparsity and normalization restrictions, that recovers most of the variability in $\mathbf{\Delta}$. This requirement leads to the penalized matrix decomposition problem (Witten et al. 2009) (see Section 7.6):

$$\underset{\mathbf{u} \in \mathbb{R}^{N^2}, \, w \in \mathbb{R}^p}{\text{maximize}} \left\{ \mathbf{u}^T \mathbf{\Delta} w \right\} \text{ subject to } \|\mathbf{u}\|_2 \leq 1, \|w\|_2 \leq 1,$$

$$\|w\|_1 \leq s, \text{ and } w \succeq 0. \tag{8.43}$$

Notice that w_j is a weight on the dissimilarity matrix for feature j. Given the optimal solution $\widehat{w}$, we rearrange the elements of $\mathbf{\Delta}w$ into a $N \times N$ matrix, and perform hierarchical clustering on this reweighted dissimilarity matrix. The result is a sparse hierarchical clustering of the data. In Figure 8.9, we see that sparse clustering isolates the informative features (bottom panel) and

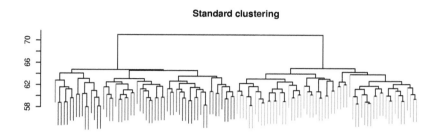

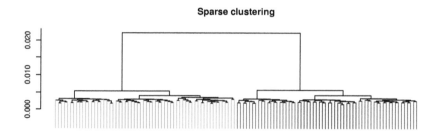

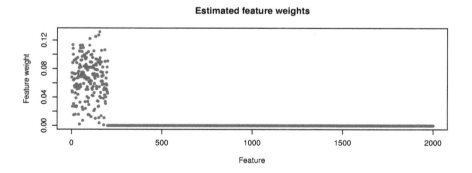

Figure 8.9 *Standard and sparse clustering applied to a simulated example. The data were generated so that the average levels of the first 200 features varied over six predefined classes, while the remaining 1800 features were noise features with the same distribution over the classes. The top two panels show the result of standard hierarchical clustering and sparse clustering, respectively. We used complete linkage in each case. The class of each sample is indicated by the color of each leaf, and was not used by the clustering procedures. The bottom panel shows the estimated weight given to each feature by the sparse clustering procedure.*

uses this information to correctly cluster the observations into the predefined groups (middle panel).

8.5.3 Sparse K-Means Clustering

Another commonly used method of clustering is called "K-means." Here we predefine the number of groups K and then try to partition the observations into K homogeneous groups. Each group is summarized by a centroid, with each observation assigned to the group with the closest centroid.

In detail, the K-means algorithm maintains a partition $\mathcal{C} = \{C_1, \ldots, C_K\}$ of the index set $\{1, 2, \ldots, N\}$, where subset C_k corresponds to those observations currently assigned to class k. It chooses these partitions by minimizing the within cluster sum of squares:

$$W(\mathcal{C}) = \sum_{k=1}^{K} \sum_{i \in C_k} \|x_i - \bar{x}_k\|_2^2. \tag{8.44}$$

Here x_i is the ith observation and $\bar{x}_k$ is a p-vector equal to the average of all observations in cluster k. The collection $\{\bar{x}_k\}_1^K$ is referred to as the *codebook* in the compression literature. The *encoder* $\tau(i)$ assigns each observation x_i to the cluster k whose centroid is closest to it. Hence $C_k = \{i : \tau(i) = k\}$. The standard algorithm for K-means clustering alternates over optimizing for $\mathcal{C}$ and $\{\bar{x}_1, \ldots, \bar{x}_K\}$, and is guaranteed to find a local minimum of $W(\mathcal{C})$. Since

$$\sum_{i, i' \in C_k} \|x_i - x_{i'}\|_2^2 = 2N_k \sum_{i \in C_k} \|x_i - \bar{x}_k\|_2^2, \tag{8.45}$$

with $N_k = |C_k|$, one can alternatively derive K-means clustering using a squared Euclidean dissimilarity matrix $D_{i,i'}$. For general dissimilarity matrices, K-*medoids* clustering is a natural generalization (Hastie et al. 2009, for example).

It might seem reasonable to define as a criterion for sparse K-means clustering the minimum of the weighted within-cluster sum of squares:

$$\underset{\mathcal{C},\, w \in \mathbb{R}^p}{\text{minimize}} \left\{ \sum_{j=1}^{p} w_j \left(\sum_{k=1}^{K} \frac{1}{N_k} \sum_{i, i' \in C_k} d_{i,i',j} \right) \right\}.$$

We still need to add constraints on w, to make the problem meaningful. Adding the constraints $\|w\|_2 \leq 1$, $\|w\|_1 \leq s$ as well as the nonnegativity constraint $w \succeq 0$ makes the problem convex in w, but leads to the pathological solution $\hat{w} = 0$. On the other hand, the triplet of constraints $\|w\|_2 \geq 1$, $\|w\|_1 \geq s$, $w \succeq 0$ would lead to a potentially useful solution, but make the problem nonconvex in w.

Witten and Tibshirani (2010) proposed a modified criterion that focuses instead on the between-cluster sum of squares:

$$\underset{\mathcal{C},\, w \in \mathbb{R}^p}{\text{maximize}} \left\{ \sum_{j=1}^{p} w_j \left(\frac{1}{N} \sum_{i=1}^{N} \sum_{i'=1}^{N} d_{i,i',j} - \sum_{k=1}^{K} \frac{1}{N_k} \sum_{i,i' \in C_k} d_{i,i',j} \right) \right\} \qquad (8.46)$$

$$\text{subject to } \|w\|_2 \le 1, \|w\|_1 \le s, w \succeq 0.$$

When $w_j = 1$ for all j, we can see from condition (8.45) that the second term is equal to $2W(\mathcal{C})$, and hence this approach is equivalent to K-means. This problem is now convex in w and generally has an interesting solution. It can be solved by a simple alternating algorithm. With $\mathcal{C} = (C_1, ..., C_K)$ fixed, the minimization over w is a convex problem, with solutions given by soft-thresholding. With w fixed, optimization with respect to $\mathcal{C}$ leads to a weighted K-means clustering algorithm. Details are given in Exercise 8.11.

8.5.4 *Convex Clustering*

The method of K-means clustering and its sparse generalization lead to problems that are biconvex but not jointly convex, and hence it is difficult to guarantee that a global solution has been attained. Here we present a different formulation of clustering that yields a convex program, and represents an interesting alternative to K-means and hierarchical clustering. Unlike the methods of the previous section which use sparsity to do feature selection, this method uses a form of sparsity to determine the number and memberships of the clusters.

In this approach, each of the N observations $x_i \in \mathbb{R}^p$ is assigned a prototype $u_i \in \mathbb{R}^p$. We then minimize the objective function

$$J(u_1, u_2, \dots, u_N) = \frac{1}{2} \sum_{i=1}^{N} \|x_i - u_i\|^2 + \lambda \sum_{i<i'} w_{ii'} \|u_i - u_{i'}\|_q \qquad (8.47)$$

for some fixed $\lambda \ge 0$, and some q-norm (typically $q = 1$ or $q = 2$). This criterion seeks prototypes that are close to the data points, but not too far from one another. The weights $w_{ii'}$ can be equal to 1, or can be a function of the distance between observations i and i'. We note that this problem is convex for $q \ge 1$. Consider for example the natural choice $q = 2$ (group lasso). Then the penalty term shrinks prototype vectors toward each other, and the distance between many pairs will be equal to zero for a sufficiently large value of λ.

Each distinct prototype $\widehat{u}_i$ in the solution represents a cluster; however, as shown in the example of Figure 8.10, we should not think of it as a typical prototype or centroid of that cluster. In this example, there are two classes each containing 50 spherical Gaussian data points, with their means separated by three units in both directions. Here we used $q = 2$ and weight function $w_{ii'} = \exp(-\|x_i - x_{i'}\|^2)$. The colors in the right panel coincide with the

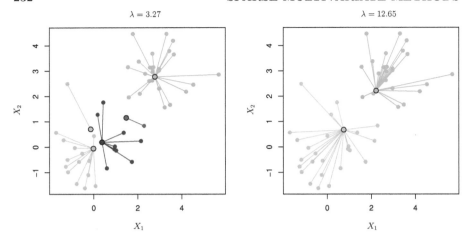

Figure 8.10 *Convex clustering applied to data generated from two spherical Gaussian populations separated by three units in each direction. We show two solutions from a path of 50 values of λ; the solution on the right was the smallest value of λ that yielded two clusters, and in this case identified the true clusters. Points x_i are associated with prototypes $\widehat{\mu}_i$ of the same color. The estimated prototypes need not be close to the centroids of their clusters.*

true clusters; further details are given in the caption. The convexity of the objective function as well as its ability to choose the number of clusters and the informative features, makes this approach attractive.

The next example is taken from Chi and Lange (2014), on the problem of clustering mammals based on their dentition. Eight different kinds of teeth were counted for each of 27 mammals: the number of top incisors, bottom incisors, top canines, bottom canines, top premolars, bottom premolars, top molars, and bottom molars. Figure 8.11 shows the resulting clustering path over λ using kernel-based weights $w_{ii'}$. For visualization, the prototypes have been projected onto the first two principal components. The continuous path of solutions provides an appealing summary of the similarity among the mammals. Both these examples were produced using the `cvxcluster` package in R (Chi and Lange 2014). The path of solutions creates a tree, which in this example is rather similar to that produced by hierarchical clustering with average linkage.

Bibliographic Notes

Jolliffe et al. (2003) proposed the original SCoTLASS criterion (8.7) for sparse PCA; the reformulation and alternating updates (8.9) were proposed by Witten et al. (2009). d'Aspremont et al. (2007) proposed the semidefinite programming relaxation (8.12) of the nonconvex SCoTLASS criterion; Amini and

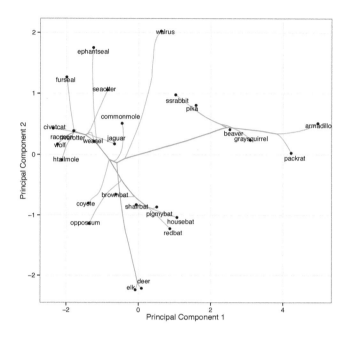

Figure 8.11 *Mammal data: path of convex clustering solutions using the criterion (8.47), and a grid of values of λ. As λ increases, the prototypes unite to form a smaller set.*

Wainwright (2009) develop some theory for the variable selection properties of this relaxation for sparse PCA. Zou et al. (2006) proposed the reconstruction-based criterion (8.13). Johnstone (2001) studied the high-dimensional asymptotics of ordinary PCA, and proposed the spiked covariance model (8.22). Johnstone and Lu (2009) and Birnbaum et al. (2013) study various types of two-stage procedures for estimating sparse principal components. Birnbaum et al. (2013) and Vu and Lei (2012) derive minimax lower bounds on ℓ_2-estimation error for sparse PCA when the eigenvectors belong ℓ_q-balls. Ma (2010, 2013) and Yuan and Zhang (2013) have studied iterative algorithms for sparse PCA based on combining the power method with soft thresholding. Berthet and Rigollet (2013) study the detection problem for high-dimensional sparse PCA, and establish a computational hardness result related to the random k-clique problem.

Olshausen and Field (1996) proposed the version of sparse coding discussed in Section 8.2.5. There is a large literature on deep learning; see Le et al. (2012) and references therein for some recent approaches.

There is a variety of papers that explore sparse canonical correlation analysis, for example Parkhomenko, Tritchler and Beyene (2009), Waaijenborg,

Versélewel de Witt Hamer and Zwinderman (2008), Hardoon and Shawe-Taylor (2011), Parkhomenko et al. (2009), Witten et al. (2009), Witten and Tibshirani (2009), and Lykou and Whittaker (2010). Dudoit, Fridlyand and Speed (2002) provide comparison of different classification methods for microarray data, including diagonal LDA. Nearest shrunken centroids was proposed by Tibshirani, Hastie, Narasimhan and Chu (2001) and Tibshirani et al. (2003). Sparse discriminant analysis from Fisher's framework and optimal scoring were explored in Trendafilov and Jolliffe (2007), Leng (2008), Clemmensen et al. (2011), and Witten and Tibshirani (2011). Minorization algorithms are discussed in Lange, Hunter and Yang (2000), Lange (2004), and Hunter and Lange (2004). Sparse hierarchical and k-means clustering are presented in Witten and Tibshirani (2010), while convex clustering was proposed by Pelckmans, De Moor and Suykens (2005) and Hocking, Vert, Bach and Joulin (2011).

Exercises

Ex. 8.1 In this exercise, we consider some elementary properties of principal component analysis.

(a) Show that the first principal component is a maximal eigenvector of the sample covariance matrix $\frac{1}{N}\mathbf{X}^T\mathbf{X}$.

(b) Suppose that the rows of $\mathbf{X}$, denoted $\{x_1, \ldots, x_N\}$ are drawn i.i.d. according to a zero-mean distribution $\mathbb{P}$, and suppose that $\mathbf{\Sigma} = \text{Cov}(x)$ has a unique maximal eigenvalue λ. Explain why, for large sample size N, one might expect $\widehat{v}$ to approach

$$v^* = \underset{\|v\|_2=1}{\arg\max} \, \text{Var}(v^T x), \quad \text{where } x \sim \mathbb{P}. \tag{8.48}$$

Ex. 8.2 Consider the principal component criterion (8.1), and the definition of the vectors v_1, v_2, and so on.

(a) Show that the principal components $\mathbf{z}_j$ are mutually uncorrelated.

(b) Show that instead of using orthogonality of the v_j to define the sequence of principal component directions, we can instead use the uncorrelatedness of the $\mathbf{z}_j$.

Ex. 8.3 Consider the reconstruction error objective (8.4) with $\mu = 0$.

(a) For fixed $\mathbf{V}_r$, show that the optimal choice of the reconstruction weights is $\lambda_i = \mathbf{V}_r^T x_i$ for each $i = 1, \ldots, N$.

(b) Use part (a) to show that the optimal $\mathbf{V}_r$ maximizes the criterion $\mathbf{V}_r^T \mathbf{X}^T \mathbf{X} \mathbf{V}$, and conclude that it can be obtained via the truncated SVD (as described in the main text).

Ex. 8.4 Consider criterion (8.8) and Algorithm 8.1 for finding a solution. By partially maximizing w.r.t. $\mathbf{u}$ with $\|\mathbf{u}\|_2 \leq 1$, show that any stationary value of v is also a stationary value for the SCoTLASS criterion (8.7).

Ex. 8.5 Consider the problem

$$\underset{\mathbf{u},v,d}{\text{minimize}} \ \|\mathbf{X} - d\mathbf{u}v^T\|_F \tag{8.49}$$
$$\text{subject to } \|\mathbf{u}\|_1 \leq c_1, \ \|v\|_1 \leq c_2, \ \|\mathbf{u}\|_2 = 1, \ \|v\|_2 = 1, \ d \geq 0,$$

where $\mathbf{X}$ is $N \times p$, and assume $1 \leq c_1 \leq \sqrt{N}$, $1 \leq c_2 \leq \sqrt{p}$. Show that a solution $\mathbf{u}_1$, v_1, d_1 is also the solution to the following problem:

$$\underset{\mathbf{u},v}{\text{maximize}} \ \mathbf{u}^T \mathbf{X} v \tag{8.50}$$
$$\text{subject to } \|\mathbf{u}\|_1 \leq c_1, \ \|v\|_1 \leq c_2, \ \|\mathbf{u}\|_2 = 1, \ \|v\|_2 = 1,$$

and $d_1 = \mathbf{u}_1^T \mathbf{X} v_1$.

Ex. 8.6 In this exercise, we explore some properties of the SCoTLASS criterion (8.7).

(a) Use the Cauchy–Schwarz inequality to show that a fixed point of Algorithm 8.1 is a local minimum of the SCoTLASS criterion (8.7).

(b) Notice that the Cauchy–Schwarz inequality implies that

$$v^T \mathbf{X}^T \mathbf{X} v \geq \frac{(v^{(m)^T} \mathbf{X}^T \mathbf{X} v)^2}{v^{(m)^T} \mathbf{X}^T \mathbf{X} v^{(m)}}, \tag{8.51}$$

and equality holds when $v = v^{(m)}$. So $\frac{(v^{(m)^T} \mathbf{X}^T \mathbf{X} v)^2}{v^{(m)^T} \mathbf{X}^T \mathbf{X} v^{(m)}}$ minorizes $v^T \mathbf{X}^T \mathbf{X} v$ at $v^{(m)}$. Hence show that the MM algorithm (Section 5.8) using this minorization function yields Algorithm 8.1.

Ex. 8.7 Show that the solution to the problem (8.15), with the additional constraint $\|\theta\|_1 \leq t$, is also a solution to the SCoTLASS problem (8.7).

Ex. 8.8 Consider the problem

$$\underset{\theta: \|\theta\|_2=1}{\text{minimize}} \sum_{i=1}^{N} \|x_i - \theta z_i\|_2^2, \tag{8.52}$$

where the vectors $\{x_i\}_{i=1}^{N}$ and θ are all p-dimensional, and the variables $\{z_i\}_{i=1}^{N}$ are scalars. Show that the optimal solution is unique and given by $\widehat{\theta} = \frac{\mathbf{X}^T \mathbf{z}}{\|\mathbf{X}^T \mathbf{z}\|_2}$.

Ex. 8.9 Show that the vectors $\mathbf{u}_k$ in (8.17) solve the multifactor sparse PCA problem (8.16).

Ex. 8.10 Consider the reconstruction criterion (8.13) for sparse principal components.

(a) With $\mathbf{V}$ fixed, derive the solution for $\boldsymbol{\Theta}$.

(b) Show that when the $\lambda_{1k} = 0$ for all k the iterations are stationary w.r.t. any set of k principal components of $\mathbf{X}$; in particular, if the algorithm is started at the largest k principal components, it will not move from them.

(c) Show under the conditions in (b) that the criterion is maximized by $\boldsymbol{\Theta} = \mathbf{V}$ and both are equal to $\mathbf{V}_k$, the matrix consisting of the largest k principal components of $\mathbf{X}$.

(d) For a solution $\mathbf{V}_k$ in (c), show that $\mathbf{V}_k\mathbf{R}$ is also a solution, for any $k \times k$ orthogonal matrix $\mathbf{R}$.

Consequently, this version of sparse principal components is similar to the Varimax method (Kaiser 1958) of rotating factors to achieve sparsity.

Ex. 8.11 *Sparse K-means clustering algorithm.* Consider the objective function (8.46).

(a) Show that with w fixed, optimization with respect to $\mathcal{C} = (C_1, ..., C_K)$ yields the problem

$$\underset{C_1,...,C_K}{\text{minimize}}\left\{\sum_{j=1}^{p} w_j \left(\sum_{k=1}^{K} \frac{1}{N_k} \sum_{i,i' \in C_k} (x_{ij} - x_{i'j})^2\right)\right\}. \tag{8.53}$$

This can be thought of as K-means clustering with weighted data. Give a sketch of its solution.

(b) With $C_1, ..., C_K$ fixed, we optimize with respect to w yielding

$$\underset{w \in \mathbb{R}^p}{\text{maximize}}\left\{\sum_{j=1}^{p} w_j \left(\frac{1}{N}\sum_{i=1}^{N}\sum_{i'=1}^{N}(x_{ij} - x_{i'j})^2 - \sum_{k=1}^{K}\frac{1}{N_k}\sum_{i,i' \in C_K}(x_{ij} - x_{i'j})^2\right)\right\}$$

such that $\|w\|_2 = 1$, $\|w\|_1 \leq s$, and $w_j \geq 0$. $\tag{8.54}$

This is a simple convex problem of the form

$$\underset{w \in \mathbb{R}^p}{\text{maximize}}\left\{w^T a\right\} \text{ such that } \|w\|_2 = 1, \|w\|_1 \leq s, \text{ and } w \succeq 0. \tag{8.55}$$

Give the details of its solution.

Ex. 8.12 Consider the optimization problem:

$$\underset{\mathbf{A},\mathbf{B} \in \mathbb{R}^{p \times m}}{\text{minimize}}\left\{\sum_{i=1}^{N}\|x_i - \mathbf{A}\mathbf{B}^T x_i\|^2\right\}, \tag{8.56}$$

where $x_i \in \mathbb{R}^p$, $i = 1, ..., N$ are the rows of $\mathbf{X}$, and $m < \min(N, p)$. Show that the solution satisfies $\hat{\mathbf{A}}\hat{\mathbf{B}}^T = \mathbf{V}_m\mathbf{V}_m^T$, where $\mathbf{V}_m$ is the matrix of the first m right-singular vectors of $\mathbf{X}$.

Ex. 8.13 Consider the sparse encoder specified in the optimization problem (8.21). Develop a simple alternating algorithm for solving this problem. Give some details for each step.

Ex. 8.14 *Canonical correlation via alternating regression:* Consider two random vectors $X \in \mathbb{R}^{m_1}$ and $Y \in \mathbb{R}^{m_2}$ with covariance matrices Σ_{11} and Σ_{22}, respectively, and cross-covariance Σ_{12}. Define $\mathbf{L} = \Sigma_{11}^{-\frac{1}{2}} \Sigma_{12} \Sigma_{22}^{-\frac{1}{2}}$. Denote by γ_i and τ_i the left and right singular vectors of $\mathbf{L}$, ordered by their singular values ρ_i.

(a) Show that the vectors β_1 and θ_1 maximizing $\mathrm{Corr}(X\beta, Y\theta)$ are given by

$$\beta_1 = \Sigma_{11}^{-\frac{1}{2}} \gamma_1$$
$$\theta_1 = \Sigma_{22}^{-\frac{1}{2}} \tau_1, \tag{8.57}$$

and the maximal correlation is ρ_1.

(b) Now consider the analogous problem based on data matrices $\mathbf{X}$ and $\mathbf{Y}$ of dimension $N \times p$ and $N \times q$, respectively, each centered to have zero-mean columns. In this setting, the canonical correlation estimates are obtained simply by replacing Σ_{11}, Σ_{22}, and Σ_{12} by their sample estimates.
Based on this formulation, show that the optimal sample canonical vectors are given by $\beta_1 = (\mathbf{X}^T \mathbf{X})^{-\frac{1}{2}} \gamma_1$ and $\theta_1 = (\mathbf{Y}^T \mathbf{Y})^{-\frac{1}{2}} \tau_1$, where γ_1 and τ_1 are the leading left and right singular vectors of the matrix

$$(\mathbf{X}^T \mathbf{X})^{-\frac{1}{2}} \mathbf{X}^T \mathbf{Y} (\mathbf{Y}^T \mathbf{Y})^{-\frac{1}{2}}.$$

(c) Denote the first canonical variates by $\mathbf{z}_1 = \mathbf{X}\beta_1$ and $\mathbf{s}_1 = \mathbf{Y}\theta_1$, both N-vectors. Let $\mathbf{H}_X = \mathbf{X}(\mathbf{X}^T \mathbf{X})^{-1} \mathbf{X}^T$ be the projection onto the column space of $\mathbf{X}$; likewise $\mathbf{H}_Y$. Show that

$$\mathbf{H}_X \mathbf{s}_1 = \rho_1 \mathbf{z}_1, \quad \text{and}$$
$$\mathbf{H}_Y \mathbf{z}_1 = \rho_1 \mathbf{s}_1.$$

Consequently, alternately regressing onto $\mathbf{X}$ and $\mathbf{Y}$ until convergence yields a solution of the maximal canonical correlation problem.

Ex. 8.15 In Exercise 8.14, we found that the leading pair of canonical variates can be found by alternating least-squares regressions. Having solved for the leading canonical variates, how could you modify this procedure to produce the second pair of canonical variates $(\mathbf{z}_2, \mathbf{s}_2)$? Propose a modification and prove that it works. Show how to extend this approach to find all subsequent pairs.

Ex. 8.16 *CCA via optimal scoring:* Given data matrices $\mathbf{X}$ and $\mathbf{Y}$ as in Exercise 8.14, both with mean-centered columns and both with full column rank, consider the problem

$$\underset{\beta, \theta}{\text{minimize}} \, \|\mathbf{Y}\theta - \mathbf{X}\beta\|_2^2 \text{ subject to } \frac{1}{N} \|\mathbf{Y}\theta\|_2 = 1. \tag{8.58}$$

(a) Characterize the solution to this problem, by first solving for β with θ fixed, and then solving for θ.

(b) Show that the optimal solution is given by $\widehat{\theta} = \theta_1$ and $\widehat{\beta} = \rho_1\beta_1$, where β_1 and θ_1 are the first pair of canonical vectors, and ρ_1 the largest canonical correlation.

(c) Show in addition that $\|\mathbf{Y}\theta_1 - \mathbf{X}\beta_1\|_2^2 = 1 - \rho_1^2$. This equivalence shows that solving the optimal scoring problem is equivalent to solving the CCA problem.

(d) Describe how to find subsequent canonical solutions, uncorrelated with the earlier solutions. Show how this can be achieved by transforming the data matrices $\mathbf{X}$ and $\mathbf{Y}$.

(e) Does the problem change if we include a constraint $\|\mathbf{X}\beta\|_2 = 1$?

Ex. 8.17 *Low-rank CCA:* Suppose that at least one of the matrices in Exercise 8.14 or 8.16 do not have full column rank. (For instance, this degeneracy will occur whenever $N < \min(p, q)$.)

(a) Show that $\rho_1 = 1$, and the CCA problem has multiple optima.

(b) Suppose that $\mathbf{Y}$ is full column rank, but $\mathbf{X}$ is not. You add a ridge constraint to (8.58), and solve

$$\underset{\beta,\theta}{\text{minimize}} \, \|\mathbf{Y}\theta - \mathbf{X}\beta\|_2^2 + \lambda\|\beta\|_2^2 \text{ subject to } \|\mathbf{Y}\theta\|_2 = 1. \qquad (8.59)$$

Is this problem degenerate? Characterize the solution.

(c) Show that the solution in (b) is equivalent to applying CCA to $\mathbf{X}$ and $\mathbf{Y}$, except that the optimal solution $\widehat{\beta}$ satisfies the normalization condition $\frac{1}{N}\widehat{\beta}^T(\mathbf{X}^T\mathbf{X} + \lambda\mathbf{I})\widehat{\beta} = 1$, corresponding to normalization by a type of ridged covariance estimate.

Ex. 8.18 For data matrices $\mathbf{X}$ and $\mathbf{Y}$ with centered columns, consider the optimization problem

$$\underset{\beta,\theta}{\text{maximize}} \left\{ \widehat{\text{Cov}}(\mathbf{X}\beta, \mathbf{Y}\theta) - \lambda_1\|\beta\|_2^2 - \lambda_2\|\beta\|_2^2 - \lambda_1'\|\theta\|_1 - \lambda_2'\|\theta\|_2 \right\}. \qquad (8.60)$$

Using the results of Exercise 8.14, outline how to solve this problem using alternating elastic-net fitting operations in place of the least-squares regressions.

Ex. 8.19 *Sparse canonical correlation analysis:* Consider the optimal scoring problem (8.58) from Exercise 8.16, but augmented with ℓ_1 constraints:

$$\underset{\beta,\theta}{\text{minimize}} \left\{ \|\mathbf{Y}\theta - \mathbf{X}\beta\|_2^2 + \lambda_1\|\beta\|_1 + \lambda_2\|\theta\|_1 \right\} \text{ subject to } \|\mathbf{X}\beta\|_2 = 1. \qquad (8.61)$$

Using the results of Exercises 8.14, 8.16, and 8.17, outline how to solve this problem using alternating elastic-net regressions in place of the least-squares regressions.

Ex. 8.20 Consider the multivariate Gaussian setup in Section 8.4.1, but assume a different covariance matrix $\mathbf{\Sigma}_k$ in each class.

(a) Show that the discriminant functions δ_k are quadratic functions of x. What can you say about the decision boundaries?

(b) Suppose that the covariance matrices $\mathbf{\Sigma}_k$ are assumed to be diagonal, meaning that the features X are conditionally independent in each class. Describe the decision boundary between class k and ℓ for this *naive Bayes classifier*.

Ex. 8.21 This exercise relates to the nearest shrunken centroids problem of Section 8.4.2. Consider the ℓ_1-regularized criterion

$$\underset{\substack{\bar{\mu}\in\mathbb{R}^p \\ \alpha_k\in\mathbb{R}^p, k=1,\ldots,p}}{\text{minimize}} \left\{ \frac{1}{2}\sum_{k=1}^{K}\sum_{i\in C_k}\sum_{j=1}^{p}\frac{(x_{ij}-\bar{\mu}_j-\alpha_{jk})^2}{s_j^2} + \lambda\sum_{k=1}^{K}\sum_{j=1}^{p}\frac{\sqrt{N_k}}{s_j}|\alpha_{jk}| \right\}.$$

$$(8.62)$$

Here we have decomposed each class mean into an overall mean plus a class-wise contrast from the overall mean, and we have weighted the penalties by the class sizes and within-class standard deviations for each feature.

(a) Show that replacing $\bar{\mu}_j$ by $\bar{x}_j$, corresponding to the overall mean for feature j, yields the shrinkage scheme (8.35a), apart from a term $1/N$ in m_k.

(b) Show that part (a) does not yield a solution to the criterion (8.33) unless we restrict $\bar{\mu}_j$ as above.

(c) A more natural criterion would add the constraints $\sum_{k=1}^{K}\alpha_{jk}=0$ for all $j=1,\ldots p$. Discuss the solution to this problem, and whether it can coincide with the solution from part (a).

Ex. 8.22 Show that the penalized Fisher's discriminant problem (8.40) and the penalized optimal scoring problem (8.42) are equivalent in the sense that any stationary point for one problem is also a stationary point for the other problem. (Clemmensen et al. 2011, Witten and Tibshirani 2011).

Chapter 9

Graphs and Model Selection

9.1   Introduction

Probabilistic graphical models provide a useful framework for building parsimonious models for high-dimensional data. They are based on an interplay between probability theory and graph theory, in which the properties of an underlying graph specify the conditional independence properties of a set of random variables. In typical applications, the structure of this graph is not known, and it is of interest to estimate it based on samples, a problem known as graphical model selection. In this chapter, we discuss a variety of methods based on ℓ_1-regularization designed for this purpose.

9.2   Basics of Graphical Models

We begin by providing a brief introduction to the basics of graphical models; for more details, we refer the reader to the references cited in the bibliographic notes at the end of the chapter. Any collection $X = (X_1, X_2, \ldots, X_p)$ of random variables can be associated with the vertex set $V = \{1, 2, \ldots, p\}$ of some underlying graph. The essential idea of a graphical model is to use the structure of the underlying graph—either its clique structure or its cut set structure—in order to constrain the distribution of the random vector X. We now make these notions more precise.

9.2.1   Factorization and Markov Properties

An ordinary graph G consists of a vertex set $V = \{1, 2, \ldots, p\}$, and an edge set $E \subset V \times V$. In this chapter, we focus exclusively on undirected graphical models, meaning that there is no distinction between an edge $(s, t) \in E$, and the edge (t, s). In contrast, *directed acyclic graphs* (DAGs) are the most popular form of graph in which the edges have directionality. In general, such directed graphs are more difficult to handle than undirected graphs, and we do not cover them here. But we do note that some methods for computation in undirected graphs can be helpful in the DAG case: see the bibliographic notes for some references.

A *graph clique* $C \subseteq V$ is a fully-connected subset of the vertex set, meaning that $(s,t) \in E$ for all $s, t \in C$. A clique is said to be maximal if it is not strictly contained within any other clique. For instance, any single vertex $\{s\}$ is itself a clique, but it is not maximal unless s is an isolated vertex (meaning that it participates in no edges). We use $\mathfrak{C}$ to denote the set of all cliques in the graph, both maximal and nonmaximal; see Figure 9.1(a) for an illustration of graph cliques.

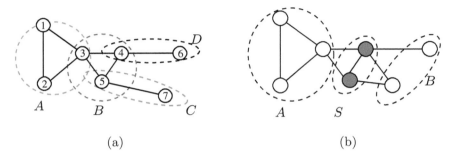

(a)   (b)

Figure 9.1 *(a) Illustration of cliques in a graph: each of the four subsets indicated are cliques. Sets A and B are 3-cliques, whereas C and D are 2-cliques, more commonly known as edges. All of these cliques are maximal. (b) Illustration of a vertex cut set S: with the vertices in S removed, the graph is broken into two subcomponents A and B.*

9.2.1.1 Factorization Property

We now describe how the clique structure of a graph can be used to constrain the probability distribution of the random vector $(X_1, \ldots, X_p)$ indexed by the graph vertices. For a given clique $C \in \mathfrak{C}$, a *compatibility function* ψ_C is a real-valued function of the subvector $x_C := (x_s, s \in C)$, taking positive real values. Given a collection of such compatibility functions, we say that the probability distribution $\mathbb{P}$ *factorizes over G* if it has the decomposition

$$\mathbb{P}(x_1, \ldots, x_p) = \frac{1}{Z} \prod_{C \in \mathfrak{C}} \psi_C(x_C). \tag{9.1}$$

Here the quantity Z, known as the partition function, is given by the sum $Z = \sum_{x \in \mathcal{X}^p} \prod_{C \in \mathfrak{C}} \psi_C(x_C)$. Thus, it ensures that $\mathbb{P}$ is properly normalized, and so defines a valid probability distribution. As a particular example, any probability distribution that factorizes over the graph in Figure 9.1(a) must have the form

$$\mathbb{P}(x_1, \ldots, x_7) = \frac{1}{Z} \psi_{123}(x_1, x_2, x_3)\, \psi_{345}(x_3, x_4, x_5)\, \psi_{46}(x_4, x_6)\, \psi_{57}(x_5, x_7),$$
$$\tag{9.2}$$

for some choice of the compatibility functions $\{\psi_{123}, \psi_{345}, \psi_{46}, \psi_{57}\}$.

A factorization of the form (9.1) is practically significant, since it can lead to substantial savings, in both storage and computation, if the clique sizes are not too large. For instance, if each variable X_s is binary, then a generic probability distribution over the vector $X \in \{-1, +1\}^p$ requires specifying $2^p - 1$ nonnegative numbers, and so grows exponentially in the graph size. On the other hand, for a clique-based factorization, the number of degrees of freedom is at most $|\mathfrak{C}| 2^c$, where c is the maximum cardinality of any clique. Thus, for the clique-based factorization, the complexity grows exponentially in the maximum clique size c, but only linearly in the number of cliques $|\mathfrak{C}|$. Luckily, many practical models of interest can be specified in terms of cliques with bounded size, in which case the clique-based representation yields substantial gains.

9.2.1.2   *Markov Property*

We now turn to an alternative way in which the graph structure can be used to constrain the distribution of X, based on its cut sets (see Figure 9.1(b)). In particular, consider a cut set S that separates the graph into disconnected components A and B, and let us introduce the symbol $\perp\!\!\!\perp$ to denote the relation *"is conditionally independent of."* With this notation, we say that the random vector X is *Markov with respect to G* if

$$X_A \perp\!\!\!\perp X_B \mid X_S \qquad\qquad \text{for all cut sets } S \subset V. \qquad (9.3)$$

The graph in Figure 9.1(b) is an example showing this conditional independence relation.

Markov chains provide a particular illustration of this property; naturally, they are based on a chain-structured graph, with edge set

$$E = \{(1, 2), (2, 3), \dots, (p - 1, p)\}.$$

In this graph, any single vertex $s \in \{2, 3, \dots, p-1\}$ forms a cut set, separating the graph into the past $P = \{1, \dots, s - 1\}$ and the future $F = \{s + 1, \dots, p\}$. For these cut sets, the Markov property (9.3) translates into the fact that, for a Markov chain, the future X_F is conditionally independent of the past X_P given the present X_s. Of course, graphs with more structure have correspondingly more complex cut sets, and thus more interesting conditional-independence properties.

9.2.1.3   *Equivalence of Factorization and Markov Properties*

A remarkable fact, known as the Hammersley–Clifford theorem, is that for any strictly positive distribution (i.e., for which $\mathbb{P}(x) > 0$ for all $x \in \mathcal{X}^p$), the two characterizations are equivalent: namely, the distribution of X factorizes according to the graph G (as in Equation (9.1)) if and only if the random vector X is Markov with respect to the graph (as in Equation (9.3)). See the bibliographic section for further discussion of this celebrated theorem.

9.2.2   Some Examples

We present some examples to provide a concrete illustration of these properties.

9.2.2.1   Discrete Graphical Models

We begin by discussing the case of a discrete graphical model, in which random variables X_s at each vertex $s \in V$ take values in a discrete space $\mathcal{X}_s$. The simplest example is the binary case, say with $\mathcal{X}_s = \{-1, +1\}$. Given a graph $G = (V, E)$, one might consider the family of probability distributions

$$\mathbb{P}_\theta(x_1, \ldots, x_p) = \exp\big\{ \sum_{s \in V} \theta_s x_s + \sum_{(s,t) \in E} \theta_{st} x_s x_t - A(\theta) \big\}, \qquad (9.4)$$

parametrized by the vector $\theta \in \mathbb{R}^{|V|+|E|}$. For later convenience, here we have introduced the notation $A(\theta) = \log Z(\theta)$, reflecting the dependence of the normalization constant on the parameter vector θ. This family of distributions is known as the *Ising model*, since it was originally used by Ising (1925) to model the behavior of magnetic materials; see the bibliographic section for further discussion. Figure 9.2 shows simulations from three different Ising models.

       (a)   (b)   (c)

Figure 9.2 *Samples generated from Ising models based on a graph with $p = 1024$ nodes. For illustrative purposes, the resulting vector $x \in \{+1, 1\}^{1024}$ is plotted as a 32×32 binary image. Panels (a) through (c) correspond to three very different distributions. The samples were drawn by running the Gibbs sampler.*

The Ising model has been used to model social networks, for example the voting behavior of politicians. In this context, the random vector $(X_1, X_2, \ldots, X_p)$ represents the set of votes cast by a set of p politicians on a particular bill. We assume that politician s provides either a "yes" vote $(X_s = +1)$ or a "no" vote $(X_s = -1)$ on the bill. With the voting results for N bills, we can make inferences on the joint distribution of X. In the factorization (9.4), a parameter $\theta_s > 0$ indicates that politician s is more likely (assuming fixed values of other politicians' votes) to vote "yes" on any given bill, with the opposite interpretation holding in the case $\theta_s < 0$. On the other hand, for any given pair s, t that are joined by an edge, a weight $\theta_{st} > 0$ means that with the behavior of all politicians held fixed, politicians s and t are more likely to share the same vote (i.e., both yes or both no) than to disagree; again, the opposite interpretation applies to the setting $\theta_{st} < 0$. See Figure 9.7 on page 257 for an application to voting-record data.

Many extensions of the Ising model are possible. First, the factorization (9.4) is limited to cliques of size at most two (i.e., edges). By allowing terms over cliques of size up to size three, one obtains the family of models

$$\mathbb{P}_\theta(x) = \exp\Big\{ \sum_{s \in V} \theta_s x_s + \sum_{(s,t) \in E} \theta_{st} x_s x_t + \sum_{(s,t,u) \in E_3} \theta_{stu} x_s x_t x_u - A(\theta) \Big\}.$$

$$(9.5)$$

where E_3 is some subset of vertex triples. This factorization can be extended up to subsets of higher order, and in the limit (where we allow an interaction term among all p variables simultaneously), it is possible to specify the distribution of any binary vector. In practice, of most interest are models based on relatively local interactions, as opposed to such a global interaction.

Another extension of the Ising model is to allow for non-binary variables, for instance $X_s \in \{0, 1, 2, \ldots, m - 1\}$ for some $m > 2$. In this case, one might consider the family of distributions

$$\mathbb{P}_\theta(x_1, \ldots, x_p) = \exp\Big\{ \sum_{s \in V} \sum_{j=1}^{m-1} \theta_{s;j} \mathbb{I}[x_s = j] + \sum_{(s,t) \in E} \theta_{st} \mathbb{I}[x_s = x_t] - A(\theta) \Big\},$$

$$(9.6)$$

where the indicator function $\mathbb{I}[x_s = j]$ takes the value 1 when $x_s = j$, and 0 otherwise. When the weight $\theta_{st} > 0$, the edge-based indicator function $\mathbb{I}[x_s = x_t]$ acts as a smoothness prior, assigning higher weight to pairs (x_s, x_t) that agree. The model (9.6) has found many uses in computer vision, for instance in image denoising and disparity computation problems.

All the models discussed here have a parallel life in the statistics and biostatistics literature, where they are referred to as log-linear models for multiway tables. However, in that setting the number of variables is typically quite small.

In Section 9.4.3 we discuss a general class of pairwise-Markov models for mixed data—allowing for both continuous and discrete variables.

9.2.2.2   Gaussian Graphical Models

Let $X \sim N(\mu, \boldsymbol{\Sigma})$ be a Gaussian distribution in p dimensions, with mean vector $\mu \in \mathbb{R}^p$ and covariance matrix $\boldsymbol{\Sigma}$:

$$\mathbb{P}_{\mu, \boldsymbol{\Sigma}}(x) = \frac{1}{(2\pi)^{\frac{p}{2}} \det[\boldsymbol{\Sigma}]^{\frac{1}{2}}} e^{-\frac{1}{2}(x-\mu)^T \boldsymbol{\Sigma}^{-1}(x-\mu)}.$$

$$(9.7)$$

If we view the multivariate Gaussian distribution as a particular type of exponential family, then (μ, Σ) are known as the mean *parameters* of the family. In order to represent the multivariate Gaussian as a graphical model, it is convenient to consider instead its parametrization in terms of the so-called canonical parameters, say a vector $\gamma \in \mathbb{R}^p$ and $\Theta \in \mathbb{R}^{p \times p}$. Any nondegenerate multivariate Gaussian—meaning whenever Σ is strictly positive definite—can be represented in the form

$$\mathbb{P}_{\gamma,\Theta}(x) = \exp\left\{ \sum_{s=1}^{p} \gamma_s x_s - \frac{1}{2} \sum_{s,t=1}^{p} \theta_{st}\, x_s x_t - A(\Theta) \right\}, \tag{9.8}$$

where $A(\Theta) = -\frac{1}{2}\log\det[\Theta/(2\pi)]$, so that $\int \mathbb{P}_{\gamma,\Theta}(x)dx = 1$. Our choice of the rescaling by $-1/2$ in the factorization (9.8) is to ensure that the matrix Θ has a concrete interpretation. In particular, with this scaling, as shown in Exercise 9.1, we have the relation $\Theta = \Sigma^{-1}$, so that Θ corresponds to the *inverse covariance, precision or concentration matrix*.

The representation (9.8) is especially convenient, because it allows us to discuss the factorization properties directly in terms of the sparsity pattern of the precision matrix Θ. In particular, whenever X factorizes according to the graph G, then based on the factorization (9.8), we must have $\Theta_{st} = 0$ for

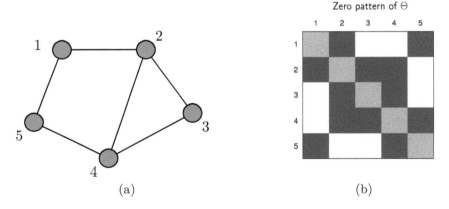

Figure 9.3 *(a) An undirected graph G on five vertices. (b) Associated sparsity pattern of the precision matrix Θ. White squares correspond to zero entries.*

any pair $(s, t) \notin E$, which sets up a correspondence between the zero pattern of Θ and the edge structure E of the underlying graph. See Figure 9.3 for an illustration of this correspondence.

9.3   Graph Selection via Penalized Likelihood

We now turn to the problem of graph selection, and the use of ℓ_1-regularized likelihood methods for solving it. The problem itself is simply stated: suppose that we are given a collection $\mathbf{X} = \{x_1, \ldots, x_N\}$ of samples from a graphical model, but the underlying graph structure is unknown. How to use the data to select the correct graph with high probability? Here we discuss the use of likelihood-based methods in conjunction with ℓ_1-regularization for this purpose. This section discusses methods based on the global likelihood function of the graphical model. In the Gaussian case, this approach leads to tractable methods for model selection based on a log-determinant convex program with ℓ_1-regularization. On the other hand, in the discrete case, this approach is computationally tractable only for relatively small graphs, or graphs with special structure.

9.3.1 Global Likelihoods for Gaussian Models

We start with model selection for the Gaussian graphical model, a problem that is also known as covariance selection. Since our primary interest is in estimating the graph structure, we assume that the distribution has mean zero, so that under the parametrization (9.8), we need only consider the symmetric precision matrix $\mathbf{\Theta} \in \mathbb{R}^{p \times p}$.

Suppose $\mathbf{X}$ represents samples from a zero-mean multivariate Gaussian with precision matrix $\mathbf{\Theta}$. Based on some straightforward algebra (see Exercise 9.2 for details), it can be seen that (up to a constant) the rescaled log-likelihood $\mathcal{L}(\mathbf{\Theta}; \mathbf{X})$ of the multivariate Gaussian takes the form

$$\begin{aligned} \mathcal{L}(\mathbf{\Theta}; \mathbf{X}) &= \frac{1}{N} \sum_{i=1}^{N} \log \mathbb{P}_{\mathbf{\Theta}}(x_i) \\ &= \log \det \mathbf{\Theta} - \text{trace}(\mathbf{S}\mathbf{\Theta}), \end{aligned} \tag{9.9}$$

where $\mathbf{S} = \frac{1}{N} \sum_{i=1}^{N} x_i x_i^T$ is the empirical covariance matrix. The log-determinant function is defined on the space of symmetric matrices as

$$\log \det(\mathbf{\Theta}) = \begin{cases} \sum_{j=1}^{p} \log(\lambda_j(\mathbf{\Theta})) & \text{if } \mathbf{\Theta} \succ \mathbf{0} \\ -\infty; & \text{otherwise,} \end{cases} \tag{9.10}$$

where $\lambda_j(\mathbf{\Theta})$ is the jth eigenvalue of $\mathbf{\Theta}$. In Exercise 9.2, we explore some additional properties of this function. The objective function (9.9) is an instance of a log-determinant program, a well-studied class of optimization problems. It is strictly concave, so that the maximum—when it is achieved—must be unique, and defines the maximum likelihood estimate $\widehat{\mathbf{\Theta}}_{\text{ML}}$, denoted MLE for short.

By classical theory, the maximum likelihood estimate $\widehat{\mathbf{\Theta}}_{\text{ML}}$ converges to the true precision matrix as the sample size N tends to infinity. Thus, at least in principle, one could use a thresholded version of $\widehat{\mathbf{\Theta}}_{\text{ML}}$ to specify an edge set, and thereby perform Gaussian graphical model selection. However, for problems frequently arising in practice, the number of nodes p may be comparable to or larger than the sample size N, in which case the MLE does not exist.

Indeed, the empirical correlation matrix $\mathbf{S}$ must be rank-degenerate whenever $N < p$, which implies that the MLE fails to exist (see Exercise 9.2(c)). Hence we must consider suitably constrained or regularized forms of the MLE. Moreover, irrespective of the sample size, we may be interested in constraining the estimated precision matrix to be sparse, and hence easier to interpret.

If we are seeking Gaussian graphical models based on relatively sparse graphs, then it is desirable to control the number of edges, which can be measured by the ℓ_0-based quantity

$$\rho_0(\mathbf{\Theta}) = \sum_{s \neq t} \mathbb{I}[\theta_{st} \neq 0], \tag{9.11}$$

where $\mathbb{I}[\theta_{st} \neq 0]$ is a 0-1-valued indicator function. Note that by construction, we have $\rho_0(\mathbf{\Theta}) = 2|E|$, where $|E|$ is the number of edges in the graph uniquely defined by $\mathbf{\Theta}$. We could then consider the optimization problem

$$\widehat{\mathbf{\Theta}} \in \underset{\substack{\mathbf{\Theta} \succeq 0 \\ \rho_0(\mathbf{\Theta}) \leq k}}{\arg\max} \left\{ \log\det(\mathbf{\Theta}) - \operatorname{trace}(\mathbf{S}\mathbf{\Theta}) \right\}. \tag{9.12}$$

Unfortunately, the ℓ_0-based constraint defines a highly nonconvex constraint set, essentially formed as the union over all $\binom{\binom{p}{2}}{k}$ possible subsets of k edges.

For this reason, it is natural to consider the convex relaxation obtained by replacing the ℓ_0 constraint with the corresponding ℓ_1-based constraint. Doing so leads to the following convex program

$$\widehat{\mathbf{\Theta}} \in \underset{\mathbf{\Theta} \succeq 0}{\arg\max} \left\{ \log\det\mathbf{\Theta} - \operatorname{trace}(\mathbf{S}\mathbf{\Theta}) - \lambda\rho_1(\mathbf{\Theta}) \right\}, \tag{9.13}$$

where $\rho_1(\mathbf{\Theta}) = \sum_{s \neq t} |\theta_{st}|$ is simply the ℓ_1-norm of the off-diagonal entries of $\mathbf{\Theta}$. The problem (9.13) can be formulated as an instance of a log-determinant program; it is thus a convex program, often referred to as the *graphical lasso*.

Since this is a convex program, one can use generic interior point methods for its solution, as in Vandenberghe, Boyd and Wu (1998). However this is not efficient for large problems. More natural are first-order block coordinate-descent approaches, introduced by d'Aspremont, Banerjee and El Ghaoui (2008) and refined by Friedman, Hastie and Tibshirani (2008). The latter authors call this the *graphical lasso* algorithm. It has a simple form which also connects it to the neighborhood-based regression approach, discussed in Section 9.4.

9.3.2   Graphical Lasso Algorithm

The subgradient equation corresponding to (9.13) is

$$\mathbf{\Theta}^{-1} - \mathbf{S} - \lambda \cdot \mathbf{\Psi} = 0, \tag{9.14}$$

with the symmetric matrix $\mathbf{\Psi}$ having diagonal elements zero, $\psi_{jk} = \operatorname{sign}(\theta_{jk})$ if $\theta_{jk} \neq 0$, else $\psi_{jk} \in [-1, 1]$ if $\theta_{jk} = 0$.

We now consider solving this system by blockwise coordinate descent. To this end we consider partitioning all the matrices into one column versus the rest; for convenience we pick the last column:

$$\mathbf{\Theta} = \begin{bmatrix} \mathbf{\Theta}_{11} & \boldsymbol{\theta}_{12} \\ \boldsymbol{\theta}_{12}^T & \theta_{22} \end{bmatrix}, \quad \mathbf{S} = \begin{bmatrix} \mathbf{S}_{11} & \mathbf{s}_{12} \\ \mathbf{s}_{12}^T & s_{22} \end{bmatrix}, \text{ etc.} \tag{9.15}$$

Denote by $\mathbf{W}$ the current working version of $\mathbf{\Theta}^{-1}$, with partitions as in (9.15). Then fixing all but the last row and column and using partitioned inverses, (9.14) leads to

$$\mathbf{W}_{11}\boldsymbol{\beta} - \mathbf{s}_{12} + \lambda \cdot \boldsymbol{\psi}_{12} = \mathbf{0}, \tag{9.16}$$

where $\boldsymbol{\beta} = -\boldsymbol{\theta}_{12}/\theta_{22}$. Here we have fixed the pth row and column: $\mathbf{W}_{11}$ is the $(p-1) \times (p-1)$ block of $\mathbf{\Theta}^{-1}$, and $\mathbf{s}_{12}$ and $\boldsymbol{\theta}_{12}$ are $p-1$ nondiagonal elements of the pth row and columns of $\mathbf{S}$ and $\mathbf{\Theta}$. Finally θ_{22} is the pth diagonal element of $\mathbf{\Theta}$. These details are derived in Exercise 9.6.[1]

It can be seen that (9.16) is equivalent to a modified version of the estimating equations for a lasso regression. Consider the usual regression setup with outcome variables $\mathbf{y}$ and predictor matrix $\mathbf{Z}$. In that problem, the lasso minimizes

$$\frac{1}{2N}(\mathbf{y} - \mathbf{Z}\boldsymbol{\beta})^T(\mathbf{y} - \mathbf{Z}\boldsymbol{\beta}) + \lambda \cdot \|\boldsymbol{\beta}\|_1 \tag{9.17}$$

The subgradient equations are

$$\frac{1}{N}\mathbf{Z}^T\mathbf{Z}\boldsymbol{\beta} - \frac{1}{N}\mathbf{Z}^T\mathbf{y} + \lambda \cdot \text{sign}(\boldsymbol{\beta}) = 0. \tag{9.18}$$

Comparing to (9.16), we see that $\frac{1}{N}\mathbf{Z}^T\mathbf{y}$ is the analog of $\mathbf{s}_{12}$, and $\frac{1}{N}\mathbf{Z}^T\mathbf{Z}$ corresponds to $\mathbf{W}_{11}$, the estimated cross-product matrix from our current model. Thus we can solve each blockwise step (9.16) using a modified algorithm for the lasso, treating each variable as the response variable and the other $p-1$ variables as the predictors. It is summarized in Algorithm 9.1.

Friedman et al. (2008) use the pathwise-coordinate-descent approach to solve the modified lasso problems at each stage, and for a decreasing series of values for λ. [This corresponds to the "covariance" version of their lasso algorithm, as implemented in the glmnet package in R and matlab (Friedman et al. 2010b).]

From Equation (9.14) we see that the diagonal elements w_{jj} of the solution matrix $\mathbf{W}$ are simply s_{jj}, and these are fixed in Step 1 of Algorithm 9.1.[2]

---

[1]On a historical note, it turns out that this algorithm is *not* block-coordinate descent on $\mathbf{\Theta}$ in (9.14) (as originally intended in Friedman et al. (2008)), but instead amounts to a block coordinate-descent step for the convex dual of problem (9.13). This is implied in Banerjee, El Ghaoui and d'Aspremont (2008), and is detailed in Mazumder and Hastie (2012). The dual variable is effectively $\mathbf{W} = \mathbf{\Theta}^{-1}$. These latter authors derive alternative coordinate descent algorithms for the primal problem (see Exercise 9.7). In some cases, this gives better numerical performance than the original graphical lasso procedure.

[2]An alternative formulation of the problem can be posed, where we penalize the diagonal of $\mathbf{\Theta}$ as well as the off-diagonal terms. Then the diagonal elements w_{jj} of the solution matrix are $s_{jj} + \lambda$, and the rest of the algorithm is unchanged.

Algorithm 9.1 GRAPHICAL LASSO.

---

1. Initialize $\mathbf{W} = \mathbf{S}$. Note that the diagonal of $\mathbf{W}$ is unchanged in what follows.

2. Repeat for $j = 1, 2, \ldots p, 1, 2, \ldots p, \ldots$ until convergence:

    (a) Partition the matrix $\mathbf{W}$ into part 1: all but the jth row and column, and part 2: the jth row and column.

    (b) Solve the estimating equations $\mathbf{W}_{11}\boldsymbol{\beta} - \mathbf{s}_{12} + \lambda \cdot \text{sign}(\boldsymbol{\beta}) = 0$ using a cyclical coordinate-descent algorithm for the modified lasso.

    (c) Update $\mathbf{w}_{12} = \mathbf{W}_{11}\hat{\boldsymbol{\beta}}$

3. In the final cycle (for each j) solve for $\hat{\boldsymbol{\theta}}_{12} = -\hat{\boldsymbol{\beta}} \cdot \hat{\theta}_{22}$, with $1/\hat{\theta}_{22} = w_{22} - \mathbf{w}_{12}^T\hat{\boldsymbol{\beta}}$.

The graphical lasso algorithm is fast, and can solve a moderately sparse problem with 1000 nodes in less than a minute. It is easy to modify the algorithm to have edge-specific penalty parameters λ_{jk}; note also that $\lambda_{jk} = \infty$ will force $\hat{\theta}_{jk}$ to be zero.

Figure 9.4 illustrates the path algorithm using a simple example. We generated 20 observations from the model of Figure 9.3, with

$$
\boldsymbol{\Theta} = \begin{bmatrix}
2 & 0.6 & 0 & 0 & 0.5 \\
0.6 & 2 & -0.4 & 0.3 & 0 \\
0 & -0.4 & 2 & -0.2 & 0 \\
0 & 0.3 & -0.2 & 2 & 0 \\
0.5 & 0 & 0 & -0.2 & 2
\end{bmatrix}. \tag{9.19}
$$

Shown are the graphical lasso estimates for a range of λ values, plotted against the ℓ_1 norm of the solution $\hat{\boldsymbol{\Theta}}_\lambda$. The true values of $\boldsymbol{\Theta}$ are indicated on the right. The solutions are spot on, but this is not surprising since the solution on the right (no regularization) is $\mathbf{S}^{-1}$ which equals $\boldsymbol{\Theta}$ (Exercise 9.3). In the right panel, we have added independent Gaussian noise to each element of the data matrix (with standard deviation 0.05). Now we see that the estimate is not nearly as accurate; in fact, the nonzero support is never recovered correctly along the path.

Some further points about the Gaussian graphical model:

- A simpler approach would be to use the observed covariance $\mathbf{S}_{11}$ in place of $\mathbf{W}_{11}$: this requires just a single pass through the predictors, carrying out a lasso regression of each variable X_j on the others. This is called *neighborhood selection* (Meinshausen and Bühlmann 2006). Like the graphical lasso algorithm, this yields a consistent estimate of the support of $\boldsymbol{\Theta}$, but is not guaranteed to produce a positive definite estimate $\hat{\boldsymbol{\Theta}}$. We discuss this in detail in Section 9.4.

- If we pre-specify the zero pattern in $\boldsymbol{\Theta}$, we can use standard linear regression in place of the lasso, leaving out the predictors that are supposed to

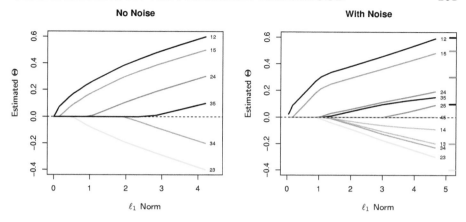

Figure 9.4 Left: *Profiles of estimates from the graphical lasso, for data simulated from the model of Figure 9.3. The actual values of Θ are achieved at the right end of the plot. Right: Same setup, except that standard Gaussian noise has been added to each column of the data. Nowhere along the path is the true edge set recovered.*

have coefficients of zero. This provides a convenient way of computing the constrained maximum likelihood estimate of Θ. Details are in Chapter 17 of Hastie et al. (2009).

9.3.3 Exploiting Block-Diagonal Structure

If the inverse covariance matrix has a block diagonal structure

$$
\Theta = \begin{bmatrix}
\Theta_{11} & 0 & \cdots & 0 \\
0 & \Theta_{22} & \cdots & 0 \\
\vdots & \vdots & \ddots & \vdots \\
0 & 0 & \cdots & \Theta_{kk}
\end{bmatrix},
\tag{9.20}
$$

for some ordering of the variables, then the graphical lasso problem can be solved separately for each block, and the solution is constructed from the individual solutions. This fact follows directly from the subgradient equations (9.14).

It turns out that there is a very simple necessary and sufficient condition for a graphical-lasso solution to have such structure (Witten, Friedman and Simon 2011, Mazumder and Hastie 2012). Let $C_1, C_2, \ldots C_K$ be a partition of the indices $1, 2, \ldots, p$ of $\mathbf{S}$ into K blocks. Then the corresponding arrangement of $\hat{\Theta}$ has this same block structure if and only if $|s_{ii'}| \leq \lambda$ for all pairs (i, i') not belonging to the same block. The proof is easy, by inspection of (9.14) and using the fact that the inverse of a block diagonal matrix has the same block-diagonal structure. This means that the elements of each block C_k are fully disconnected from elements of all other blocks.

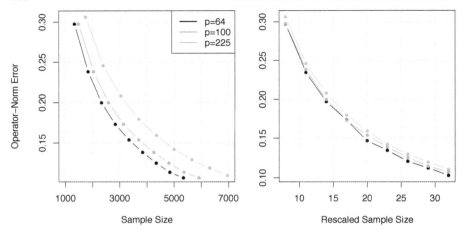

Figure 9.5 *Plots of the operator-norm error* $\|\widehat{\Theta} - \Theta^*\|_2$ *between the graphical lasso estimate* $\widehat{\Theta}$ *and the true inverse covariance matrix. Left: plotted versus the raw sample size* N*, for three different graph sizes* $p \in \{64, 100, 225\}$*. Note how the curves shift to the right as the graph size* p *increases, reflecting the fact that larger graphs require more samples for consistent estimation. Right: the same operator-norm error curves plotted versus the* rescaled sample size $\frac{N}{d^2 \log p}$ *for three different graph sizes* $p \in \{64, 100, 225\}$*. As predicted by theory, the curves are now quite well-aligned.*

This fact can be exploited to provide a substantial speedup in computation, by first identifying the disconnected components and then solving the subproblems in each block. The number of blocks is monotone in λ. This means, for example, that solutions can be found for very large problems (that perhaps could not be solved in general), as long as λ is sufficiently large.

9.3.4   Theoretical Guarantees for the Graphical Lasso

In order to explore the behavior of graphical lasso or log-determinant method, we carried out a series of simulations. For a given graph type with p nodes and specified covariance matrix, we generated a set of N zero-mean multivariate Gaussian samples, and used them to form the empirical covariance matrix $\mathbf{S}$. We solved the graphical lasso program using the regularization parameter $\lambda_N = 2\sqrt{\frac{\log p}{N}}$ (Ravikumar, Wainwright, Raskutti and Yu 2011, as suggested by theory) and plot the operator norm error $\|\widehat{\Theta} - \Theta^*\|_2$ versus the sample size N. Figure 9.5 (left) shows such plots for three different dimensions $p \in \{64, 100, 225\}$ of two-dimensional grid graphs, in which each node has degree four. For each graph size, we generated an inverse covariance matrix $\Theta^* \in$

$\mathbb{R}^{p \times p}$ with entries

$$\theta^*_{st} = \begin{cases} 1 & \text{if } s = t, \\ 0.2 & \text{if } |s - t| = 1, \text{ and} \\ 0 & \text{otherwise} \end{cases}$$

The plots in the left panel show that the graphical lasso is a consistent procedure for estimating the inverse covariance Θ^* in operator norm, since the error curves converge to zero as N increases. Comparing the curves for different graph sizes, we see a rightward-upward shift in the error curves, reflecting the fact that larger graphs require more samples to achieve the same error tolerance.

It is known that the solution $\widehat{\Theta}$ to the graphical lasso satisfies, with high probability, the error bound

$$\|\widehat{\Theta} - \Theta^*\|_2 \lesssim \sqrt{\frac{d^2 \log p}{N}}, \tag{9.21}$$

where d is the maximum degree of any node in the graph, and $\lesssim$ denotes inequality up to constant terms. (See the bibliographic notes on page 261 for details). If this theoretical prediction were sharp, one would expect that the same error curves—if replotted versus the *rescaled sample size* $\frac{N}{d^2 \log p}$—should be relatively well-aligned. The right panel in Figure 9.5 provides empirical confirmation of this prediction. We note that there are also theoretical results that guarantee that, as long as $N = \Omega(d^2 \log p)$, the support set $\widehat{\Theta}$ of the graphical lasso estimate coincides with the support set of Θ^*. Thus, the graphical lasso also succeeds in recovering the true graph structure.

9.3.5 Global Likelihood for Discrete Models

In principle, one could imagine adopting the ℓ_1-regularized version of the global likelihood to the problem of graph selection in models with discrete variables. A major challenge here is that the partition functions A in (9.4)–(9.6)—in sharp contrast to the multivariate Gaussian case—are difficult to compute in general. For instance, in the case of the Ising model (9.4), it takes the form

$$A(\theta) = \log \left[\sum_{x \in \{-1, +1\}^p} \exp \left\{ \sum_{s \in V} \theta_s x_s + \sum_{(s,t) \in E} \theta_{st} x_s x_t \right\} \right].$$

Thus, a brute force approach is intractable for large p, since it involves a summation over 2^p terms. With the exception of some special cases, computing the value of $A(\theta)$ is computationally intractable in general.

There are various approaches that can be taken to approximate or bound the partition function (see the bibliographic discussion for some references). However, these methods are somewhat off the main course of development in this chapter. Instead, the next section moves away from global likelihoods toward the idea of conditional or pseudo-likelihoods. These approaches can be used both for Gaussian and discrete variable models, and are computationally tractable, meaning polynomial-time in both sample size and graph size in either case.

9.4 Graph Selection via Conditional Inference

An alternative approach to graph selection is based on the idea of neighborhood-based likelihood, or products of such quantities that are known as pseudo-likelihoods. Both of these methods focus on conditional distributions, which for many situations are tractable.

For a given vertex $s \in V$, we use

$$X_{\setminus\{s\}} = \{X_t, t \in V \setminus \{s\}\} \in \mathbb{R}^{p-1}.$$

to denote the collection of all other random variables in the graph. Now consider the distribution of X_s given the random vector $X_{\setminus\{s\}}$. By the conditional

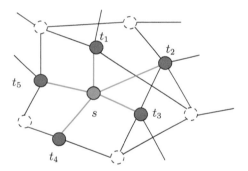

Figure 9.6 *The dark blue vertices form the neighborhood set $\mathcal{N}(s)$ of vertex s (drawn in red); the set $\mathcal{N}^+(s)$ is given by the union $\mathcal{N}(s) \cup \{s\}$. Note that $\mathcal{N}(s)$ is a cut set in the graph that separates $\{s\}$ from $V \setminus \mathcal{N}^+(s)$. Consequently, the variable X_s is conditionally independent of $X_{V \setminus \mathcal{N}^+(s)}$ given the variables $X_{\mathcal{N}(s)}$ in the neighborhood set. This conditional independence implies that the optimal predictor of X_s based on all other variables in the graph depends only on $X_{\mathcal{N}(s)}$.*

independence properties of any undirected graphical model (see Section 9.2), the only relevant variables to this conditioning are those in the *neighborhood set*

$$\mathcal{N}(s) = \{t \in V \mid (s,t) \in E\}. \tag{9.22}$$

Indeed, as shown in Figure 9.6, the set $\mathcal{N}(s)$ is a cut set that separates $\{s\}$ from the remaining vertices $V \setminus \mathcal{N}^+(s)$, where we define $\mathcal{N}^+(s) = \mathcal{N}(s) \cup$

$\{s\}$. Consequently, we are guaranteed that the variable X_s is conditionally independent of $X_{V \setminus \mathcal{N}^+(s)}$ given the variables $X_{\mathcal{N}(s)}$, or equivalently that

$$(X_s \mid X_{\setminus \{s\}}) \stackrel{d}{=} (X_s \mid X_{\mathcal{N}(s)}). \tag{9.23}$$

How can this conditional independence (CI) property be exploited in the context of graph selection? If we consider the problem of predicting the value of X_s based on $X_{V \setminus \{s\}}$, then by the CI property, the best predictor can be specified as a function of only $X_{\mathcal{N}(s)}$. Consequently, the problem of finding the neighborhood can be tackled by solving a prediction problem, as detailed next.

9.4.1 Neighborhood-Based Likelihood for Gaussians

Let us first develop this approach for the multivariate Gaussian. In this case, the conditional distribution of X_s given $X_{\setminus \{s\}}$ is also Gaussian, so that X_s can be decomposed into the best linear prediction based on $X_{\setminus \{s\}}$ and an error term—namely

$$X_s = X_{\setminus \{s\}}^T \beta^s + W_{\setminus \{s\}}. \tag{9.24}$$

In this decomposition, $W_{\setminus \{s\}}$ is zero-mean Gaussian variable with $\mathrm{Var}(W_{\setminus \{s\}}) = \mathrm{Var}(X_s \mid X_{\setminus \{s\}})$, corresponding to the prediction error, and is independent of $X_{\setminus \{s\}}$. So the dependence is captured entirely by the linear regression coefficients β^s, which are a scalar multiple of θ^s, the corresponding subvector of Θ in (9.8) in Section 9.2.2 (Exercise 9.4).

The decomposition (9.24) shows that in the multivariate Gaussian case, the prediction problem reduces to a linear regression of X_s on $X_{\setminus \{s\}}$. The key property here—as shown in Exercise 9.4—is that the regression vector β^s satisfies $\mathrm{supp}(\beta^s) = \mathcal{N}(s)$. If the graph is relatively sparse—meaning that the degree $|\mathcal{N}(s)|$ of node s is small relative to p—then it is natural to consider estimating β^s via the lasso. This leads to the following neighborhood-based approach to Gaussian graphical model selection, based on a set of samples $\mathbf{X} = \{x_1, \dots, x_N\}$.

In step 1(a), $x_{i, V \setminus \{s\}}$ represents the $(p-1)$ dimensional subvector of the p-vector x_i, omitting the sth component. To clarify step 2, the AND rule declares that edge (s, t) belongs to the estimated edge set $\widehat{E}$ if and only $s \in \widehat{\mathcal{N}}(t)$ *and* $t \in \widehat{\mathcal{N}}(s)$. On the other hand, the OR rule is less conservative, allowing $(s, t) \in \widehat{E}$ if either $s \in \widehat{\mathcal{N}}(t)$ *or* $t \in \widehat{\mathcal{N}}(s)$.

An advantage of neighborhood models is speed. Many efficient implementations of the lasso are available, and the p regression problems can be solved independently of each other, and hence in parallel.

The AND/OR rules can be avoided by using a joint estimation approach using the pseudo-likelihood, which is essentially the sum of the log-likelihoods in (9.25). In this case we would enforce the symmetry of Θ. While more elegant, this does incur a small additional computational cost (Friedman, Hastie

---

Algorithm 9.2 NEIGHBORHOOD-BASED GRAPH SELECTION FOR GAUSSIAN
GRAPHICAL MODELS.

---

1. For each vertex $s = 1, 2, \ldots, p$:

   (a) Apply the lasso to solve the neighborhood prediction problem:

   $$\widehat{\beta}^s \in \arg\min_{\beta^s \in \mathbb{R}^{p-1}} \left\{ \frac{1}{2N} \sum_{i=1}^{N} (x_{is} - x_{i,V\setminus\{s\}}^T \beta^s)^2 + \lambda\|\beta^s\|_1 \right\}. \tag{9.25}$$

   (b) Compute the estimate $\widehat{\mathcal{N}}(s) = \text{supp}(\widehat{\beta}^s)$ of the neighborhood set $\mathcal{N}(s)$.

2. Combine the neighborhood estimates $\{\widehat{\mathcal{N}}(s), s \in V\}$ via the AND or OR rule to form a graph estimate $\widehat{G} = (V, \widehat{E})$.

---

and Tibshirani 2010a). It also produces an estimate $\widehat{\Theta}$, and not just its graph. We discuss such approaches in more generality in Section 9.4.3.

9.4.2 Neighborhood-Based Likelihood for Discrete Models

The idea of a neighborhood-based likelihood is not limited to Gaussian models, but can also be applied to other types of graphical models that can be written in the exponential family form. In fact, given that global likelihood calculations are intractable for discrete graphical models, the neighborhood-based approach is especially attractive in the discrete case, at least from the computational perspective.

The simplest type of discrete graphical model is the Ising model (9.4), used to model a collection of variables $(X_1, X_2, \ldots, X_p) \in \{-1, +1\}^p$ that interact in a pairwise manner. In this case, as we explore in Exercise 9.5, the conditional log-odds for the probability of X_s given $X_{V\setminus\{s\}}$ takes the form[3]

$$
\begin{aligned}
\eta_{\theta^s}(X_{\setminus\{s\}}) &= \log\left[\frac{\mathbb{P}(X_s = +1 \mid X_{\setminus\{s\}})}{\mathbb{P}(X_s = -1 \mid X_{\setminus\{s\}})}\right] \\
&= 2\theta_s + \sum_{t \in V\setminus\{s\}} 2\theta_{st}X_t,
\end{aligned}
\tag{9.26}
$$

with $\theta^s = [\theta_s, \{\theta_{st}\}_{t \in V\setminus\{s\}}]$. Consequently, the neighborhood-based approach for the Ising model has the same form as Algorithm 9.2, with the ordinary lasso in step 1(a) replaced by the lasso logistic regression problem

$$\widehat{\theta}^s \in \arg\min_{\theta^s \in \mathbb{R}^p} \left\{ \frac{1}{N} \sum_{i=1}^{N} \ell[x_{is}, \eta_{\theta^s}(x_{i,\setminus\{s\}})] + \lambda \sum_{t \in V\setminus\{s\}} |\theta_{st}| \right\}, \tag{9.27}$$

---

[3]the factor of 2 comes from the particular response coding +1/-1 rather than the traditional 0/1.

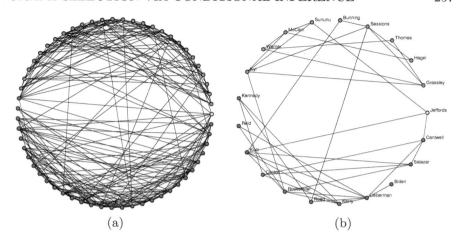

(a) (b)

Figure 9.7 *Politician networks estimated from voting records of U.S. Senate (2004–2006). A total of $N = 546$ votes were collected for each of $p = 100$ senators, with $X_s = +1$ (respectively $X_s = -1$) meaning that senator s voted "yes" (respectively "no"). A pairwise graphical model was fit to the dataset using the neighborhood-based logistic regression approach. (a) A subgraph of 55 senators from the fitted graph, with Democratic/Republican/Independent senators coded as blue/red/yellow nodes, respectively. Note that the subgraph shows a strong bipartite tendency with clustering within party lines. A few senators show cross-party connections. (b) A smaller subgraph of the same social network.*

where ℓ is the negative log-likelihood for the binomial distribution. This problem is again a convex program, and any algorithm for ℓ_1-penalized logistic regression can be used, such as the coordinate-descent procedure discussed in Chapter 5. As in the Gaussian case, rules have to be used to enforce symmetry in the edge calls.

Hoefling and Tibshirani (2009) present a pseudo-likelihood method for this problem which imposes the symmetry. It can be thought of as intermediate to the exact, but computationally intractable, global likelihood approach and the neighborhood-based method described above. We cover their approach in Section 9.4.3.

Figure 9.7 shows the results of fitting an Ising model to represent the politician social network for the U.S. Senate, reconstructed on the basis of voting records. Details are given in the figure caption; overall we see strong clustering within each party.

The neighborhood-based approach for graphical model selection can be shown to be consistent under relatively mild conditions on the sample size (see the bibliographic section for references and further discussion). In the case of the Ising model, let $\widehat{G}$ denote the output of the neighborhood-based approach using logistic regression. It is known that $N = \Omega(d^2 \log p)$ samples

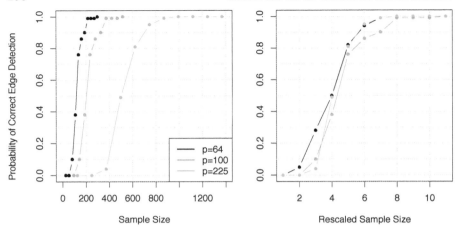

Figure 9.8 *Plots of the probability* $\mathbb{P}[\widehat{G} = G]$ *of correctly recovering the graph versus the sample size. The neighborhood-based logistic regression procedure was applied to recover a star-shaped graph (central hub with spokes) with p vertices and hub degree* $d = \lceil 0.1p \rceil$. *(a) Plotted versus the raw sample size N. As would be expected, larger graphs require more samples for consistent estimation. (b) The same simulation results plotted versus the rescaled sample size* $\frac{N}{d \log p}$.

are sufficient to ensure that $\widehat{G} = G$ with high probability. Figure 9.8 illustrates the sufficiency of this condition in practice, when the method is applied to a star-shaped graph with p nodes, in which a central hub node is connected to $d = \lceil 0.1p \rceil$ spoke nodes. (The remaining nodes in the graph are disconnected from the hub-spoke subgraph.) We implemented the neighborhood-based logistic-regression method for graphs with $p \in \{64, 100, 225\}$ nodes, using the AND rule to combine the neighborhoods so as to form the graph estimate. In panel (a) of Figure 9.8, we plot the probability $\mathbb{P}[\widehat{G} = G]$ of correctly recovering the unknown graph versus the sample size N, with each curve corresponding to a different graph size as labelled. Each of these plots show that the method is model-selection consistent, in that as the sample size N increases, the probability $\mathbb{P}[\widehat{G} = G]$ converges to one. Naturally, the transition from failure to success occurs later (at larger sample sizes) for larger graphs, reflecting that the problem is more difficult. Panel (b) shows the same simulation results plotted versus the rescaled sample size $\frac{N}{d \log p}$; on this new axis, all three curves are now well-aligned. This simulation confirms that the theoretical scaling $N = \Omega(d^2 \log p)$ is sufficient to ensure successful graph recovery, but not necessary for this class of graphs. However, there are other classes of graphs for which this scaling is both sufficient and necessary (see the bibliographic section for details).

9.4.3   Pseudo-Likelihood for Mixed Models

So far we have covered graphical models for all continuous variables (the Gaussian graphical model), and models for all binary variables (the Ising model). These do not cover other frequently occurring situations:

- discrete variables with more than two states;
- mixed data types: i.e., some continuous and some discrete.

In this section, we extend the models covered so far to include these cases, and demonstrate an approach for inference based on the pseudo-likelihood.

A simple generalization of the Gaussian and Ising models is the pairwise Markov random field model. For convenience in notation, we denote the p continuous variables by X and the q discrete variables by Y. The density $\mathbb{P}_\Omega(x, y)$ is proportional to

$$\exp\left\{ \sum_{s=1}^{p} \gamma_s x_s - \frac{1}{2} \sum_{s=1}^{p} \sum_{t=1}^{p} \theta_{st} x_s x_t + \sum_{s=1}^{p} \sum_{j=1}^{q} \rho_{sj}[y_j] x_s + \sum_{j=1}^{q} \sum_{r=1}^{q} \psi_{jr}[y_j, y_r] \right\}.$$

(9.28)

The first two terms are as in the Gaussian graphical model (9.8). The term ρ_{sj} represents an edge between continuous X_s and discrete Y_j. If Y_j has L_j possible states or levels, then ρ_{sj} is a vector of L_j parameters, and $\rho_{sj}[y_j]$ references the y_j^{th} value. Likewise ψ_{jr} will be an $L_j \times L_r$ matrix representing an edge between discrete Y_j and Y_r, and $\psi_{jr}[y_j, y_r]$ references the element in row y_j and column y_r. The terms ψ_{jj} will be diagonal, and represent the *node potentials* (they correspond to the θ_s in the Ising model (9.4) on page 244). The matrix Ω represents the entire collection of parameters. Needless to say, the partition function is typically intractable, except in very low-dimensional cases.

Here the pseudo-likelihood is attractive: it is the product of the $p + q$ conditional likelihoods, and each of these are simple (Exercise 9.8), depending on the type of response:

Continuous: The conditional distribution for each of the p continuous variables is Gaussian, with mean linear in the conditioning variables.

$$\mathbb{P}(X_s | X_{\backslash\{s\}}, Y; \Omega) = \left(\frac{\theta_{ss}}{2\pi}\right)^{\frac{1}{2}} e^{-\frac{\theta_{ss}}{2}\left(X_s - \frac{\gamma_s + \sum_j \rho_{sj}[Y_j] - \sum_{t\neq s}\theta_{st}X_t}{\theta_{ss}}\right)^2}$$

(9.29)

The contributions of the discrete conditioning variables on the right-hand side are different additive constants, as for qualitative factors in linear regression models; i.e., a constant for each level, determined by the ρ_{sj}.

Discrete: The conditional distribution for each of the q discrete variables is multinomial, with log-odds linear in the conditioning variables.

$$\mathbb{P}(Y_j | X, Y_{\backslash\{j\}}; \Omega) = \frac{e^{\psi_{jj}[Y_j, Y_j] + \sum_s \rho_{sj}[Y_j]X_s + \sum_{r\neq j}\psi[Y_j, Y_r]}}{\sum_{\ell=1}^{L_j} e^{\psi_{jj}[\ell, \ell] + \sum_s \rho_{sj}[\ell]X_s + \sum_{r\neq j}\psi[\ell, Y_r]}}$$

(9.30)

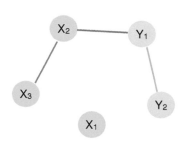
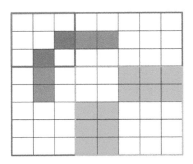

Figure 9.9 *Mixed graphical model, with three continuous and two discrete variables.*
Y_1 *has two states, and* Y_3 *has three. The diagram on the right shows the sets of parameters associated with each edge. The group lasso treats these as sets, and includes or excludes them all together.*

With this notation, the pseudo-log-likelihood is defined to be

$$\ell^p(\mathbf{\Omega}; \mathbf{X}, \mathbf{Y}) = \sum_{i=1}^{N} \left[\sum_{s=1}^{p} \log \mathbb{P}(x_{is} | x_{i \setminus \{s\}}, y_i; \mathbf{\Omega}) \sum_{j=1}^{q} \log \mathbb{P}(y_{ij} | x_i, y_{i \setminus \{j\}}; \mathbf{\Omega}) \right].$$
(9.31)

One can show that (9.31) is a concave function of $\mathbf{\Omega}$. Notice that each of the parameters appears twice: once when one of the indices corresponds to the response, and again when that index refers to a conditioning variable.

The edge parameters are now of three kinds: scalars θ_{st}, vectors ρ_{sj}, and matrices ψ_{jr}. Figure 9.9 illustrates with a small example. Lee and Hastie (2014) use the group-lasso to select these different parameter types; they propose optimizing the penalized pseudo-log-likelihood

$$\ell^p(\mathbf{\Omega}; \mathbf{X}, \mathbf{Y}) - \lambda \left(\sum_{s=1}^{p} \sum_{t=1}^{s-1} |\theta_{st}| + \sum_{s=1}^{p} \sum_{j=1}^{q} \|\rho_{sj}\|_2 + \sum_{j=1}^{q} \sum_{r=1}^{j-1} \|\psi_{jr}\|_2 \right)$$
(9.32)

w.r.t. the parameters $\mathbf{\Omega}$. Notice that not all the parameters are penalized. In particular the diagonal of $\mathbf{\Theta}$ is left alone (as in the graphical lasso algorithm), as are each of the node-potentials ψ_{jj}. This imposes interesting constraints on some of the parameter estimates, which are explored in Exercise 9.9.

With only continuous variables, this is exactly the penalized pseudo-likelihood for the Gaussian graphical model. For all binary variables, one can show that this is equivalent to the lasso-penalized pseudo-likelihood for the Ising model (see Hoefling and Tibshirani (2009) and Exercise 9.10).

Block coordinate descent is attractive here, since each of the components has well-studied solutions. However, the parameters are shared, and so care must be taken to respect this symmetry. Lee and Hastie (2014) use a proximal-Newton algorithm.

9.5   Graphical Models with Hidden Variables

Chandrasekaran et al. (2012) propose a method for undirected graphical models, in which some of the variables are unobserved (or "latent"). Suppose for example that we are modelling stock prices in a certain sector, and they are heavily dependent on the price of energy, the latter being unmeasured in our data. Then the concentration matrix of the stock prices will not look sparse in our data, but may instead be sparse if we could condition on energy price.

Let the covariance matrix of all variables—observed and unobserved—be $\mathbf{\Sigma}$. The sub-block of $\mathbf{\Sigma}$ corresponding to the observed variables is $\mathbf{\Sigma}_O$. Let $\mathbf{K} = \mathbf{\Sigma}^{-1}$ be the concentration matrix for the set of observed and hidden variables, with sub-matrices $\mathbf{K}_O$, $\mathbf{K}_{O,H}$, $\mathbf{K}_{H,O}$ and $\mathbf{K}_H$. These capture the dependencies among the observed variables, between the observed and hidden variables, and among the hidden variables, respectively. Making use of the partitioned-inverse formulas, we get the following expression for the concentration matrix of the observed variables:

$$\tilde{\mathbf{K}}_O = \mathbf{\Sigma}_O^{-1} = \mathbf{K}_O - \mathbf{K}_{O,H}\mathbf{K}_H^{-1}\mathbf{K}_{H,O} \tag{9.33}$$

Here $\mathbf{K}_O$ is the concentration matrix of the conditional statistics of the observed variables given the hidden variables. Now $\tilde{\mathbf{K}}_O$ may not be sparse, but if the graphical model for all variables (observed and hidden) has few edges then $\mathbf{K}_O$ will be sparse.

Motivated by the form (9.33), letting $\mathbf{K}_O = \mathbf{\Theta}$ the idea is to write

$$\tilde{\mathbf{K}}_O = \mathbf{\Theta} - \mathbf{L}, \tag{9.34}$$

where $\mathbf{L}$ is assumed to be low rank, with the rank at most the number of hidden variables. We then solve the problem

$$\underset{\mathbf{\Theta},\mathbf{L}}{\text{minimize}} \left\{ \text{trace}[\mathbf{S}(\mathbf{\Theta} - \mathbf{L})] - \log[\det(\mathbf{\Theta} - \mathbf{L})] + \lambda\|\mathbf{\Theta}\|_1 + \text{trace}(\mathbf{L}) \right\} \tag{9.35}$$

over the set $\{\mathbf{\Theta} - \mathbf{L} \succ \mathbf{0}, \mathbf{L} \succeq \mathbf{0}\}$. Like the graphical lasso, this is again a convex problem. This relates to the "sparse plus low rank" idea discussed in Mazumder et al. (2010) and Chandrasekaran et al. (2011). Ma, Xue and Zou (2013) propose first-order alternating direction-of-multipliers (ADMM) techniques for this problem, and compare them to second order methods. Some details are given in Exercise 9.11.

Bibliographic Notes

Detailed discussion of graphical models can be found in Whittaker (1990), Lauritzen (1996), Cox and Wermuth (1996), Edwards (2000), Pearl (2000), Anderson (2003), and Koller and Friedman (2009).

The Hammersley–Clifford theorem was first announced in the unpublished note of Hammersley and Clifford (1971). Independent proofs were given by Besag (1974) and Grimmett (1973), the latter proof using the Möebius inversion formula. See Clifford (1990) for some historical discussion and context of the result.

Welsh (1993) discusses the computational intractability of evaluating the partition function for general discrete graphical models. For graphs with special structure, exact computation of the cumulant function is possible in polynomial time. Examples include graphs with low tree width, for which the junction tree algorithm can be applied (Lauritzen and Spiegelhalter 1988, Lauritzen 1996), and certain classes of planar models (Kastelyn 1963, Fisher 1966). For other special cases, there are rapidly mixing Markov chains that can be used to obtain good approximations to the cumulant function (Jerrum and Sinclair 1993, Jerrum and Sinclair 1996, for example, and the references therein). A complementary approach is provided by the class of variational methods, which provide approximations to the cumulant generating function (e.g., see the monograph by Wainwright and Jordan (2008) and references therein). Examples include the mean-field algorithm, the sum-product or belief-propagation algorithm, expectation propagation, as well as various other convex relaxations. For certain graphs, particularly those that are "locally tree-like," there are various kinds of asymptotic exactness results (e.g., see the book by Mézard and Montanari (2008) and references therein).

Gaussian graphical models are used for modelling gene expression data (Dobra, Hans, Jones, Nevins, Yao and West 2004), and other genomic and proteomic assays. The Ising model (9.4) was first proposed in the context of statistical physics by Ising (1925). In more recent work, it and related models have been used as simple models for binary images (Geman and Geman 1984, Greig, Porteous and Seheuly 1989, Winkler 1995), voting behavior in politicians (Banerjee et al. 2008), citation network analysis (Zhao, Levina and Zhu 2011).

Some of the methods discussed in this chapter for undirected models can be used to aid in the more difficult model search for directed graphical models; see for example Schmidt, Niculescu-Mizil and Murphy (2007). The paper by Vandenberghe et al. (1998) provides an introduction to the problem of determinant maximization with constraints; the Gaussian MLE (with or without regularization) is a special case of this class of problems. Yuan and Lin (2007a) proposed the use of ℓ_1-regularization in conjunction with the Gaussian (log-determinant) likelihood for the covariance-selection problem, and used interior point methods (Vandenberghe et al. 1998) to solve it. d'Aspremont et al. (2008) and Friedman et al. (2008) develop faster coordinate descent algorithms for solving the graphical lasso (9.13), based on solving a sequence of subproblems. Mazumder and Hastie (2012) offer variants on these algorithms with better convergence properties. Witten et al. (2011) and Mazumder and Hastie (2012) show how to exploit block-diagonal structure in $\mathbf{S}$ in computing graphical-lasso solutions. Rothman, Bickel, Levina and Zhu (2008) established consistency of the estimator in Frobenius norm, whereas Ravikumar et al. (2011) provide some results on model selection consistency as well as rates for operator norm. In particular, they proved the operator-norm bound (9.21) illustrated in Figure 9.5.

The idea of pseudo-likelihood itself is quite old, dating back to the seminal work of Besag (1975). Meinshausen and Bühlmann (2006) were the first to propose and develop the lasso-based neighborhood selection for Gaussian graphical models, and to derive consistency results under high-dimensional scaling; see also the papers by Zhao and Yu (2006) and Wainwright (2009) for related results on static graphs. Zhou, Lafferty and Wasserman (2008) consider the problem of tracking a time-varying sequence of Gaussian graphical models.

Ravikumar, Wainwright and Lafferty (2010) proposed ℓ_1-regularized logistic regression for model selection in discrete binary graphical models, and showed that it is model-selection consistent under the scaling $N = \Omega(d^3 \log p)$. Subsequent analysis by Bento and Montanari (2009) improved this scaling to $N = \Omega(d^2 \log p)$ for Ising models below the phase transition. Koh et al. (2007) develop an interior-point algorithm suitable for large-scale ℓ_1-regularized logistic regression. Instead of solving separate a logistic regression problem at each node, Hoefling and Tibshirani (2009) propose minimization of the ℓ_1-regularized pseudo-likelihood, and derive efficient algorithms for it; see also Friedman et al. (2010a). Santhanam and Wainwright (2008) derive information-theoretic lower bounds on Ising model selection, showing that no method can succeed more than half the time if $N = \mathcal{O}(d^2 \log p)$. This shows that the neighborhood approach is an optimal procedure up to constant factors.

Cheng, Levina and Zhu (2013) and Lee and Hastie (2014) discuss mixed graphical models, involving both continuous and discrete variables. Kalisch and Bühlmann (2007) show that a variant of the PC algorithm can be used for high-dimensional model selection in directed graphs.

A different kind of graphical model is the *covariance graph* or *relevance network*, in which vertices are connected by bidirectional edges if the covariance (rather than the partial covariance) between the corresponding variables is nonzero. These are popular in genomics; see for example Butte, Tamayo, Slonim, Golub and Kohane (2000). The negative log-likelihood from these models is not convex, making the computations more challenging (Chaudhuri, Drton and Richardson 2007). Recent progress on this problem has been made by Bien and Tibshirani (2011) and Wang (2014). The latter paper derives a blockwise coordinate descent algorithm analogous to the to the graphical lasso procedure. Some theoretical study of the estimation of large covariance matrices is given by Bickel and Levina (2008) and El Karoui (2008).

Exercises

Ex. 9.1 The most familiar parametrization of the multivariate Gaussian is in terms of its mean vector $\mu \in \mathbb{R}^p$ and covariance matrix $\mathbf{\Sigma} \in \mathbb{R}^{p \times p}$. Assuming that the distribution is nondegenerate (i.e., $\mathbf{\Sigma}$ is strictly positive definite), show that the canonical parameters $(\gamma, \Theta) \in \mathbb{R}^p \times \mathcal{S}_+^p$ from the factorization (9.8) are related by

$$\mu = -\Theta^{-1}\gamma, \quad \text{and} \quad \Sigma = \Theta^{-1}. \tag{9.36}$$

Ex. 9.2 Let $\{x_1, \ldots, x_N\}$ be N i.i.d. samples from a Gaussian graphical model, and let $\mathcal{L}(\Theta; \mathbf{X}) = \frac{1}{N} \sum_{i=1}^{N} \log \mathbb{P}_\Theta(x_i)$ denote the rescaled log-likelihood of the sample.

(a) Show that
$$\mathcal{L}(\Theta; \mathbf{X}) = \log \det \Theta - \text{trace}(\mathbf{S}\Theta) + C,$$

where $\mathbf{S} = \frac{1}{N} \sum_{i=1}^{N} x_i x_i^T$ is the empirical covariance matrix, and C is a constant independent of Θ.

(b) Show that the function $f(\Theta) = -\log \det \Theta$ is a strictly convex function on the cone of positive definite matrices. Prove that $\nabla f(\Theta) = \Theta^{-1}$ for any $\Theta \in \mathcal{S}_+^p$.

(c) The (unregularized) Gaussian MLE is given by

$$\widehat{\Theta} \in \underset{\Theta \in \mathcal{S}_+^p}{\arg\max} \left\{ \log \det \Theta - \text{trace}(\mathbf{S}\Theta) \right\},$$

when this maximum is attained. Assuming that the maximum is attained, show that $\widehat{\Theta} = \mathbf{S}^{-1}$. Discuss what happens when $N < p$.

(d) Now consider the graphical lasso (9.13), based on augmenting the rescaled log-likelihood with an ℓ_1-regularizer. Derive the Karush–Kuhn–Tucker equations that any primal-optimal pair $(\widehat{\Theta}, \widehat{\mathbf{W}}) \in \mathcal{S}_+^p \times \mathbb{R}^{p \times p}$ must satisfy.

(e) Derive the dual program associated with the graphical lasso. Can you generalize your result to regularization with any ℓ_q-norm, for $q \in [1, \infty]$?

Ex. 9.3 Show that if $\mathbf{S}$ is positive definite, the graphical lasso algorithm with $\lambda = 0$ computes $\widehat{\Theta} = \mathbf{S}^{-1}$.

Ex. 9.4 In this exercise, we explore properties of jointly Gaussian random vectors that guarantee Fisher consistency of the neighborhood-based lasso approach to covariance selection. Let $(X_1, X_2, \ldots, X_p)$ be a zero-mean jointly Gaussian random vector with positive definite covariance matrix Σ. Letting $T = \{2, 3, \ldots, p\}$, consider the conditioned random variable $Z = (X_1 \mid X_T)$.

(a) Show that there is a vector $\theta \in \mathbb{R}^{p-1}$ such that

$$Z = \theta^T X_T + W,$$

where W is a zero-mean Gaussian variable independent of X_T. *Hint:* consider the best linear predictor of X_1 given X_T.

(b) Show that $\theta = \Sigma_{TT}^{-1} \Sigma_{T1}$, where $\Sigma_{T1} \in \mathbb{R}^{p-1}$ is the vector of covariances between X_1 and X_T.

(c) Show that $\theta_j = 0$ if and only if $j \notin \mathcal{N}(1)$. *Hint:* The following elementary fact could be useful: let $\mathbf{A}$ be an invertible matrix, given in the block-partitioned form

$$\mathbf{A} = \begin{bmatrix} \mathbf{A}_{11} & \mathbf{A}_{12} \\ \mathbf{A}_{21} & \mathbf{A}_{22} \end{bmatrix}.$$

Then letting $\mathbf{B} = \mathbf{A}^{-1}$, we have $\mathbf{B}_{12} = \mathbf{A}_{11}^{-1}\mathbf{A}_{12}\left[\mathbf{A}_{21}\mathbf{A}_{11}^{-1}\mathbf{A}_{12} - \mathbf{A}_{22}\right]^{-1}$ (Horn and Johnson 1985, for example).

Ex. 9.5 Consider the neighborhood-based likelihood approach for selection of Ising models.

(a) Derive the conditional distribution $\mathbb{P}(x_s \mid x_{V\setminus\{s\}}; \theta)$, and show how the neighborhood-prediction reduces to logistic regression.

(b) Verify that the method is Fisher-consistent, meaning that the true conditional distribution is the population minimizer.

Ex. 9.6 Here we show how, in expression (9.14), we can solve for $\boldsymbol{\Theta}$ and its inverse $\mathbf{W} = \boldsymbol{\Theta}^{-1}$ one row and column at a time. For simplicity let's focus on the last row and column. Then the upper right block of Equation (9.14) can be written as

$$\mathbf{w}_{12} - \mathbf{s}_{12} - \lambda \cdot \text{sign}(\boldsymbol{\theta}_{12}) = 0. \tag{9.37}$$

Here we have partitioned the matrices into two parts: part 1 being the first $p-1$ rows and columns, and part 2 the pth row and column. With $\mathbf{W}$ and its inverse $\boldsymbol{\Theta}$ partitioned in a similar fashion

$$\begin{pmatrix} \mathbf{W}_{11} & \mathbf{w}_{12} \\ \mathbf{w}_{12}^T & w_{22} \end{pmatrix} \begin{pmatrix} \boldsymbol{\Theta}_{11} & \boldsymbol{\theta}_{12} \\ \boldsymbol{\theta}_{12}^T & \theta_{22} \end{pmatrix} = \begin{pmatrix} \mathbf{I} & \mathbf{0} \\ \mathbf{0}^T & 1 \end{pmatrix}, \tag{9.38}$$

show that

$$\mathbf{w}_{12} = -\mathbf{W}_{11}\boldsymbol{\theta}_{12}/\theta_{22} \tag{9.39}$$
$$= \mathbf{W}_{11}\boldsymbol{\beta} \tag{9.40}$$

where $\boldsymbol{\beta} = -\boldsymbol{\theta}_{12}/\theta_{22}$. This is obtained from the formula for the inverse of a partitioned inverse of a matrix (Horn and Johnson 1985, for example). Substituting (9.40) into (9.37) gives

$$\mathbf{W}_{11}\boldsymbol{\beta} - \mathbf{s}_{12} + \lambda \cdot \text{sign}(\boldsymbol{\beta}) = 0. \tag{9.41}$$

Ex. 9.7 With the partitioning as in (9.38), write down the expressions for the partitioned inverses of each matrix in terms of the other. Show that since $\mathbf{W}_{11}$ depends on $\boldsymbol{\theta}_{12}$, we are not really holding $\mathbf{W}_{11}$ *fixed* as assumed in the graphical lasso Algorithm 9.1.

(a) Show that as an alternative we can write (9.37) as

$$\boldsymbol{\Theta}_{11}^{-1}\boldsymbol{\theta}_{12}w_{22} + \mathbf{s}_{12} + \lambda\,\text{sign}(\boldsymbol{\theta}_{12}) = \mathbf{0}. \tag{9.42}$$

(b) Show how to use the solution $\boldsymbol{\theta}_{12}$ to update the current version of $\mathbf{W}$ and $\widehat{\boldsymbol{\Theta}}$ in $O(p^2)$ operations.

(c) Likewise, show how to move to a new block of equations in $O(p^2)$ operations.

(d) You have derived a *primal* graphical lasso algorithm. Write it down in algorithmic form, as in Algorithm 9.1

(Mazumder and Hastie 2012)

Ex. 9.8 Derive the conditional distributions (9.29) and (9.30) for the mixed graphical model.

Ex. 9.9 Close inspection of the pairwise Markov random field model (9.28) will show that it is overparametrized with respect to the discrete potentials ρ_{sj} and ψ_{jr}. This exercise shows that this aliasing is resolved by the quadratic penalties in the penalized pseudo-likelihood, in the form of "sum-to-zero" constraints familiar in regression and ANOVA modelling.

Consider the penalized pseudo log-likelihood (9.32), with $\lambda > 0$.

(a) Since the γ_s are not penalized, show that the solution $\hat{\rho}_{sj}$ for any s and j satisfies

$$\sum_{\ell=1}^{L_j} \hat{\rho}_{sj}[\ell] = 0.$$

(b) Since the (diagonal) matrices ψ_{jj} are not penalized, show that the solution $\hat{\psi}_{jr}$ for any $j \neq r$ satisfies

$$\sum_{\ell=1}^{L_j} \hat{\psi}_{jr}[\ell, m] \;=\; 0, \; m = 1, \dots, L_r; \tag{9.43}$$

$$\sum_{m=1}^{L_r} \hat{\psi}_{jr}[\ell, m] \;=\; 0, \; \ell = 1, \dots, L_j. \tag{9.44}$$

Ex. 9.10 Consider the pairwise Markov random field model with only binary discrete variables. This appears to be different from the Ising model, since we have four parameters per edge. Use Exercise 9.9 to show that with the quadratic constraints in (9.32), it is exactly equivalent to a lasso-penalized pseudo log-likelihood for the Ising model.

Ex. 9.11 Consider the objective function (9.35) for the graphical model that allows for latent variables. Defining a new variable $\mathbf{R} = \boldsymbol{\Theta} - \mathbf{L}$, derive the details of the steps of an ADMM algorithm for solving (9.35) using the augmented Lagrangian

$$L_\mu(\mathbf{R}, \boldsymbol{\Theta}_0, \mathbf{L}, \boldsymbol{\Gamma}) \;=\; \operatorname{trace}(\mathbf{SR}) - \log \det \mathbf{R} + \lambda \|\boldsymbol{\Theta}\|_1 + \beta \cdot \operatorname{trace}(\mathbf{L})$$

$$+ I(\mathbf{L} \succeq \mathbf{0}) - \operatorname{trace}[\boldsymbol{\Gamma}(\mathbf{R} - \boldsymbol{\Theta} + \mathbf{L})] + \frac{1}{2\mu} \|\mathbf{R} - \boldsymbol{\Theta} + \mathbf{L}\|_F^2$$

successively over $\mathbf{R}, \boldsymbol{\Theta}, \mathbf{L}$ and $\boldsymbol{\Gamma}$ (Ma et al. 2013).

Chapter 10

Signal Approximation and Compressed Sensing

10.1 Introduction

In this chapter, we discuss applications of ℓ_1-based relaxation to problems of signal recovery and approximation. Our focus is the role played by sparsity in signal representation and approximation, and the use of ℓ_1-methods for exploiting this sparsity for solving problems like signal denoising, compression, and approximation. We begin by illustrating that many classes of "natural" signals are sparse when represented in suitable bases, such as those afforded by wavelets and other multiscale transforms. We illustrate how such sparsity can be exploited for compression and denoising in orthogonal bases. Next we discuss the problem of signal approximation in overcomplete bases, and the role of ℓ_1-relaxation in finding near-optimal approximations. Finally, we discuss the method of compressed sensing for recovering sparse signals. It is a combination of two ideas: taking measurements of signals via random projections, and solving a lasso-type problem for reconstruction.

10.2 Signals and Sparse Representations

Let us begin by providing some background on the role of sparse representations in signal processing. To be clear, our use of the term "signal" is general, including (among other examples) data such as sea water levels, seismic recordings, medical time series, audio recordings, photographic images, video data, and financial data. In all cases, we represent the signal by a vector $\theta^* \in \mathbb{R}^p$. (For two-dimensional signals such as images, the reader should think about a vectorized form of the image.)

10.2.1 Orthogonal Bases

In signal processing, it is frequently useful to represent signals in different types of bases. Examples include Fourier representations, useful for extracting periodic structure in time series, and multiscale representations such as wavelets. Such representations are described by a collection of vectors

$\{\psi_j\}_{j=1}^p$ that form an orthonormal basis of $\mathbb{R}^p$. If we define the $p \times p$ matrix $\mathbf{\Psi} := \begin{bmatrix} \psi_1 & \psi_2 & \dots & \psi_p \end{bmatrix}$, then the orthonormality condition guarantees that $\mathbf{\Psi}^T \mathbf{\Psi} = I_{p \times p}$. Given an orthonormal basis, any signal $\theta^* \in \mathbb{R}^p$ can be expanded in the form

$$\theta^* := \sum_{j=1}^p \beta_j^* \psi_j, \tag{10.1}$$

where the jth *basis coefficient* $\beta_j^* := \langle \theta^*, \psi_j \rangle = \sum_{i=1}^p \theta_i^* \psi_{ij}$ is obtained by projecting the signal onto the jth basis vector ψ_j. Equivalently, we can write the transformation from signal $\theta^* \in \mathbb{R}^p$ to basis coefficient vector $\beta^* \in \mathbb{R}^p$ as the matrix-vector product $\beta^* = \mathbf{\Psi}^T \theta^*$.

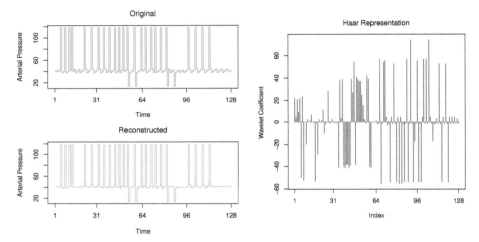

Figure 10.1 *Illustration of sparsity in time series data. Left, top panel: Signal $\theta^* \in \mathbb{R}^p$ of arterial pressure versus time over $p = 128$ points. Left, bottom panel: Reconstruction $\widehat{\theta}^{128}$ based on retaining the largest (in absolute amplitude) $k = 64$ coefficients from the Haar basis. Right: Haar basis coefficients $\beta^* = \mathbf{\Psi}^T \theta^*$ of the signal.*

To give a simple example, consider the following matrix

$$\mathbf{\Psi} := \begin{bmatrix} \frac{1}{2} & \frac{1}{2} & \frac{1}{\sqrt{2}} & 0 \\ \frac{1}{2} & \frac{1}{2} & \frac{-1}{\sqrt{2}} & 0 \\ \frac{1}{2} & -\frac{1}{2} & 0 & \frac{1}{\sqrt{2}} \\ \frac{1}{2} & -\frac{1}{2} & 0 & \frac{-1}{\sqrt{2}} \end{bmatrix}. \tag{10.2}$$

It is an orthonormal matrix, satisfying $\mathbf{\Psi}^T \mathbf{\Psi} = \mathbf{I}_{4 \times 4}$, and corresponds to a two-level Haar transform for signal length $p = 4$. For any given signal $\theta^* \in \mathbb{R}^4$, the Haar basis coefficients $\beta^* = \mathbf{\Psi}^T \theta^*$ have a natural interpretation. The first coefficient $\beta_1^* = \langle \psi_1, \theta^* \rangle = \frac{1}{2} \sum_{j=1}^4 \theta_j^*$ is a rescaled version of the averaged signal. The second column ψ_2 is a differencing operator on the full signal, whereas the third and fourth columns are local differencing operators on each half of the signal. This Haar transform is the simplest example of a wavelet transform.

An important fact is that many signal classes, while not sparse in the canonical basis, become sparse when represented in a different orthogonal basis. Figure 10.1 provides an illustration of this phenomenon for some medical time series data. The top-left panel shows $p = 128$ samples of arterial pressure from a patient, showing that the signal θ^* itself is not at all sparse. The right panel shows the Haar coefficient representation $\beta^* = \mathbf{\Psi}^T \theta^*$ of the signal; note how in contrast it is relatively sparse. Finally, the bottom-left panel shows a reconstruction $\widehat{\theta}$ of the original signal, based on discarding half of the Haar coefficients. Although not a perfect reconstruction, it captures the dominant features of the time series.

Figure 10.2 provides a second illustration of this sparsity phenomenon, this time for the class of photographic images and two-dimensional wavelet transforms. Panel (a) shows a 512×512 portion of the "Boats" image; in our framework, we view this two-dimensional image as a vector in $p = 512^2 = 262,144$ dimensions. Shown in panel (b) is the form of a particular two-dimensional wavelet; as can be discerned from the shape, it is designed to extract diagonally oriented structure at a particular scale. Taking inner products with this wavelet over all spatial positions of the image (a procedure known as convolution) yields a collection of wavelet coefficients at all spatial positions of the image. These coefficients are then sub-sampled, depending on the scale of the wavelet. Then we reconstruct the image from these coefficients. Doing so at multiple scales (three in this illustration) and orientations (four in this illustration) yields the multiscale pyramid shown in panel (c). Once again, although the original image is not a sparse signal, its representation in this multiscale basis is very sparse, with many coefficients either zero or very close to zero. As a demonstration of this sparsity, panel (d) shows a histogram of one of the wavelet coefficients, obtained by pooling its values over all spatial positions of the image. This histogram is plotted on the log scale, and the sharp peak around zero reveals the sparsity of the coefficient distribution.

10.2.2 Approximation in Orthogonal Bases

The goal of signal compression is to represent the signal $\theta^* \in \mathbb{R}^p$, typically in an approximate manner, using some number $k \ll p$ of coefficients much smaller than the ambient dimension. In the setting of orthogonal bases, one method for doing so is based on using only a sparse subset of the orthogonal vectors $\{\psi_j\}_{j=1}^p$. In particular, for an integer $k \in \{1, 2, \ldots, p\}$ that characterizes the

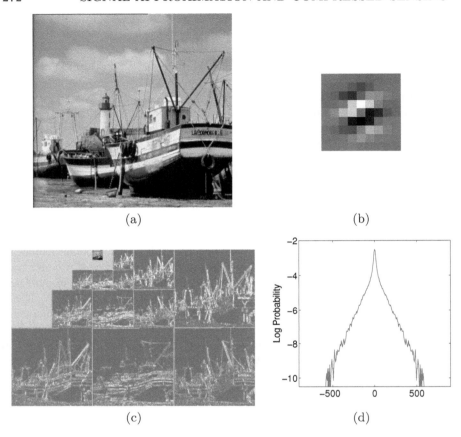

Figure 10.2 *Sparsity in wavelet-based representations of natural images. (a) "Boats" image. (b) Basis vector of a multiscale pyramid transform, drawn here as a 2-dimensional image. (c) Three levels of a multiscale representation of "Boats" image with four different orientations at each scale. (d) Log histogram of the amplitudes of a wavelet coefficient from a fixed scale and orientation, pooled over all pixels within the image. Note that the majority of coefficients are close to zero, with relatively few large in absolute value.*

approximation accuracy, let us consider reconstructions of the form

$$\boldsymbol{\Psi}\beta \;=\; \sum_{j=1}^{p} \beta_j \psi_j, \qquad \text{such that } \|\beta\|_0 := \sum_{j=1}^{p} \mathbb{I}[\beta_j \neq 0] \;\leq\; k. \qquad (10.3)$$

Here we have introduced the ℓ_0-"norm," which simply counts the number of nonzero elements in the vector $\beta \in \mathbb{R}^p$. We then consider the problem of

(a) (b)

Figure 10.3 *Illustration of image compression based on wavelet thresholding. (a) Zoomed portion of the original "Boats" image from Figure 10.2(a). (b) Reconstruction based on retaining 5% of the wavelet coefficients largest in absolute magnitude. Note that the distortion is quite small, and concentrated mainly on the fine-scale features of the image.*

optimal k-sparse approximation—namely, to compute

$$\widehat{\beta}^k \in \arg\min_{\beta \in \mathbb{R}^p} \|\theta^* - \Psi\beta\|_2^2 \qquad \text{such that } \|\beta\|_0 \leq k. \tag{10.4}$$

Given the optimal solution $\widehat{\beta}^k$ of this problem, the reconstruction

$$\theta^k := \sum_{j=1}^p \widehat{\beta}_j^k \psi_j \tag{10.5}$$

defines the best least-squares approximation to θ^* based on k terms. Figure 10.3 illustrates the idea.

Note that the problem (10.4) is nonconvex and combinatorial, due to the ℓ_0-norm constraint. Despite this fact, it is actually very easy to solve in this particular case, essentially due to the structure afforded by orthonormal transforms. In particular, suppose that we order the vector $\beta^* \in \mathbb{R}^p$ of basis coefficients in terms of their absolute values, thereby defining the order statistics

$$|\beta_{(1)}^*| \geq |\beta_{(2)}^*| \geq \ldots \geq |\beta_{(p)}^*|. \tag{10.6}$$

Then for any given integer $k \in \{1, 2, \ldots, p\}$, it can be shown that the optimal k-term approximation is given by

$$\widehat{\theta}^k := \sum_{j=1}^k \beta_{(j)}^* \psi_{\sigma(j)}, \tag{10.7}$$

where $\sigma(j)$ denotes the basis vector associated with the jth order statistic. In words, we retain only the basis vectors associated with the largest k coefficients in absolute value.

In summary, then, we have the following simple algorithm for computing optimal k-term approximations in an orthogonal basis:

1. Compute the basis coefficients $\beta_j^* = \langle \theta^*, \psi_j \rangle$ for $j = 1, 2, \ldots, p$. In matrix-vector notation, compute the vector $\beta^* = \mathbf{\Psi}^T \theta^*$.

2. Sort the coefficients in terms of absolute values as in (10.6), and extract the top k coefficients.

3. Compute the best k-term approximation $\widehat{\theta}^k$ as in (10.7).

For any orthogonal basis, the computational complexity of this procedure is at most $\mathcal{O}(p^2)$, with the $\mathcal{O}(p \log p)$ complexity of sorting in step 2 dominated by the complexity of computing the basis coefficients in step 1. An attractive feature of many orthogonal representations, including Fourier bases and discrete wavelets, is that the basis coefficients can be computed in time $\mathcal{O}(p \log p)$.

As discussed previously, Figure 10.1 provides one illustration of signal approximation within the Haar wavelet basis. In particular, the bottom-left panel shows the approximated signal $\widehat{\theta}^{64}$, based on retaining only half of the Haar wavelet coefficients ($k/p = 64/128 = 0.5$).

10.2.3 Reconstruction in Overcomplete Bases

Orthonormal bases, though useful in many ways, have a number of shortcomings. In particular, there is a limited class of signals that have sparse representations in any given orthonormal basis. For instance, Fourier bases are particularly well-suited to reconstructing signals with a globally periodic structure; in contrast, the Haar basis with its localized basis vectors is rather poor at capturing this kind of structure. On the other hand, the Haar basis excels at capturing step discontinuities, whereas such jumps have very nonsparse representations in the Fourier basis.

Based on this intuition, it is relatively straightforward to construct signals that are in some sense "simple," but fail to have sparse representations in a classical orthonormal basis. As an illustration, panel (a) of Figure 10.4 shows a signal $\theta^* \in \mathbb{R}^{128}$ that contains a mixture of both some globally periodic components, and some rapid (nearly discontinuous) transitions. As shown in panel (b), its Haar coefficients $\beta^* = \mathbf{\Psi}^T \theta^*$ are relatively dense, because many basis vectors are required to reconstruct the globally periodic portion of the signal. Similarly, as shown in panel (c), its representation $\alpha^* = \mathbf{\Phi}^T \theta^*$ in the discrete cosine basis (a type of Fourier representation) is also relatively dense. Due to this lack of sparsity, neither basis alone will provide a good sparse approximation to the original signal.

However, suppose that we allow the reconstruction to use subsets of vectors from *both bases* simultaneously; in this case, it might be possible to obtain a significantly more accurate, or even exact, sparse approximation. To set up

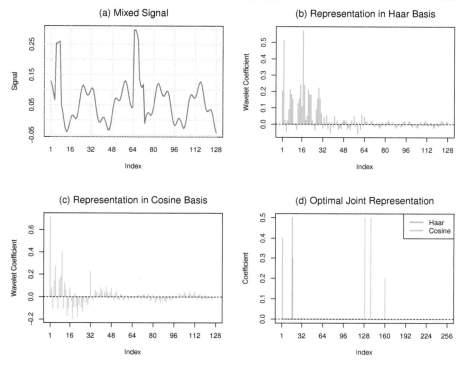

Figure 10.4 *(a) Original signal $\theta^* \in \mathbb{R}^p$ with $p = 128$. (b) Representation $\mathbf{\Psi}^T \theta^*$ in the Haar basis. (c) Representation $\mathbf{\Phi}^T \theta^*$ in the discrete cosine basis. (d) Coefficients $(\widehat{\alpha}, \widehat{\beta}) \in \mathbb{R}^p \times \mathbb{R}^p$ of the optimally sparse joint representation obtained by solving basis pursuit linear program (10.11).*

the problem more precisely, given a pair of orthonormal bases $\{\psi_j\}_{j=1}^p$ and $\{\phi_j\}_{j=1}^p$, let us consider reconstructions of the form

$$\underbrace{\sum_{j=1}^p \alpha_j \phi_j}_{\mathbf{\Phi}\alpha} + \underbrace{\sum_{j=1}^p \beta_j \psi_j}_{\mathbf{\Psi}\beta} \quad \text{such that } \|\alpha\|_0 + \|\beta\|_0 \leq k, \qquad (10.8)$$

and the associated optimization problem

$$\underset{(\alpha,\beta)\in\mathbb{R}^p\times\mathbb{R}^p}{\text{minimize}} \|\theta^* - \mathbf{\Phi}\alpha - \mathbf{\Psi}\beta\|_2^2 \quad \text{such that } \|\alpha\|_0 + \|\beta\|_0 \leq k. \qquad (10.9)$$

Despite its superficial similarity to our earlier k-term approximation problem (10.5), the optimization problem (10.9) is actually very difficult to solve. Unlike the earlier case, we are now working in an *overcomplete basis* described by the union of the two bases $\mathbf{\Phi}$ and $\mathbf{\Psi}$.

Nonetheless, we can resort to our usual relaxation of the ℓ_0-"norm," and consider the following convex program

$$\underset{(\alpha,\beta)\in\mathbb{R}^p\times\mathbb{R}^p}{\text{minimize}} \ \|\theta^* - \mathbf{\Phi}\alpha - \mathbf{\Psi}\beta\|_2^2 \qquad \text{such that } \|\alpha\|_1 + \|\beta\|_1 \leq R, \qquad (10.10)$$

where $R > 0$ is a user-defined radius. This program is a constrained version of the lasso program, also referred to as the relaxed basis-pursuit program. When seeking a perfect reconstruction, we can also consider the even simpler problem

$$\underset{(\alpha,\beta)\in\mathbb{R}^p\times\mathbb{R}^p}{\text{minimize}} \ \|\alpha\|_1 + \|\beta\|_1 \qquad \text{such that } \theta^* = \begin{bmatrix} \mathbf{\Phi} & \mathbf{\Psi} \end{bmatrix} \begin{bmatrix} \alpha \\ \beta \end{bmatrix}. \qquad (10.11)$$

This problem is a linear program (LP), often referred to as the basis-pursuit linear program.

Returning to the example discussed in Figure 10.4, panel (d) shows the optimal coefficients $(\widehat{\alpha}, \widehat{\beta}) \in \mathbb{R}^p \times \mathbb{R}^p$ obtained by solving the basis pursuit LP (10.11). We thus find that the original signal in panel (a) can be generated by an extremely sparse combination, with only six nonzero coefficients, in the overcomplete basis formed by combining the Haar and discrete cosine representations. In fact, this is the sparsest possible representation of the signal, so that in this case, solving the basis pursuit LP (10.11) is equivalent to solving the ℓ_0-constrained problem (10.9).

Naturally, the reader might wonder about the generality of this phenomenon—namely, when does the solution to the basis pursuit LP coincide with the computationally difficult ℓ_0-problem (10.9)? As it turns out, the answer to this question depends on the degree of incoherence between the two bases $\mathbf{\Phi}$ and $\mathbf{\Psi}$, as we explore at more length in Section 10.4.

10.3 Random Projection and Approximation

In the previous sections, we discussed approximating a signal by computing its projection onto each of a fixed set of basis functions. We now turn to the use of random projections in signal approximation. This allows one to use a smaller number of (random) basis functions than is required under a fixed basis. We will combine this with an ℓ_1-penalty on the coefficient of each projection, leading to the idea of *compressed sensing*.

A random projection of a signal θ^* is a measurement of the form

$$y_i = \langle z_i, \theta^* \rangle = \sum_{j=1}^{p} z_{ij}\theta_j^*, \qquad (10.12)$$

where $z_i \in \mathbb{R}^p$ is a random vector. The idea of using random projections for dimensionality reduction and approximation is an old one, dating back (at least) to classical work on metric embedding and spherical sections of convex bodies (see the bibliographic section for more details). We begin by describing a classical use of random projection, namely for embedding data while preserving distances between points, and then move on to discuss compressed sensing, which combines random projections with ℓ_1-relaxation.

10.3.1 Johnson–Lindenstrauss Approximation

As one application of random projection, let us consider how they can be used to approximate a finite collection of vectors, say representing some dataset. The technique that we describe is often known as Johnson–Lindenstrauss embedding, based on the authors who pioneered its use in studying the more general problem of metric embedding (see the bibliographic section for more details). Suppose that we are given M data points $\{u_1, \ldots, u_M\}$ lying in $\mathbb{R}^p$. If the data dimension p is large, then it might be too expensive to store the dataset. In this setting, one approach is to design a dimension-reducing mapping $F : \mathbb{R}^p \to \mathbb{R}^N$ with $N \ll p$ that preserves some "essential" features of the dataset, and then store only the projected dataset $\{F(u_1), \ldots, F(u_M)\}$. For example, since many algorithms operate on datasets by computing pairwise distances, we might be interested in a mapping F with the guarantee that for some tolerance $\delta \in (0, 1)$, we have

$$(1-\delta)\|u_i - u_{i'}\|_2^2 \leq \|F(u_i) - F(u_{i'})\|_2^2 \leq (1+\delta)\|u_i - u_{i'}\|_2^2 \text{ for all pairs } i \neq i'. \tag{10.13}$$

Of course, this is always possible if the projected dimension N is large enough, but the goal is to do it with relatively small N.

As shown in the seminal work of Johnson and Lindenstrauss, random projections provide one method for designing such approximate distance-preserving embeddings. The construction is straightforward:

(a) Form a random matrix $\mathbf{Z} \in \mathbb{R}^{N \times p}$ with each $Z_{ij} \sim N(0, 1)$, i.i.d., and define the linear mapping $F : \mathbb{R}^p \to \mathbb{R}^N$ via

$$F(u) := \frac{1}{\sqrt{N}} \mathbf{Z} u. \tag{10.14}$$

(b) Compute the projected dataset $\{F(u_1), F(u_2), \ldots, F(u_M)\}$.

An interesting question is the following: for a given tolerance $\delta \in (0, 1)$ and number of data points M, how large should we choose the projected dimension N to ensure that approximate distance-preserving property (10.13) holds with high probability? In Exercises 10.1 and 10.2, we show that this property holds with high probability as long as $N > \frac{c}{\delta^2} \log M$ for some universal constant c. Thus, the dependence on the number M of data points scales logarithmically, and hence is very mild.

As a particular example, suppose that our goal is to obtain a compressed representation of all Boolean vectors $u \in \{-1, 0, 1\}^p$ that are k-sparse.[1] By

---

[1] A vector $u \in \mathbb{R}^p$ is k-sparse if only $k \leq p$ elements are nonzero.

a simple counting argument, there are $M = 2^k \binom{p}{k}$ such vectors. Noting that $\log M \leq k \log \left(\frac{e^2 p}{k} \right)$, we see that a projection dimension $N > \frac{c}{\delta^2} k \log \left(\frac{e^2 p}{k} \right)$ suffices to preserve pairwise distances up to δ-accuracy between all k-sparse Boolean vectors. This example provides a natural segue to the method of compressed sensing, which combines random projections with ℓ_1-relaxation.

10.3.2 Compressed Sensing

Compressed sensing is a combination of random projection and ℓ_1-regularization that was introduced in independent work by Candes and Tao (2005) and Donoho (2006); since this pioneering work, an extensive literature on the topic has developed, with numerous applications including medical imaging and single-pixel cameras, among others. In this section, we provide a brief introduction to the basic ideas.

The motivation for compressed sensing is the inherent wastefulness of the standard method for compressing signals in an orthogonal basis. As described in Section 10.2.2, this approach involves first computing the full vector $\beta^* \in \mathbb{R}^p$ of basis coefficients (step 1 on page 274), and then discarding a large fraction of them in order to obtain the k-sparse approximation $\widehat{\theta}^k$ of the underlying signal θ^* (step 2). Given that we end up discarding most of the basis coefficients, is it really necessary to compute all of them? Of course, if one knew *a priori* which subset of k coefficients were to be retained for the sparse approximation $\widehat{\theta}^k$, then one could simply compute this subset of basis coefficients. We refer to this approach as the oracle technique. Of course, it is unimplementable in practice, since we don't know *a priori* which coefficients are the most relevant for a given signal.

The power of compressed sensing is that it enables one to mimic the behavior of the oracle with very little computational overhead. It combines random projection with ℓ_1-minimization in the following way. Instead of pre-computing all of the basis coefficients $\beta^* = \mathbf{\Psi}^T \theta^*$, suppose that we compute some number N of random projections, say of the form $y_i = \langle z_i, \theta^* \rangle$, for $i = 1, 2, \ldots, N$. We are free to choose the form of the random projection vectors $z_i \in \mathbb{R}^p$, and we discuss a number of reasonable choices shortly.

Thus, the setup of our problem is as follows: we are given an N-vector $\mathbf{y}$ of random projections of the signal θ^*. Also known to us is the $N \times p$ random matrix $\mathbf{Z}$ with i^{th} row z_i, used to compute the random projections; we refer to $\mathbf{Z}$ as the *design matrix* or measurement matrix. The observation vector $\mathbf{y}$ and design matrix $\mathbf{Z}$ are linked to the unknown signal $\theta^* \in \mathbb{R}^N$ by the matrix-vector equation $\mathbf{y} = \mathbf{Z}\theta^*$, and our goal is to recover (exactly or approximately) the signal $\theta^* \in \mathbb{R}^p$. See Figure 10.5(a) for an illustration of this setup.

At first sight, the problem seems very simple, since determining θ^* amounts to solving a linear system. However, for this method to be cheaper than the standard approach (and therefore of practical interest), it is essential that the number of projections (or sample size) N be much smaller than the ambient dimension p. For this reason, the linear system $\mathbf{y} = \mathbf{Z}\theta^*$ is highly under-

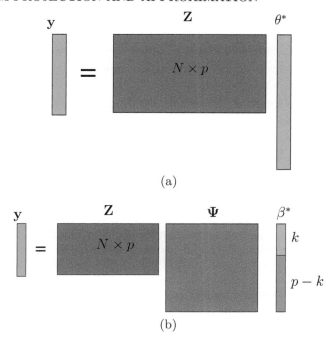

(a)

(b)

Figure 10.5 *(a) An under-determined linear system* $\mathbf{y} = \mathbf{Z}\theta^*$*: Each row* z_i *of the* $N \times p$ *measurement matrix* $\mathbf{Z}$ *defines the random projection* $y_i = \langle z_i, \theta^* \rangle$. *The signal* $\theta^* \in \mathbb{R}^p$ *need not be sparse in the canonical basis. (b) Equivalent representation of the linear system: Basis coefficients* $\beta^* = \mathbf{\Psi}^T \theta^*$ *are assumed to be* k*-sparse. This transformation defines an equivalent linear system* $\mathbf{y} = \widetilde{\mathbf{Z}}\beta^*$ *with sparsity that can be exploited.*

determined: there are many signals θ that are consistent with the observed random projections.

However, if we also have the additional side-information that $\mathbf{\Psi}^T \theta^*$ is sparse, then it could be possible to recover θ^* exactly, even though the linear system on its own is under-determined. In an ideal world, we would like to exploit this sparsity by solving the ℓ_0-based problem

$$\underset{\theta \in \mathbb{R}^p}{\text{minimize}} \, \|\mathbf{\Psi}^T \theta\|_0 \qquad \text{such that } \mathbf{y} = \mathbf{Z}\theta. \qquad (10.15)$$

The ℓ_0-problem is combinatorial, and known to be computationally intractable (NP-hard) in general; thus, we are led to consider the ℓ_1-relaxation

$$\underset{\theta \in \mathbb{R}^p}{\text{minimize}} \, \|\mathbf{\Psi}^T \theta\|_1 \qquad \text{such that } \mathbf{y} = \mathbf{Z}\theta. \qquad (10.16)$$

Equivalently, we can write this problem in terms of the transform coefficient vector $\beta \in \mathbb{R}^p$, namely as

$$\underset{\beta \in \mathbb{R}^p}{\text{minimize}} \, \|\beta\|_1 \qquad \text{such that } \mathbf{y} = \widetilde{\mathbf{Z}}\beta, \qquad (10.17)$$

where we have defined the transformed matrix $\widetilde{\mathbf{Z}} := \mathbf{Z}\mathbf{\Psi} \in \mathbb{R}^{N \times p}$. See Figure 10.5(b) for an illustration of this transformed linear system.

In summary, then, the method of compressed sensing operates as follows:

1. For a given sample size N, compute the random projections $y_i = \langle z_i, \theta^* \rangle$ for $i = 1, 2, \ldots, N$.

2. Estimate the signal θ^* by solving the linear program (10.16) to obtain $\widehat{\theta}$. (Equivalently, solve the linear program (10.17) to obtain $\widehat{\beta}$, and set $\widehat{\theta} = \mathbf{\Psi}\widehat{\beta}$.)

To be clear, we have actually described a family of procedures, depending on our choice of the random projection vectors $\{z_i\}_{i=1}^{N}$, or equivalently the transformed design matrix $\mathbf{Z}$. A variety of different design matrices $\mathbf{Z}$ have been studied for the purposes of compressed sensing. Perhaps the simplest choice is to choose its entries $z_{ij} \sim N(0,1)$ in an i.i.d. manner, leading to a standard Gaussian random matrix. Other choices of matrices for compressed sensing include random Bernoulli matrices formed with i.i.d. entries drawn as $z_{ij} \in \{-1, +1\}$ with equal probability, as well as random submatrices of Fourier matrices.

When can compressed sensing succeed using a number of projections N less than the signal dimension p? As we discuss in Section 10.4.2, it is sufficient that the columns of the transformed design matrix $\widetilde{\mathbf{Z}}$ be "incoherent", and there are different measures of such incoherence. The simplest measure of incoherence is pairwise, based on the inner products between the columns of $\widetilde{\mathbf{Z}}$. A more sophisticated notion of incoherence is the restricted isometry property (RIP), based on looking on the conditioning of submatrices of $\widetilde{\mathbf{Z}}$ consisting of up to k columns. An important fact is that the random design matrices discussed above satisfy RIP with high probability using a relatively small number of projections N. For instance, for the standard Gaussian or Bernoulli cases, it can be shown that RIP holds with high probability with as few as $N = \Omega\big(k \log \frac{p}{k}\big)$ samples, where $k < p$ is the sparsity of the basis coefficient vector β^*. Note that any method—even the unimplementable oracle that already knew the support of β^*—would require at least $N = k$ random projections for exact recovery. Thus, compressed sensing incurs a multiplicative overhead of only $\mathcal{O}\big(\log(p/k)\big)$ relative to oracle performance.

10.4   Equivalence between ℓ_0 and ℓ_1 Recovery

Thus far, we have discussed a number of applications of ℓ_1-norm regularization in signal processing, including sparse approximation in overcomplete bases (Section 10.2.3), and compressed sensing (Section 10.3.2). In both cases, the ℓ_1-norm is introduced as a computationally tractable surrogate to optimization problems involving the intractable ℓ_0-"norm." Up to this point, we have not addressed in any depth an important question: when is solving the ℓ_1-relaxation equivalent to solving the original ℓ_0-problem?

More precisely, given an observation vector $\mathbf{y} \in \mathbb{R}^p$ and a design matrix

$\mathbf{X} \in \mathbb{R}^{N \times p}$, let us consider the two problems

$$\underset{\beta \in \mathbb{R}^p}{\text{minimize}} \, \|\beta\|_0 \qquad \text{such that } \mathbf{X}\beta = \mathbf{y}, \tag{10.18}$$

and

$$\underset{\beta \in \mathbb{R}^p}{\text{minimize}} \, \|\beta\|_1 \qquad \text{such that } \mathbf{X}\beta = \mathbf{y}. \tag{10.19}$$

This setup includes as a special case the problem of sparse approximation in an overcomplete basis, as discussed in Section 10.2.3; in this case, the observation $\mathbf{y}$ is equal to the signal θ^* to be approximated, and the design matrix $\mathbf{X} = \begin{bmatrix} \mathbf{\Phi} & \mathbf{\Psi} \end{bmatrix}$. It also includes the case of compressed sensing, where $\mathbf{X}$ is the transformed version of the random projection matrix (namely, $\widetilde{\mathbf{Z}}$ in our earlier notation).

10.4.1   Restricted Nullspace Property

Suppose that the ℓ_0-based problem (10.18) has a unique optimal solution, say $\beta^* \in \mathbb{R}^p$. Our interest is in understanding when β^* is also the unique optimal solution of the ℓ_1-based problem (10.19), in which case we say that the basis pursuit LP is *equivalent* to ℓ_0-recovery. Remarkably, there exists a very simple necessary and sufficient condition on the design matrix $\mathbf{X}$ for this equivalence to hold. For a given subset $S \subseteq \{1, 2, \ldots, p\}$, it is stated in terms of the set

$$\mathbb{C}(S) := \big\{\beta \in \mathbb{R}^p \mid \|\beta_{S^c}\|_1 \le \|\beta_S\|_1\big\}. \tag{10.20}$$

The set $\mathbb{C}(S)$ is a cone, containing all vectors that are supported on S, and other vectors as well. Roughly, it corresponds to the cone of vectors that have most of their mass allocated to S. Given a matrix $\mathbf{X} \in \mathbb{R}^{N \times p}$, its nullspace is given by $\text{null}(\mathbf{X}) = \{\beta \in \mathbb{R}^p \mid \mathbf{X}\beta = \mathbf{0}\}$.

Definition 10.1. Restricted nullspace property. For a given subset $S \subseteq \{1, 2, \ldots, p\}$, we say that the design matrix $\mathbf{X} \in \mathbb{R}^{N \times p}$ satisfies the *restricted nullspace property* over S, denoted by $\text{RN}(S)$, if

$$\text{null}(\mathbf{X}) \cap \mathbb{C}(S) = \{0\}. \tag{10.21}$$

In words, the $\text{RN}(S)$ property holds when the only element of the cone $\mathbb{C}(S)$ that lies within the nullspace of $\mathbf{X}$ is the all-zeroes vector. The following theorem highlights the significance of this property:

Theorem 10.1. ℓ_0 and ℓ_1 equivalence. Suppose that $\beta^* \in \mathbb{R}^p$ is the unique solution to the ℓ_0 problem (10.18), and has support S. Then the basis pursuit relaxation (10.19) has a unique solution equal to β^* if and only if $\mathbf{X}$ satisfies the $\text{RN}(S)$ property.

The proof of Theorem 10.1 is relatively short, and is provided in Section 10.4.3.

Since the subset S is not known in advance—indeed, it is usually what we are trying to determine—it is natural to seek matrices that satisfy a uniform version of the restricted nullspace property. For instance, we say that the uniform RN property of order k holds if $\mathrm{RN}(S)$ holds for all subsets of size at most k. In this case, we are guaranteed that the ℓ_1-relaxation succeeds for any vector supported on any subset of size at most k.

10.4.2 Sufficient Conditions for Restricted Nullspace

Of course, in order for Theorem 10.1 to be useful in practice, we need to verify the restricted nullspace property. A line of work has developed various conditions for certifying the uniform RN property. The simplest and historically earliest condition is based on the *pairwise incoherence*

$$\nu(\mathbf{X}) := \max_{\substack{j,j'=1,2,\ldots,p \\ j \neq j'}} \frac{|\langle \mathbf{x}_j, \mathbf{x}_{j'} \rangle|}{\|\mathbf{x}_j\|_2 \|\mathbf{x}_{j'}\|_2}. \tag{10.22}$$

For centered $\mathbf{x}_j$ this is the maximal absolute pairwise correlation. When $\mathbf{X}$ is rescaled to have unit-norm columns, an equivalent representation is given by $\nu(\mathbf{X}) = \max_{j \neq j'} |\langle \mathbf{x}_j, \mathbf{x}_{j'} \rangle|$, which illustrates that the pairwise incoherence measures how close the Gram matrix $\mathbf{X}^T \mathbf{X}$ is to the p-dimensional identity matrix in an element-wise sense.

The following result shows that having a low pairwise incoherence is sufficient to guarantee exactness of the basis pursuit LP:

Proposition 10.1. Pairwise incoherence implies RN. Suppose that for some integer $k \in \{1, 2, \ldots, p\}$, the pairwise incoherence satisfies the bound $\nu(\mathbf{X}) < \frac{1}{3k}$. Then $\mathbf{X}$ satisfies the uniform RN property of order k, and hence, the basis pursuit LP is exact for all vectors with support at most k.

See Section 10.4.3 for the proof of this claim.

An attractive feature of the pairwise incoherence is that it is easily computed; in particular, in $\mathcal{O}(Np^2)$ time. A disadvantage is that it provides very conservative bounds that do not always capture the actual performance of ℓ_1-relaxation in practice. For instance, consider the matrix $\mathbf{X} = \begin{bmatrix} \mathbf{\Phi} & \mathbf{\Psi} \end{bmatrix}$, as arises in the overcomplete basis problem (10.11). We can numerically compute the incoherence, say for the discrete cosine and Haar bases in dimension $p = 128$, as illustrated in Figure 10.4. We find that Proposition 10.1 guarantees exact recovery of all signals with sparsity $k = 1$, whereas in practice, the ℓ_1-relaxation works for much larger values of k.

For random design matrices, such as those that arise in compressed sensing, one can use probabilistic methods to bound the incoherence. For instance, consider a random matrix $\mathbf{X} \in \mathbb{R}^{N \times p}$ with i.i.d. $N(0, 1/N)$ entries. Here we have rescaled the variance so that the columns of $\mathbf{X}$ have expected norm equal to one. For such a matrix, one can show that $\nu(\mathbf{X}) \precsim \sqrt{\frac{\log p}{N}}$ with high probability as (N, p) tend to infinity (see Exercise 10.5). Combined with Proposition 10.1, we conclude that the ℓ_1-relaxation (10.16) will exactly recover all signals with sparsity at most k as long as the number of projections scales as $N \succsim k^2 \log p$.

In fact, for random designs and compressed sensing, this scaling can be sharpened using the *restricted isometry property* (RIP). Recall that the incoherence condition (10.22) is a measure of the orthonormality of pairs of columns of the design matrix $\mathbf{X}$. The notion of restricted isometry is to constrain much larger submatrices of $\mathbf{X}$ to have nearly orthogonal columns.

Definition 10.2. Restricted isometry property. For a tolerance $\delta \in (0,1)$ and integer $k \in \{1, 2, \ldots, p\}$, we say that RIP$(k, \delta)$ holds if

$$\|\mathbf{X}_S^T \mathbf{X}_S - \mathbf{I}_{k \times k}\|_{\mathrm{op}} \le \delta \tag{10.23}$$

for all subsets $S \subset \{1, 2, \ldots, p\}$ of cardinality k.

We recall here that $\|\cdot\|_{\mathrm{op}}$ denotes the operator norm, or maximal singular value of a matrix. Due to the symmetry of $\mathbf{X}_S^T \mathbf{X}_S$, we have the equivalent representation

$$\|\mathbf{X}_S^T \mathbf{X}_S - \mathbf{I}_{k \times k}\|_{\mathrm{op}} = \sup_{\|u\|_2 = 1} \left| u^T \left(\mathbf{X}_S^T \mathbf{X}_S - \mathbf{I}_{k \times k} \right) u \right| = \sup_{\|u\|_2 = 1} \left| \|\mathbf{X}_S u\|_2^2 - 1 \right|.$$

Thus, we see that RIP(k, δ) holds if and only if for all subsets S of cardinality k, we have

$$\frac{\|\mathbf{X}_S u\|_2^2}{\|u\|_2^2} \in [1 - \delta, 1 + \delta] \qquad \text{for all } u \in \mathbb{R}^k \backslash \{0\},$$

hence the terminology of restricted isometry.

The following result shows that RIP is a sufficient condition for the restricted nullspace to hold:

Proposition 10.2. RIP implies restricted nullspace. If RIP$(2k, \delta)$ holds with $\delta < 1/3$, then the uniform RN property of order k holds, and hence the ℓ_1-relaxation is exact for all vectors supported on at most k elements.

We work through the proof of a slightly weaker version of this claim in Exercise 10.8. Observe that the RIP$(2k, \delta)$ condition imposes constraints on a huge number of submatrices, namely $\binom{p}{2k}$ in total. On the other hand, as opposed to the pairwise incoherence condition, the actual RIP constant δ has no dependence on k.

From known results in random matrix theory, various choices of random projection matrices $\mathbf{X}$ satisfy RIP with high probability as long as $N \succsim k \log \frac{ep}{k}$. Among other matrix ensembles, this statement applies to a standard Gaussian random matrix $\mathbf{X}$ with i.i.d. $N(0, \frac{1}{N})$ entries; see Exercise 10.6 for details. Thus, we see that the RIP-based approach provides a certificate for exact recovery based on far fewer samples than pairwise incoherence, which as previously discussed, provides guarantees when $N \succsim k^2 \log p$. On the other hand, a major drawback of RIP is that—in sharp contrast to the pairwise incoherence—it is very difficult to verify in practice due to the number $\binom{p}{2k}$ of submatrices.

We conclude the chapter by providing proofs of the claims given in the preceding section.

10.4.3.1   Proof of Theorem 10.1

First, suppose that $\mathbf{X}$ satisfies the RN(S) property. Let $\widehat{\beta} \in \mathbb{R}^p$ be any optimal solution to the basis pursuit LP (10.19), and define the error vector $\Delta := \widehat{\beta} - \beta^*$. Our goal is to show that $\Delta = 0$, and in order to do so, it suffices to show that $\Delta \in \text{null}(\mathbf{X}) \cap \mathbb{C}(S)$. On the one hand, since β^* and $\widehat{\beta}$ are optimal (and hence feasible) solutions to the ℓ_0 and ℓ_1 problems, respectively, we are guaranteed that $\mathbf{X}\beta^* = \mathbf{y} = \mathbf{X}\widehat{\beta}$, showing that $\mathbf{X}\Delta = 0$. On the other hand, since β^* is also feasible for the ℓ_1-based problem (10.19), the optimality of $\widehat{\beta}$ implies that $\|\widehat{\beta}\|_1 \le \|\beta^*\|_1 = \|\beta_S^*\|_1$. Writing $\widehat{\beta} = \beta^* + \Delta$, we have

$$\|\beta_S^*\|_1 \ge \|\widehat{\beta}\|_1 = \|\beta_S^* + \Delta_S\|_1 + \|\Delta_{S^c}\|_1$$
$$\ge \|\beta_S^*\|_1 - \|\Delta_S\|_1 + \|\Delta_{S^c}\|_1,$$

where the final bound follows by triangle inequality. Rearranging terms, we find that $\Delta \in \mathbb{C}(S)$; since $\mathbf{X}$ satisfies the RN(S) condition by assumption, we conclude that $\Delta = 0$ as required.

We lead the reader through a proof of the converse in Exercise 10.4.

10.4.3.2   Proof of Proposition 10.1

We may assume without loss of generality (rescaling as needed) that $\|\mathbf{x}_j\|_2 = 1$ for all $j = 1, 2, \ldots, p$. To simplify notation, let us assume an incoherence condition of the form $\nu(\mathbf{X}) < \frac{\delta}{k}$ for some $\delta > 0$, and verify the sufficiency of $\delta = 1/3$ in the course of the argument.

For an arbitrary subset S of cardinality k, suppose that $\beta \in \mathbb{C}(S) \backslash \{0\}$. It suffices to show that $\|\mathbf{X}\beta\|_2^2 > 0$, and so we begin with the lower bound

$$\|\mathbf{X}\beta\|_2^2 \ge \|\mathbf{X}_S\beta_S\|_2^2 + 2\beta_S^T \mathbf{X}_S^T \mathbf{X}_{S^c}\beta_{S^c}. \tag{10.24}$$

On one hand, we have

$$2\left|\beta_S^T \mathbf{X}_S^T \mathbf{X}_{S^c}\beta_{S^c}\right| \le 2 \sum_{i \in S} \sum_{j \in S^c} |\beta_i| \cdot |\beta_j| \cdot |\langle \mathbf{x}_i, \mathbf{x}_j \rangle|$$
$$\overset{(i)}{\le} 2\|\beta_S\|_1 \|\beta_{S^c}\|_1 \, \nu(\mathbf{X})$$
$$\overset{(ii)}{\le} \frac{2\delta\|\beta_S\|_1^2}{k}$$
$$\overset{(iii)}{\le} 2\,\delta\,\|\beta_S\|_2^2,$$

where inequality (i) uses the definition (10.22) of the pairwise incoherence;

inequality (ii) exploits the assumed bound on $\nu(\mathbf{X})$ combined with the fact that $\beta \in \mathbb{C}(S)$; and inequality (iii) uses the fact that $\|\beta_S\|_1^2 \leq k\|\beta_S\|_2^2$, by Cauchy–Schwarz, since the cardinality of S is at most k. Consequently, we have established that

$$\|\mathbf{X}\beta\|_2^2 \geq \|\mathbf{X}_S\beta_S\|_2^2 - 2\delta\|\beta_S\|_2^2. \tag{10.25}$$

In order to complete the proof, it remains to lower bound $\|\mathbf{X}_S\beta_S\|_2^2$. Letting $\|\cdot\|_{\mathrm{op}}$ denote the operator norm (maximum singular value) of a matrix, we have

$$\|\mathbf{X}_S^T\mathbf{X}_S - \mathbf{I}_{k\times k}\|_{\mathrm{op}} \leq \max_{i\in S} \sum_{j\in S\setminus\{i\}} |\langle x_i,\, x_j\rangle| \leq k\frac{\delta}{k} = \delta.$$

Consequently, $\|\mathbf{X}_S\beta_S\|_2^2 \geq (1-\delta)\|\beta_S\|_2^2$, and combined with the bound (10.25), we conclude that $\|\mathbf{X}\beta\|_2^2 > (1 - 3\delta)\|\beta_S\|_2^2$, so that $\delta = 1/3$ is sufficient as claimed.

Bibliographic Notes

There is an extensive literature on the sparsity of images and other signal classes when represented in wavelet and other multiscale bases (Field 1987, Ruderman 1994, Wainwright, Simoncelli and Willsky 2001, Simoncelli 2005). Sparse approximation in overcomplete bases is discussed in various papers (Donoho and Stark 1989, Chen et al. 1998, Donoho and Huo 2001, Elad and Bruckstein 2002, Feuer and Nemirovski 2003). The multiscale basis illustrated in Figure 10.2 is known as the steerable pyramid (Simoncelli and Freeman 1995). Random projection is a widely used technique in computer science and numerical linear algebra (Vempala 2004, Mahoney 2011, Pilanci and Wainwright 2014, e.g.). Johnson and Lindenstrauss (1984) proved the lemma that now bears their name in the context of establishing the existence of metric embeddings, using random projection as a proof technique. Compressed sensing was introduced independently by Candès, Romberg and Tao (2006) and Donoho (2006). Lustig, Donoho, Santos and Pauly (2008) discuss the applications of compressed sensing to medical imaging, whereas Candès and Wakin (2008) discuss various applications in signal processing.

The restricted nullspace property is discussed in Donoho and Huo (2001), Feuer and Nemirovski (2003), and Cohen, Dahmen and DeVore (2009). Various authors (Donoho and Huo 2001, Elad and Bruckstein 2002, Feuer and Nemirovski 2003) have studied the pairwise incoherence of overcomplete bases and other design matrices, as a sufficient condition for the restricted nullspace property. Candès and Tao (2005) introduced the restricted isometry property as a milder sufficient condition for the restricted nullspace property. For random matrices with i.i.d. sub-Gaussian rows, it follows from a combination of union bound and standard results in random matrix theory (Davidson and Szarek 2001, Vershynin 2012) that a sample size $N > ck\log\left(\frac{ep}{k}\right)$ suffices to ensure that the RIP is satisfied with high probability. Baraniuk, Davenport, DeVore and Wakin (2008) point out connections between the RIP and the Johnson–Lindenstrauss lemma; see also Exercise 10.6 for some related calculations. Krahmer and Ward (2011) establish a partial converse, showing that restricted isometry can be used to establish Johnson–Lindenstrauss type guarantees.

Exercises

Ex. 10.1 *Chi-squared concentration.* If $Y_1, \ldots, Y_N$ are i.i.d $\mathcal{N}(0,1)$ variates, then the variable $Z := \sum_{i=1}^{N} Y_i^2$ has a chi-squared distribution with N degrees of freedom. (In short, we write $Z \sim \chi_N^2$.)

(a) Show that for all $\lambda \in [-\infty, 1/2)$, we have

$$\mathbb{E}[\exp(\lambda(Z - d))] = \left[\frac{e^{-\lambda}}{\sqrt{1 - 2\lambda}}\right]^N. \tag{10.26}$$

(b) Use the bound (10.26) to show that

$$\mathbb{P}\big[|Z - N| \geq tN\big] \leq 2e^{-\frac{Nt^2}{32}} \qquad \text{for all } t \in (0, 1/2). \tag{10.27}$$

(The constants in this tail bound are not sharp, and can be improved.)

Ex. 10.2 *Johnson–Lindenstrauss approximation.* Recall from Section 10.3.1 the problem of distance-preserving embedding.

(a) Show that for any vector u with unit Euclidean norm, the random variable $N\|F(u)\|_2^2$ follows a χ^2-squared distribution with N degrees of freedom.

(b) For any $\delta \in (0, 1)$, define the event

$$\mathcal{E}(\delta) := \left\{\frac{\|F(u_i) - F(u_j)\|_2^2}{\|u_i - u_j\|_2^2} \in [1 - \delta, 1 + \delta] \qquad \text{for all pairs } i \neq j.\right\}$$

Use the results of Exercise 10.1 and the union bound to show that

$$\mathbb{P}\big[\mathcal{E}(\delta)\big] \geq 1 - 2e^{-N}.$$

as long as $N > \frac{64}{\delta^2} \log M$.

Ex. 10.3 For a given compact set $\mathcal{A} \subset \mathbb{R}^p$, an ϵ-covering set is a subset $\{u_1, \ldots, u_M\}$ of elements of $\mathcal{A}$ with the property for any $u \in \mathcal{A}$, there is some index $j \in \{1, \ldots, M\}$ such that $\|u - u_j\|_2 \leq \epsilon$. A ϵ-packing set is a subset $\{v^1, \ldots, v^{M'}\}$ of elements of $\mathcal{A}$ such that such that $\|v^i - v^j\|_2 > \epsilon$ for all pairs $i \neq j$ in $\{1, \ldots, M'\}$. We use $M(\epsilon)$ to denote the size of the largest ϵ-packing, and $N(\epsilon)$ to denote the size of the smallest ϵ-covering.

(a) Show that $M(2\epsilon) \leq N(\epsilon)$.

(b) Show that $N(\epsilon) \le M(\epsilon)$.

(c) Consider the Euclidean ball $\mathbb{B}_2(1) = \{u \in \mathbb{R}^p \mid \|u\|_2 = 1\}$. For each $\epsilon \in (0,1)$, show that there exists an ϵ-covering set with at most $M = (c/\epsilon)^p$ elements, for some universal constant $c > 0$. (*Hint:* Use part (b) and consider the volumes of Euclidean balls in p-dimensions.)

Ex. 10.4 In this exercise, we work through the proof of the converse of Theorem 10.1, in particular showing that if the ℓ_1-relaxation has a unique optimal solution, equal to the ℓ_0-solution, for all S-sparse vectors, then the set $\text{null}(\mathbf{X})\backslash\{0\}$ has no intersection with $\mathbb{C}(S)$.

(a) For a given vector $\beta^* \in \text{null}(\mathbf{X})\backslash\{0\}$, consider the basis-pursuit problem

$$\underset{\beta \in \mathbb{R}^p}{\text{minimize}} \, \|\beta\|_1 \quad \text{such that} \quad \mathbf{X}\beta = \mathbf{X}\begin{bmatrix} \beta_S^* \\ 0 \end{bmatrix}.$$

What is the link between its unique optimal solution $\widehat{\beta}$ and the vector

$$\begin{bmatrix} 0 \\ -\beta_{S^c}^* \end{bmatrix}?$$

(b) Use part (a) to show that $\beta^* \notin \mathbb{C}(S)$.

Ex. 10.5 Let $\mathbf{X} \in \mathbb{R}^{N \times p}$ be a random matrix with i.i.d. $\mathcal{N}(0, 1/N)$ entries. Show that it satisfies the pairwise incoherence condition (10.22) as long as $N > ck^2 \log p$ for a universal constant c. (*Hint:* The result of Exercise 10.1 may be useful.)

Ex. 10.6 Let $\mathbf{X} \in \mathbb{R}^{N \times p}$ be a random matrix with i.i.d. $\mathcal{N}(0, 1/N)$ entries. In this exercise, we show that the restricted isometry property (RIP) holds with high probability as long as $N > ck \log(ep/k)$ for a sufficiently large constant $c > 0$.

(a) Explain why it is sufficient to prove that there are constants c_1, c_2 such that

$$\|\mathbf{X}_S^T \mathbf{X}_S - \mathbf{I}_{2k \times 2k}\|_{\text{op}} \le t \tag{10.28}$$

with probability at least $1 - c_1 e^{-c_2 N t^2}$, for any fixed subset S of cardinality $2k$, and any $t \in (0, 1)$.

(b) Let $\mathbb{B}_2(1; S) = \{u \in \mathbb{R}^p \mid \|u\|_2 = 1 \text{ and } u_{S^c} = 0\}$ denote the intersection of the Euclidean ball with the subspace of vectors supported on a given subset S. Let $\{u_1, \ldots, u_M\}$ be an ϵ-covering of the set $\mathbb{B}_2(1; S)$, as previously defined in Exercise 10.3. Show that the bound (10.28) is implied by a bound of the form

$$\max_{j=1,\ldots,M} \left| \|\mathbf{X}u_j\|_2^2 - 1 \right| \le \epsilon,$$

with probability at least $1 - c_3 e^{c_4 N \epsilon^2}$, for any $\epsilon \in (0, 1)$.

(c) Use part (b) and Exercise 10.3 to complete the proof.

Ex. 10.7 ℓ_0 *and* ℓ_1*-balls.* In this exercise, we consider the relationship between ℓ_0 and ℓ_1-balls, and prove a containment property related to the success of ℓ_1-relaxation. For an integer $r \in \{1, \ldots, p\}$, consider the following two subsets:

$$
\mathbb{L}_0(r) := \mathbb{B}_2(1) \cap \mathbb{B}_0(r) = \{\theta \in \mathbb{R}^p \mid \|\theta\|_2 \le 1, \text{ and } \|\theta\|_0 \le r\},
$$
$$
\mathbb{L}_1(r) := \mathbb{B}_2(1) \cap \mathbb{B}_1(\sqrt{r}) = \{\theta \in \mathbb{R}^p \mid \|\theta\|_2 \le 1, \text{ and } \|\theta\|_1 \le \sqrt{r}\}.
$$

Let $\overline{\mathrm{conv}}$ denote the closure of the convex hull (when applied to a set).

(a) Prove that $\overline{\mathrm{conv}}\big(\mathbb{L}_0(r)\big) \subseteq \mathbb{L}_1(r)$.

(b) Prove that $\mathbb{L}_1(r) \subseteq 2\,\overline{\mathrm{conv}}\big(\mathbb{L}_0(r)\big)$.

(*Hint:* Part (b) is a more challenging problem: you may find it useful to consider the support functions of the two sets.)

Ex. 10.8 In this exercise, we work through a proof of (a slightly weaker version of) Proposition 10.2.

(a) For any subset S of cardinality k, the set $\mathbb{C}(S) \cap \mathbb{B}_2(1)$ is contained within the set $\mathbb{L}_1(r)$ with $r = 4k$.

(b) Now show that if $\mathrm{RIP}(8k, \delta)$ holds with $\delta < 1/4$, then the restricted nullspace property holds. (*Hint:* Part (b) of Exercise 10.7 could be useful.)

Chapter 11

Theoretical Results for the Lasso

In this chapter, we turn our attention to some theoretical results concerning the behavior of the lasso. We provide non-asymptotic bounds for the ℓ_2 and prediction error of the lasso, as well as its performance in recovering the support set of the unknown regression vector.

11.1 Introduction

Consider the standard linear regression model in matrix-vector form

$$\mathbf{y} = \mathbf{X}\beta^* + \mathbf{w}, \tag{11.1}$$

where $\mathbf{X} \in \mathbb{R}^{N \times p}$ is the model (design) matrix, $\mathbf{w} \in \mathbb{R}^N$ is a vector of noise variables, and $\beta^* \in \mathbb{R}^p$ is the unknown coefficient vector. In this chapter, we develop some theoretical guarantees for both the constrained form of the lasso

$$\underset{\|\beta\|_1 \leq R}{\text{minimize}} \|\mathbf{y} - \mathbf{X}\beta\|_2^2, \tag{11.2}$$

as well as for its Lagrangian version

$$\underset{\beta \in \mathbb{R}^p}{\text{minimize}} \left\{ \frac{1}{2N} \|\mathbf{y} - \mathbf{X}\beta\|_2^2 + \lambda_N \|\beta\|_1 \right\}. \tag{11.3}$$

As we have discussed previously, by Lagrangian duality, there is a correspondence between these two families of quadratic programs, where λ_N can be interpreted as the Lagrange multiplier associated with the constraint $\|\beta\|_1 \leq R$.

11.1.1 Types of Loss Functions

Given a lasso estimate $\widehat{\beta} \in \mathbb{R}^p$, we can assess its quality in various ways. In some settings, we are interested in the predictive performance of $\widehat{\beta}$, so that we might compute a *prediction loss function* of the form

$$\mathcal{L}_{\text{pred}}(\widehat{\beta}; \beta^*) = \frac{1}{N} \|\mathbf{X}\widehat{\beta} - \mathbf{X}\beta^*\|_2^2, \tag{11.4}$$

corresponding to the mean-squared error of $\widehat{\beta}$ over the given samples of $\mathbf{X}$. In other applications—among them medical imaging, remote sensing, and compressed sensing—the unknown vector β^* is of primary interest, so that it is most appropriate to consider loss functions such as the ℓ_2-error

$$\mathcal{L}_2(\widehat{\beta}; \beta^*) = \|\widehat{\beta} - \beta^*\|_2^2, \tag{11.5}$$

which we refer to as a *parameter estimation loss*. Finally, we might actually be interested in variable selection or *support recovery*, and so use the loss function

$$\mathcal{L}_{\mathrm{vs}}(\widehat{\beta}; \beta^*) = \begin{cases} 0 & \text{if } \mathrm{sign}(\widehat{\beta}_i) = \mathrm{sign}(\beta_i^*) \text{ for all } i = 1, \ldots, p, \\ 1 & \text{otherwise.} \end{cases} \tag{11.6}$$

This assesses whether or not the estimate $\widehat{\beta}$ shares the same signed support as β^*. In this chapter, we provide theoretical results for all three types of losses.

11.1.2   Types of Sparsity Models

A classical analysis of a method such as the lasso would fix the number of covariates p, and then take the sample size N to infinity. Although this type of analysis is certainly useful in some regimes, there are many settings in which the number of covariates p may be of the same order, or substantially larger than the sample size N. Examples include microarray gene expression analysis, which might involve $N = 100$ observations of $p = 10,000$ genes, or social networks, in which one makes relatively few observations of a large number of individuals. In such settings, it is doubtful whether theoretical results based on "fixed p, large N" scaling would provide useful guidance to practitioners.

Accordingly, our aim in this chapter is to develop theory that is applicable to the high-dimensional regime, meaning that it allows for the scaling $p \gg N$. Of course, if the model lacks any additional structure, then there is no hope of recovering useful information about a p-dimensional vector with limited samples. Indeed, whenever $N < p$, the linear model (11.1) is unidentifiable; for instance, it is impossible to distinguish between the models $\beta^* = 0$ and $\beta^* = \Delta$, where $\Delta \in \mathbb{R}^p$ is any element of the $p - N$-dimensional nullspace of $\mathbf{X}$.

For this reason, it is necessary to impose additional constraints on the unknown regression vector $\beta^* \in \mathbb{R}^p$, and here we focus on various types of sparsity constraints. The first setting is that of *hard sparsity*, in which we assume that β^* has at most $k \le p$ nonzero entries. For such hard-sparse models, it makes sense to consider the prediction and ℓ_2-norm losses as well as the variable selection loss (11.6). Assuming that the model is exactly supported on k coefficients may be overly restrictive, so that we also consider the case of *weakly sparse* models, meaning that β^* can be closely approximated by vectors with few nonzero entries. For instance, one way of formalizing this

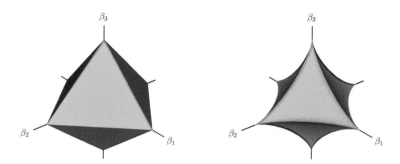

Figure 11.1 *Left: For $q = 1$, the set $\mathbb{B}(R_q)$ corresponds to the ℓ_1-ball, which is a convex set. Right: Setting $q = 0.75$ yields a nonconvex set, with spikes along the coordinate axes.*

notion is by defining, for a parameter $q \in [0, 1]$ and radius $R_q > 0$, the set

$$\mathbb{B}(R_q) = \big\{ \beta \in \mathbb{R}^p \mid \sum_{j=1}^{p} |\beta_i|^q \leq R_q \big\}. \tag{11.7}$$

This set is known as the ℓ_q-"ball" of radius[1] R_q; as illustrated in Figure 11.1, for $q \in [0, 1)$, it is not a ball in the strict sense of the word, since it is a nonconvex set. In the special case $q = 0$, imposing the constraint $\beta^* \in \mathbb{B}(R_0)$ is equivalent to requiring that β^* has at most $k = R_0$ nonzero entries.

11.2   Bounds on Lasso ℓ_2-Error

We begin by developing some results on the ℓ_2-norm loss (11.5) between a lasso solution $\widehat{\beta}$ and the true regression vector β^*. We focus on the case when β^* is k-sparse, meaning that its entries are nonzero on a subset $S(\beta^*) \subset \{1, 2, \ldots, p\}$ of cardinality $k = |S(\beta^*)|$. In the exercises, we work through some extensions to the case of weakly-sparse coefficient vectors.

11.2.1   *Strong Convexity in the Classical Setting*

We begin by developing some conditions on the model matrix $\mathbf{X}$ that are needed to establish bounds on ℓ_2-error. In order to provide some intuition for these conditions, we begin by considering one route for proving ℓ_2-consistency in the classical setting (i.e., p fixed, N tending to infinity). Suppose that we estimate some parameter vector β^* by minimizing a data-dependent objective

---

[1]Strictly speaking, the radius would be $R_q^{\frac{1}{q}}$, but we take this liberty so as to simplify notation.

function $f_N(\beta)$ over some constraint set. (For instance, the lasso minimizes the least-squares loss $f_N(\beta) = \frac{1}{N}\|\mathbf{y} - \mathbf{X}\beta\|_2^2$ subject to an ℓ_1-constraint.) Let us suppose that the difference in function values $\Delta f_N = |f_N(\widehat{\beta}) - f_N(\beta^*)|$ converges to zero as the sample size N increases. The key question is the following: what additional conditions are needed to ensure that the ℓ_2-norm of the parameter vector difference $\Delta\beta = \|\widehat{\beta} - \beta^*\|_2$ also converges to zero?

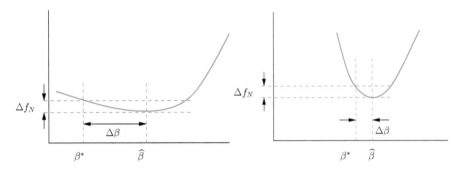

Figure 11.2 *Relation between differences in objective function values and differences in parameter values. Left: the function f_N is relatively "flat" around its optimum $\widehat{\beta}$, so that a small function difference $\Delta f_N = |f_N(\widehat{\beta}) - f_N(\beta^*)|$ does not imply that $\Delta\beta = \|\widehat{\beta} - \beta^*\|_2$ is small. Right: the function f_N is strongly curved around its optimum, so that a small difference Δf_N in function values translates into a small difference in parameter values.*

To understand the issues involved, suppose that for some N, the objective function f_N takes the form shown in Figure 11.2(a). Due to the relative "flatness" of the objective function around its minimum $\widehat{\beta}$, we see that the difference $\Delta f_N = |f_N(\widehat{\beta}) - f_N(\beta^*)|$ in function values is quite small while at the same time the difference $\Delta\beta = \|\widehat{\beta} - \beta^*\|_2$ in parameter values is relatively large. In contrast, Figure 11.2(b) shows a more desirable situation, in which the objective function has a high degree of curvature around its minimum $\widehat{\beta}$. In this case, a bound on the function difference $\Delta f_N = |f_N(\widehat{\beta}) - f_N(\beta^*)|$ translates directly into a bound on $\Delta\beta = \|\widehat{\beta} - \beta^*\|_2$.

How do we formalize the intuition captured by Figure 11.2? A natural way to specify that a function is suitably "curved" is via the notion of strong convexity. More specifically, given a differentiable function $f : \mathbb{R}^p \to \mathbb{R}$, we say that it is *strongly convex* with parameter $\gamma > 0$ at $\theta \in \mathbb{R}^p$ if the inequality

$$f(\theta') - f(\theta) \geq \nabla f(\theta)^T (\theta' - \theta) + \frac{\gamma}{2}\|\theta' - \theta\|_2^2 \qquad (11.8)$$

hold for all $\theta' \in \mathbb{R}^p$. Note that this notion is a strengthening of ordinary convexity, which corresponds to the case $\gamma = 0$. When the function f is twice continuously differentiable, an alternative characterization of strong convexity

is in terms of the Hessian $\nabla^2 f$: in particular, the function f is strongly convex
with parameter γ around $\beta^* \in \mathbb{R}^p$ if and only if the minimum eigenvalue of the
Hessian matrix $\nabla^2 f(\beta)$ is at least γ for all vectors β in a neighborhood of β^*.
If f is the negative log-likelihood under a parametric model, then $\nabla^2 f(\beta^*)$ is
the observed *Fisher information* matrix, so that strong convexity corresponds
to a uniform lower bound on the Fisher information in all directions.

11.2.2   *Restricted Eigenvalues for Regression*

Let us now return to the high-dimensional setting, in which the number of
parameters p might be larger than N. It is clear that the least-squares objective
function $f_N(\beta) = \frac{1}{2N}\|\mathbf{y} - \mathbf{X}\beta\|_2^2$ is always convex; under what additional
conditions is it also strongly convex? A straightforward calculation yields that
$\nabla^2 f(\beta) = \mathbf{X}^T\mathbf{X}/N$ for all $\beta \in \mathbb{R}^p$. Thus, the least-squares loss is strongly
convex if and only if the eigenvalues of the $p \times p$ positive semidefinite matrix
$\mathbf{X}^T\mathbf{X}$ are uniformly bounded away from zero. However, it is easy to see that
any matrix of the form $\mathbf{X}^T\mathbf{X}$ has rank at most $\min\{N, p\}$, so it is always
rank-deficient—and hence *not* strongly convex—whenever $N < p$. Figure 11.3
illustrates the situation.

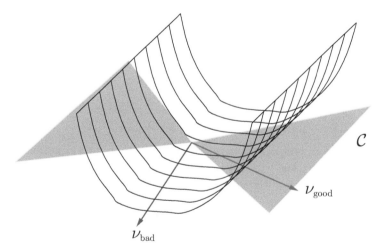

Figure 11.3 *A convex loss function in high-dimensional settings (with $p \gg N$) cannot be strongly convex; rather, it will be curved in some directions but flat in others.
As shown in Lemma 11.1, the lasso error $\widehat{\nu} = \widehat{\beta} - \beta^*$ must lie in a restricted subset $\mathcal{C}$
of $\mathbb{R}^p$. For this reason, it is only necessary that the loss function be curved in certain
directions of space.*

For this reason, we need to relax our notion of strong convexity. It turns
out, as will be clarified by the analysis below, that it is only necessary to
impose a type of strong convexity condition for some subset $\mathcal{C} \subset \mathbb{R}^p$ of possible

perturbation vectors $\nu \in \mathbb{R}^p$. In particular, we say that a function f satisfies *restricted strong convexity* at β^* with respect to $\mathcal{C}$ if there is a constant $\gamma > 0$ such that

$$\frac{\nu^T \nabla^2 f(\beta) \nu}{\|\nu\|_2^2} \geq \gamma \text{ for all nonzero } \nu \in \mathcal{C}, \tag{11.9}$$

and for all $\beta \in \mathbb{R}^p$ in a neighborhood of β^*. In the specific case of linear regression, this notion is equivalent to lower bounding the *restricted eigenvalues* of the model matrix—in particular, requiring that

$$\frac{\frac{1}{N} \nu^T \mathbf{X}^T \mathbf{X} \nu}{\|\nu\|_2^2} \geq \gamma \text{ for all nonzero } \nu \in \mathcal{C}. \tag{11.10}$$

What constraint sets $\mathcal{C}$ are relevant? Suppose that the parameter vector β^* is sparse—say supported on the subset $S = S(\beta^*)$. Defining the lasso error $\widehat{\nu} = \widehat{\beta} - \beta^*$, let $\widehat{\nu}_S \in \mathbb{R}^{|S|}$ denote the subvector indexed by elements of S, with $\widehat{\nu}_{S^c}$ defined in an analogous manner. For appropriate choices of the ℓ_1-ball radius—or equivalently, of the regularization parameter λ_N—it turns out that the lasso error satisfies a *cone constraint* of the form

$$\|\widehat{\nu}_{S^c}\|_1 \leq \alpha \|\widehat{\nu}_S\|_1, \tag{11.11}$$

for some constant $\alpha \geq 1$. This fact is easiest to see for the lasso in its constrained version. Indeed, assuming that we solve the constrained lasso (11.2) with ball radius $R = \|\beta^*\|_1$, then since $\widehat{\beta}$ is feasible for the program, we have

$$\begin{aligned}
R = \|\beta_S^*\|_1 &\geq \|\beta^* + \widehat{\nu}\|_1 \\
&= \|\beta_S^* + \widehat{\nu}_S\|_1 + \|\widehat{\nu}_{S^c}\|_1 \\
&\geq \|\beta_S^*\|_1 - \|\widehat{\nu}_S\|_1 + \|\widehat{\nu}_{S^c}\|_1.
\end{aligned}$$

Rearranging this inequality, we see that the bound (11.11) holds with $\alpha = 1$. If we instead solve the regularized version (11.3) of the lasso with a "suitable" choice of λ_N, then it turns out that the error satisfies the constraint

$$\|\widehat{\nu}_{S^c}\|_1 \leq 3\|\widehat{\nu}_S\|_1. \tag{11.12}$$

(We establish this fact during the proof of Theorem 11.1 to follow.) Thus, in either its constrained or regularized form, the lasso error is restricted to a set of the form

$$\mathcal{C}(S; \alpha) := \{\nu \in \mathbb{R}^p \mid \|\nu_{S^c}\|_1 \leq \alpha \|\nu_S\|_1\}, \tag{11.13}$$

for some parameter $\alpha \geq 1$; see Figure 11.3 for an illustration.

11.2.3   A Basic Consistency Result

With this intuition in place, we now state a result that provides a bound on the lasso error $\|\widehat{\beta} - \beta^*\|_2$, based on the linear observation model $\mathbf{y} = \mathbf{X}\beta^* + \mathbf{w}$, where β^* is k-sparse, supported on the subset S.

Theorem 11.1. Suppose that the model matrix $\mathbf{X}$ satisfies the restricted eigenvalue bound (11.10) with parameter $\gamma > 0$ over $\mathcal{C}(S; 3)$.

(a) Then any estimate $\widehat{\beta}$ based on the constrained lasso (11.2) with $R = \|\beta^*\|_1$ satisfies the bound

$$\|\widehat{\beta} - \beta^*\|_2 \leq \frac{4}{\gamma} \sqrt{\frac{k}{N}} \, \|\frac{\mathbf{X}^T \mathbf{w}}{\sqrt{N}}\|_\infty. \tag{11.14a}$$

(b) Given a regularization parameter $\lambda_N \geq 2\|\mathbf{X}^T \mathbf{w}\|_\infty / N > 0$, any estimate $\widehat{\beta}$ from the regularized lasso (11.3) satisfies the bound

$$\|\widehat{\beta} - \beta^*\|_2 \leq \frac{3}{\gamma} \sqrt{\frac{k}{N}} \, \sqrt{N} \, \lambda_N. \tag{11.14b}$$

Before proving these results, let us discuss the different factors in the bounds (11.14a) and (11.14b), and then illustrate them with some examples. First, it is important to note that these results are deterministic, and apply to any set of linear regression equations with a given observed noise vector $\mathbf{w}$. Below we obtain results for specific statistical models, as determined by assumptions on the noise vector $\mathbf{w}$ and/or the model matrix. These assumptions will affect the rate through the restricted eigenvalue constant γ, and the terms $\|\mathbf{X}^T \mathbf{w}\|_\infty$ and λ_N in the two bounds. Based on our earlier discussion of the role of strong convexity, it is natural that lasso ℓ_2-error is inversely proportional to the restricted eigenvalue constant $\gamma > 0$. The second term $\sqrt{k/N}$ is also to be expected, since we are trying to estimate unknown regression vector with k unknown entries based on N samples. As we have discussed, the final term in both bounds, involving either $\|\mathbf{X}^T \mathbf{w}\|_\infty$ or λ_N, reflects the interaction of the observation noise $\mathbf{w}$ with the model matrix $\mathbf{X}$.

It is instructive to consider the consequences of Theorem 11.1 for some linear regression models that are commonly used and studied.

Example 11.1. Classical linear Gaussian model. We begin with the classical linear Gaussian model, for which the observation noise $\mathbf{w} \in \mathbb{R}^N$ is Gaussian, with i.i.d. $N(0, \sigma^2)$ entries. Let us view the design matrix $\mathbf{X}$ as fixed, with columns $\{\mathbf{x}_1, \ldots, \mathbf{x}_p\}$. For any given column $j \in \{1, \ldots, p\}$, a simple calculation shows that the random variable $\mathbf{x}_j^T \mathbf{w}/N$ is distributed as $N(0, \frac{\sigma^2}{N} \cdot \frac{\|\mathbf{x}_j\|_2^2}{N})$. Consequently, if the columns of the design matrix $\mathbf{X}$ are normalized (meaning $\|\mathbf{x}_j\|_2 / \sqrt{N} \leq 1$ for all $j = 1, \ldots, p$), then this variable is stochastically dominated by a $N(0, \frac{\sigma^2}{N})$ variable, so that we have the Gaussian tail bound

$$\mathbb{P}\left[\frac{|\mathbf{x}_j^T \mathbf{w}|}{N} \geq t\right] \leq 2e^{-\frac{Nt^2}{2\sigma^2}}.$$

Since $\frac{\|\mathbf{X}^T \mathbf{w}\|_\infty}{N}$ corresponds to the maximum over p such variables, the union bound yields

$$\mathbb{P}\left[\frac{\|\mathbf{X}^T \mathbf{w}\|_\infty}{N} \geq t\right] \leq 2e^{-\frac{Nt^2}{2\sigma^2} + \log p} = 2e^{-\frac{1}{2}(\tau - 2)\log p},$$

where the second equality follows by setting $t = \sigma \sqrt{\frac{\tau \log p}{N}}$ for some $\tau > 2$. Consequently, we conclude that the lasso error satisfies the bound

$$\|\widehat{\beta} - \beta^*\|_2 \leq \frac{c\sigma}{\gamma} \sqrt{\frac{\tau k \log p}{N}}. \tag{11.15}$$

probability at least $1 - 2e^{-\frac{1}{2}(\tau-2)\log p}$. This calculation has also given us a choice of the regularization parameter λ_N that is valid for the Lagrangian lasso in Theorem 11.1(b). In particular, from our calculations, setting $\lambda_N = 2\sigma \sqrt{\tau \frac{\log p}{N}}$ for some $\tau > 2$ will be a valid choice with the same high probability.

It should also be noted that the rate (11.15) is intuitively reasonable. Indeed, if support set $S(\beta^*)$ were known, then estimation of β^* would require approximating a total of k parameters—namely, the elements β_i^* for all $i \in S(\beta^*)$. Even with knowledge of the support set, since the model has k free parameters, no method can achieve squared ℓ_2-error that decays more quickly than $\frac{k}{N}$. Thus, apart from the logarithmic factor, the lasso rate matches the best possible that one could achieve, even if the subset $S(\beta^*)$ were known *a priori*. In fact, the rate (11.15)—including the logarithmic factor—is known to be minimax optimal, meaning that it cannot be substantially improved upon by any estimator. See the bibliographic section for further discussion.

Example 11.2. Compressed sensing. In the domain of compressed sensing (Chapter 10), the design matrix $\mathbf{X}$ can be chosen by the user, and one standard choice is to form a random matrix with i.i.d. $N(0, 1)$ entries, and model the noise vector $\mathbf{w} \in \mathbb{R}^N$ as deterministic, say with bounded entries ($\|\mathbf{w}\|_\infty \leq \sigma$.) Under these assumptions, each variable $\frac{1}{N}\mathbf{x}_j^T\mathbf{w}$ is a zero-mean Gaussian with variance at most $\frac{\sigma^2}{N}$. Thus, by following the same argument as in the preceding example, we conclude that the lasso error will again obey the bound (11.15) with high probability for this set-up.

By a more refined argument, one can derive a strengthening of the error bound (11.15), namely:

$$\|\widehat{\beta} - \beta^*\|_2 \leq c\sigma \sqrt{\frac{k \log(ep/k)}{N}}. \tag{11.16}$$

where $e \approx 2.71828$, and c is a universal constant. This bound suggests that the sample size N should satisfy the lower bound

$$N \geq k \log(ep/k) \tag{11.17}$$

in order for the lasso to have small error.

Following Donoho and Tanner (2009), let us consider the ratios $\rho = k/N$ and $\alpha = N/p$, in which form the bound (11.17) can be rewritten as

$$\rho(1 - \log(\rho\alpha)) \leq 1. \tag{11.18}$$

In order to study the accuracy of this prediction, we generated random ensembles of the linear regression problem in dimension $p = 200$ and sample sizes N ranging from 10 and 200, where each feature $x_{ij} \sim N(0, 1)$ was generated independently. Given this random design, we then generated outcomes from a linear model $y_i = \nu\langle x_i, \beta^* \rangle + \sigma w_i$ where $w_i \sim N(0, 1)$ and $\sigma = 4$. For a given sparsity level k, we chose a random subset S of size k, and for each $j \in S$, we generated $\beta_j^* \sim N(0, 1)$ independently at random. In all cases, the pre-factor ν was chosen for each N and k, so that the signal-to-noise ratio was approximately equal to 10. We then solved the Lagrangian lasso using the regularization parameter $\lambda_N = 2\sigma\sqrt{3\frac{\log \frac{ep}{k}}{N}}$. Figure 11.4 is a heatmap of the median of the Euclidean error $\|\widehat{\beta} - \beta^*\|_2$ over 10 realizations, with the boundary (11.18) super-imposed. We see that there is a fairly sharp change at the theoretical boundary, indicating that more samples are needed when the underlying model is more dense.

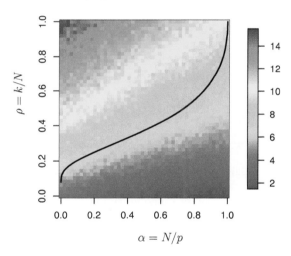

Figure 11.4 *Simulation experiment: median of the error $\|\widehat{\beta} - \beta^*\|_2$ over 10 realizations, with the boundary (11.18) super-imposed.*

Equipped with this intuition, we now turn to the proof of Theorem 11.1.

Proof of Theorem 11.1: The proof is very straightforward for the constrained lasso bound (11.14a), and requires a bit more work for the regularized lasso bound (11.14b).

Constrained Lasso. In this case, since β^* is feasible and $\widehat{\beta}$ is optimal, we have the inequality $\|\mathbf{y} - \mathbf{X}\widehat{\beta}\|_2^2 \leq \|\mathbf{y} - \mathbf{X}\beta^*\|_2^2$. Defining the error vector $\widehat{\nu} := \widehat{\beta} - \beta^*$, substituting in the relation $\mathbf{y} = \mathbf{X}\beta^* + \mathbf{w}$, and performing some algebra yields the *basic inequality*

$$\frac{\|\mathbf{X}\widehat{\nu}\|_2^2}{2N} \leq \frac{\mathbf{w}^T\mathbf{X}\widehat{\nu}}{N}. \tag{11.19}$$

Applying a version of Hölder's inequality to the right-hand side yields the upper bound $\frac{1}{N}|\mathbf{w}^T\mathbf{X}\widehat{\nu}| \leq \frac{1}{N}\|\mathbf{X}^T\mathbf{w}\|_\infty \|\widehat{\nu}\|_1$. As shown in Chapter 10, the inequality $\|\widehat{\beta}\|_1 \leq R = \|\beta^*\|_1$ implies that $\widehat{\nu} \in \mathcal{C}(S; 1)$, whence we have

$$\|\widehat{\nu}\|_1 = \|\widehat{\nu}_S\|_1 + \|\widehat{\nu}_{S^c}\|_1 \leq 2\|\widehat{\nu}_S\|_1 \leq 2\sqrt{k}\|\widehat{\nu}\|_2.$$

On the other hand, applying the restricted eigenvalue condition (11.10) to the left-hand side of the inequality (11.19) yields $\frac{1}{N}\|\mathbf{X}\widehat{\nu}\|_2^2 \geq \gamma\|\widehat{\nu}\|_2^2$. Putting together the pieces yields the claimed bound (11.14a).

Lagrangian Lasso. Define the function

$$G(\nu) := \frac{1}{2N}\|\mathbf{y} - \mathbf{X}(\beta^* + \nu)\|_2^2 + \lambda_N\|\beta^* + \nu\|_1. \tag{11.20}$$

Noting that $\widehat{\nu} := \widehat{\beta} - \beta^*$ minimizes G by construction, we have $G(\widehat{\nu}) \leq G(0)$. Some algebra yields the *modified basic inequality*

$$\frac{\|\mathbf{X}\widehat{\nu}\|_2^2}{2N} \leq \frac{\mathbf{w}^T\mathbf{X}\widehat{\nu}}{N} + \lambda_N\{\|\beta^*\|_1 - \|\beta^* + \widehat{\nu}\|_1\}. \tag{11.21}$$

Now since $\beta^*_{S^c} = 0$, we have $\|\beta^*\|_1 = \|\beta^*_S\|_1$, and

$$\|\beta^* + \widehat{\nu}\|_1 = \|\beta^*_S + \widehat{\nu}_S\|_1 + \|\widehat{\nu}_{S^c}\|_1 \geq \|\beta^*_S\|_1 - \|\widehat{\nu}_S\|_1 + \|\widehat{\nu}_{S^c}\|_1.$$

Substituting these relations into inequality (11.21) yields

$$\begin{aligned}\frac{\|\mathbf{X}\widehat{\nu}\|_2^2}{2N} &\leq \frac{\mathbf{w}^T\mathbf{X}\widehat{\nu}}{N} + \lambda_N\left\{\|\widehat{\nu}_S\|_1 - \|\widehat{\nu}_{S^c}\|_1\right\} \\ &\leq \frac{\|\mathbf{X}^T\mathbf{w}\|_\infty}{N}\|\widehat{\nu}\|_1 + \lambda_N\left\{\|\widehat{\nu}_S\|_1 - \|\widehat{\nu}_{S^c}\|_1\right\},\end{aligned} \tag{11.22}$$

where the second step follows by applying Hölder's inequality with ℓ_1 and ℓ_∞ norms. Since $\frac{1}{N}\|\mathbf{X}^T\mathbf{w}\|_\infty \leq \frac{\lambda_N}{2}$ by assumption, we have

$$\frac{\|\mathbf{X}\widehat{\nu}\|_2^2}{2N} \leq \frac{\lambda_N}{2}\{\|\widehat{\nu}_S\|_1 + \|\widehat{\nu}_{S^c}\|_1\} + \lambda_N\{\|\widehat{\nu}_S\|_1 - \|\widehat{\nu}_{S^c}\|_1\} \leq \frac{3}{2}\sqrt{k}\lambda_N\|\widehat{\nu}\|_2, \tag{11.23}$$

where the final step uses the fact that $\|\widehat{\nu}_S\|_1 \leq \sqrt{k}\|\widehat{\nu}\|_2$.

In order to complete the proof, we require the following auxiliary result:

Lemma 11.1. Suppose that $\lambda_N \geq 2\|\frac{\mathbf{X}^T\mathbf{w}}{N}\|_\infty > 0$. Then the error $\widehat{\nu} := \widehat{\beta} - \beta^*$ associated with any lasso solution $\widehat{\beta}$ belongs to the cone set $\mathcal{C}(S; 3)$.

Taking this claim as given for the moment, let us complete the proof of the bound (11.14b). Lemma 11.1 allows us to apply the γ-RE condition (11.10) to $\widehat{\nu}$, which ensures that $\frac{1}{N}\|\mathbf{X}\widehat{\nu}\|_2^2 \geq \gamma\|\widehat{\nu}\|_2^2$. Combining this lower bound with our earlier inequality (11.23) yields

$$\frac{\gamma}{2}\|\widehat{\nu}\|_2^2 \leq \frac{3}{2}\lambda_N\sqrt{k}\,\|\widehat{\nu}\|_2,$$

and rearranging yields the bound (11.14b).

It remains to prove Lemma 11.1. Since $\frac{\|\mathbf{X}^T\mathbf{w}\|_\infty}{N} \leq \frac{\lambda_N}{2}$, inequality (11.22) implies that

$$0 \leq \frac{\lambda_N}{2}\|\widehat{\nu}\|_1 + \lambda_N\left\{\|\widehat{\nu}_S\|_1 - \|\widehat{\nu}_{S^c}\|_1\right\},$$

Rearranging and then dividing out by $\lambda_N > 0$ yields that $\|\widehat{\nu}_{S^c}\|_1 \leq 3\|\widehat{\nu}_S\|_1$ as claimed.

$\square$

Some extensions. As stated, Theorem 11.1 applies to regression models in which β^* has at most k nonzero entries, an assumption that we referred to as hard sparsity. However, a similar type of analysis can be performed for weakly sparse models, say with β^* belonging to the ℓ_q-ball $\mathbb{B}_q(R_q)$ previously defined in Equation (11.7). Under a similar set of assumptions, it can be shown that the lasso error will satisfy the bound

$$\|\widehat{\beta} - \beta^*\|_2^2 \leq c\,R_q\left(\frac{\sigma^2\,\log p}{N}\right)^{1-q/2} \tag{11.24}$$

with high probability. We work through portions of this derivation in Exercise 11.3. In the special case $q = 0$, assuming that β^* belongs to the ℓ_0 ball is equivalent to the assumption of hard sparsity (with radius $R_0 = k$), so that this rate (11.24) is equivalent to our previous result (11.16) derived as a consequence of Theorem 11.1. Otherwise, note that the rate slows down as the weak sparsity parameter q increases away from zero toward one, reflecting the fact that we are imposing weaker conditions on the true regression vector β^*. The rate (11.24) is known to be minimax-optimal over the ℓ_q-ball, meaning that no other estimator can achieve a substantially smaller ℓ_2-error; see the bibliographic section for further discussion.

11.3 Bounds on Prediction Error

Thus far, we have studied the performance of the lasso in recovering the true regression vector, as assessed by the Euclidean error $\|\widehat{\beta} - \beta^*\|_2$. In other settings, it may suffice to obtain an estimate $\widehat{\beta}$ that has a relatively low (in-sample) prediction error $\mathcal{L}_{\mathrm{pred}}(\widehat{\beta}, \beta^*) = \frac{1}{N}\|\mathbf{X}(\widehat{\beta} - \beta^*)\|_2^2$. In this section, we develop some theoretical guarantees on this form of loss. For concreteness, we focus on the Lagrangian lasso (11.3), although analogous results can be derived for other forms of the lasso.

Theorem 11.2. Consider the Lagrangian lasso with a regularization parameter $\lambda_N \geq \frac{2}{N}\|\mathbf{X}^T\mathbf{w}\|_\infty$.

(a) If $\|\beta^*\|_1 \leq R_1$, then any optimal solution $\widehat{\beta}$ satisfies

$$\frac{\|\mathbf{X}(\widehat{\beta} - \beta^*)\|_2^2}{N} \leq 12\,R_1\,\lambda_N. \tag{11.25a}$$

(b) If β^* is supported on a subset S, and the design matrix $\mathbf{X}$ satisfies the γ-RE condition (11.10) over $\mathcal{C}(S;3)$, then any optimal solution $\widehat{\beta}$ satisfies

$$\frac{\|\mathbf{X}(\widehat{\beta} - \beta^*)\|_2^2}{N} \leq \frac{9}{\gamma}\,|S|\,\lambda_N^2. \tag{11.25b}$$

As we have discussed, for various statistical models, the choice $\lambda_N = c\,\sigma\,\sqrt{\frac{\log p}{N}}$ is valid for Theorem 11.2 with high probability, so the two bounds take the form

$$\frac{\|\mathbf{X}(\widehat{\beta} - \beta^*)\|_2^2}{N} \leq c_1\,\sigma\,R_1\,\sqrt{\frac{\log p}{N}}, \quad \text{and} \tag{11.26a}$$

$$\frac{\|\mathbf{X}(\widehat{\beta} - \beta^*)\|_2^2}{N} \leq c_2\,\frac{\sigma^2}{\gamma}\,\frac{|S|\,\log p}{N}. \tag{11.26b}$$

The bound (11.26a), which depends on the ℓ_1-ball radius R_1, is known as the "slow rate" for the lasso, since the squared prediction error decays as $1/\sqrt{N}$. On the other hand, the bound (11.26b) is known as the "fast rate," since it decays as $1/N$. Note that the latter is based on much stronger assumptions: namely, the hard sparsity condition that β^* is supported on a small subset S, and more disconcertingly, the γ-RE condition on the design matrix $\mathbf{X}$. In principle, prediction performance should not require an RE condition, so that one might suspect that this requirement is an artifact of our proof technique. Remarkably, as we discuss in the bibliographic section, this dependence turns out to be unavoidable for any polynomial-time method.

Proof of Theorem 11.2: The proofs of both claims are relatively straightforward given our development thus far.

Proof of bound (11.25a): Beginning with the modified basic inequality (11.21), we have

$$\begin{aligned}
0 &\leq \frac{\|\mathbf{X}^T\mathbf{w}\|_\infty}{N}\,\|\widehat{\nu}\|_1 + \lambda_N\big\{\|\beta^*\|_1 - \|\beta^* + \widehat{\nu}\|_1\big\} \\
&\leq \big\{\frac{\|\mathbf{X}^T\mathbf{w}\|_\infty}{N} - \lambda_N\big\}\|\widehat{\nu}\|_1 + 2\lambda_N\|\beta^*\|_1 \\
&\overset{(i)}{\leq} \frac{1}{2}\lambda_N\big\{-\|\widehat{\nu}\|_1 + 4\|\beta^*\|_1\big\},
\end{aligned}$$

where step (i) uses the fact that $\frac{1}{N}\|\mathbf{X}^T\mathbf{w}\|_\infty \leq \frac{\lambda_N}{2}$ by assumption. Putting together the pieces, we conclude that $\|\widehat{\nu}\|_1 \leq 4\|\beta^*\|_1 \leq 4R_1$. Returning again to the modified basic inequality (11.21), we have

$$\frac{\|\mathbf{X}\widehat{\nu}\|_2^2}{2N} \leq \left\{\frac{\|\mathbf{X}^T\mathbf{w}\|_\infty}{N} + \lambda_N\right\}\|\widehat{\nu}\|_1 \leq 6\lambda_N R_1,$$

which establishes the claim (11.25a).

Proof of bound (11.25b):   Given the stated choice of λ_N, inequality (11.23) holds, whence

$$\frac{\|\mathbf{X}\widehat{\nu}\|_2^2}{2N} \leq 3\sqrt{k}\lambda_N\|\widehat{\nu}\|_2.$$

By Lemma 11.1, the error vector $\widehat{\nu}$ belongs to the cone $\mathcal{C}(S; 3)$, so that the γ-RE condition guarantees that $\|\widehat{\nu}\|_2^2 \leq \frac{1}{N\gamma}\|\mathbf{X}\widehat{\nu}\|_2^2$. Putting together the pieces yields the claim (11.25b).   □

11.4   Support Recovery in Linear Regression

Thus far, we have focused on bounds on either the ℓ_2-error or the prediction error associated with a lasso solution. In other settings, we are interested in a somewhat more refined question, namely whether or not a lasso estimate $\widehat{\beta}$ has nonzero entries in the same positions as the true regression vector β^*. More precisely, suppose that the true regression vector β^* is k-sparse, meaning that it is supported on a subset $S = S(\beta^*)$ of cardinality $k = |S|$. In such a setting, a natural goal is to correctly identify the subset S of relevant variables. In terms of the lasso, we ask the following question: given an optimal lasso solution $\widehat{\beta}$, when is its support set—denoted by $\widehat{S} = S(\widehat{\beta})$—exactly equal to the true support S? We refer to this property as *variable selection consistency* or *sparsistency*.

Note that it is possible for the ℓ_2 error $\|\widehat{\beta} - \beta^*\|_2$ to be quite small even if $\widehat{\beta}$ and β^* have different supports, as long as $\widehat{\beta}$ is nonzero for all "suitably large" entries of β^*, and not "too large" in positions where β^* is zero. Similarly, it is possible for the prediction error $\|\mathbf{X}(\widehat{\beta} - \beta^*)\|_2/\sqrt{N}$ to be small even when $\widehat{\beta}$ and β^* have very different supports. On the other hand, as we discuss in the sequel, given an estimate $\widehat{\beta}$ that correctly recovers the support of β^*, we can estimate β^* very well—both in ℓ_2-norm and the prediction semi-norm— simply by performing an ordinary least-squares regression restricted to this subset.

11.4.1   Variable-Selection Consistency for the Lasso

We begin by addressing the issue of variable selection in the context of deterministic design matrices $\mathbf{X}$. It turns out that variable selection requires a condition related to but distinct from the restricted eigenvalue condition (11.10).

In particular, we assume a condition known either as *mutual incoherence* or *irrepresentability*: there must exist some $\gamma > 0$ such that

$$\max_{j \in S^c} \|(\mathbf{X}_S^T \mathbf{X}_S)^{-1} \mathbf{X}_S^T \mathbf{x}_j\|_1 \leq 1 - \gamma. \tag{11.27}$$

To interpret this condition, note that the submatrix $\mathbf{X}_S \in \mathbb{R}^{N \times k}$ corresponds to the subset of covariates that are in the support set. For each index j in the complementary set S^c, the k-vector $(\mathbf{X}_S^T \mathbf{X}_S)^{-1} \mathbf{X}_S^T \mathbf{x}_j$ is the regression coefficient of $\mathbf{x}_j$ on $\mathbf{X}_S$; this vector is a measure of how well the column $\mathbf{x}_j$ aligns with the columns of the submatrix $\mathbf{X}_S$. In the most desirable case, the columns $\{\mathbf{x}_j, j \in S^c\}$ would all be orthogonal to the columns of $\mathbf{X}_S$, and we would be guaranteed that $\gamma = 1$. Of course, in the high-dimensional setting ($p \gg N$), this complete orthogonality is not possible, but we can still hope for a type of "near orthogonality" to hold.

In addition to this incoherence condition, we also assume that the design matrix has normalized columns

$$\max_{j=1,\ldots,p} \|\mathbf{x}_j\|_2 / \sqrt{N} \leq K_{\text{clm}}. \tag{11.28}$$

For example, we can take $\|\mathbf{x}_j\|_2 = \sqrt{N}$ and $K_{\text{clm}} = 1$. Further we assume that the submatrix $\mathbf{X}_S$ is well-behaved in the sense that

$$\lambda_{\min}(\mathbf{X}_S^T \mathbf{X}_S / N) \geq C_{\min}. \tag{11.29}$$

Note that if this condition were violated, then the columns of $\mathbf{X}_S$ would be linearly dependent, and it would be impossible to estimate β^* even in the "oracle case" when the support set S were known.

The following result applies to the regularized lasso (11.3) when applied to an instance the linear observation model (11.1) such that the true parameter β^* has support size k.

Theorem 11.3. Suppose that the design matrix $\mathbf{X}$ satisfies the mutual incoherence condition (11.27) with parameter $\gamma > 0$, and the column normalization condition (11.28) and the eigenvalue condition (11.29) both hold. For a noise vector $\mathbf{w} \in \mathbb{R}^N$ with i.i.d. $N(0, \sigma^2)$ entries, consider the regularized lasso program (11.3) with

$$\lambda_N \geq \frac{8 K_{\text{clm}} \sigma}{\gamma} \sqrt{\frac{\log p}{N}}. \tag{11.30}$$

Then with probability greater than $1 - c_1 e^{-c_2 N \lambda_N^2}$, the lasso has the following properties:

(a) *Uniqueness:* The optimal solution $\widehat{\beta}$ is unique.

(b) *No false inclusion:* The unique optimal solution has its support $S(\widehat{\beta})$ contained within the true support $S(\beta^*)$.

(c) ℓ_∞-bounds: The error $\widehat{\beta} - \beta^*$ satisfies the ℓ_∞ bound

$$\|\widehat{\beta}_S - \beta_S^*\|_\infty \le \lambda_N \underbrace{\left[\frac{4\sigma}{\sqrt{C_{\min}}} + \left\|(\mathbf{X}_S^T\mathbf{X}_S/N)^{-1}\right\|_\infty\right]}_{B(\lambda_N, \sigma; \mathbf{X})}, \qquad (11.31)$$

where for a matrix $\mathbf{A}$, its ∞-norm is given by $\|\mathbf{A}\|_\infty = \max_{\|u\|_\infty = 1} \|\mathbf{A}u\|_\infty$.

(d) No false exclusion: The lasso solution includes all indices $j \in S(\beta^*)$ such that $|\beta_j^*| > B(\lambda_N, \sigma; \mathbf{X})$, and hence is variable selection consistent as long as $\min_{j \in S} |\beta_j^*| > B(\lambda_N, \sigma; \mathbf{X})$.

Before proving this result, let us try to interpret its main claims. First, the uniqueness claim in part (a) is not trivial in the high-dimensional setting, because as discussed previously, although the lasso objective is convex, it can never be strictly convex when $p > N$. The uniqueness claim is important, because it allows us to talk unambiguously about the support of the lasso estimate $\widehat{\beta}$. Part (b) guarantees that the lasso does not falsely include variables that are not in the support of β^*, or equivalently that $\widehat{\beta}_{S^c} = 0$, whereas part (c) guarantees that $\widehat{\beta}_S$ is uniformly close to β_S^* in the ℓ_∞-norm. Finally, part (d) is a consequence of this uniform norm bound: as long as the minimum value of $|\beta_j^*|$ over indices $j \in S$ is not too small, then the lasso is variable-selection consistent in the full sense.

11.4.1.1   Some Numerical Studies

In order to learn more about the impact of these results in practice, we ran a few small simulation studies. We first explore the impact of the irrepresentability condition (11.27). We fixed the sample size to $N = 1000$, and for a range of problem dimensions p, we generated p i.i.d standard Gaussian variates, with a fraction $f = k/p$ of them being in the support set S. For correlations ρ ranging over the interval $[0, 0.6]$, for each $j \in S$ we randomly chose a predictor $\ell \in S^c$, and set $\mathbf{x}_\ell \leftarrow \mathbf{x}_\ell + c \cdot \mathbf{x}_j$ with c chosen so that $\mathrm{corr}(\mathbf{x}_j, \mathbf{x}_\ell) = \rho$. Figure 11.5 shows the average value of $1 - \gamma$, the value of the irrepresentability condition (11.27), over five realizations. We see for example with $\rho = 0$, we fall into the "good" region $1 - \gamma < 1$ when $p \le 1000$ and there is $f \le 2\%$ sparsity or $p \le 500$ with $f \le 5\%$ sparsity. However the maximum size of p and sparsity level f decrease as the correlation ρ increases.

We also ran a small simulation study to examine the false discovery and false exclusion rates for a lasso regression. We set $N = 1000$ and $p = 500$ with $k = 15$ predictors in S having nonzero coefficients. The data matrices $\mathbf{X}_S$ and $\mathbf{X}_{S^c}$ were generated as above, with different values for the correlations ρ. We then generated a response $\mathbf{y}$ according to $\mathbf{y} = \mathbf{X}_S\beta_S + \mathbf{w}$, with the elements of $\mathbf{w}$ i.i.d. $N(0, 1)$.

We tried two different values for the nonzero regression coefficients in β_S: all 0.25 or all 0.15, with randomly selected signs. These result in "effect sizes"

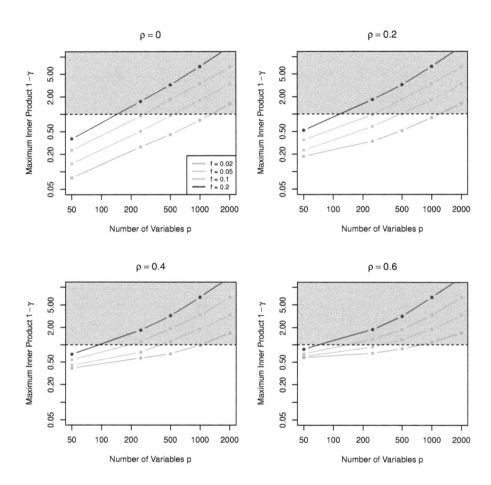

Figure 11.5 *Irrepresentability condition in practice. Each plot shows values for*
$1 - \gamma$ *in (11.27) for simulated Gaussian data. Values less than one are good, and the*
smaller the better. The sample size $N = 1000$ is fixed, and the number of predictors
p varies along the horizontal axis. The fraction $f = k/p$ of true nonzero coefficients
(the sparsity level) varies within each panel, and finally, the correlation between each
true predictor and its null predictor partner (as described in the text) varies across
the four panels. A horizontal broken line is drawn at $1 - \gamma = 1$, below which the
irrepresentability condition holds. Each point is a mean of $1-\gamma$ over five simulations;
the standard errors of the means are small, averaging about 0.03.

(absolute standardized regression coefficients) for the 15 true predictors of 7.9 and 4.7, respectively. Finally, we chose λ_N in an "optimal" way in each run: we used the value yielding the correct number of nonzero coefficients (15).

The top row of Figure 11.6 shows the results. In the top left panel (the best case), the average false discovery and false exclusion probabilities are zero until ρ is greater than about 0.6. After that point, the lasso starts to include false variables and exclude good ones, due to the high correlation between signal and noise variables. The value γ from the irrepresentability condition is also shown, and drops below zero at around the value $\rho = 0.6$. (Hence the condition holds below correlation 0.6.) In the top right panel, we see error rates increase overall, even for small ρ. Here the effect size is modestly reduced from 7.9 to 4.7, which is the cause of the increase.

The lower panel of Figure 11.6 shows the results when the sample size N is reduced to 200 ($p < N$) and the size k of the support set is increased to 25. The values used for the nonzero regression coefficients were 5.0 and 0.5, yielding effect sizes of about 71 and 7, respectively. The irrepresentability condition and other assumptions of the theorem do not hold. Now the error rates are 15% or more irrespective of ρ, and recovery of the true support set seems unrealistic in this scenario.

11.4.2 Proof of Theorem 11.3

We begin by developing the necessary and sufficient conditions for optimality in the lasso. A minor complication arises because the ℓ_1-norm is not differentiable, due to its sharp point at the origin. Instead, we need to work in terms of the subdifferential of the ℓ_1-norm. Here we provide a very brief introduction; see Chapter 5 for further details. Given a convex function $f : \mathbb{R}^p \to \mathbb{R}$, we say that $z \in \mathbb{R}^p$ is a subgradient at β, denoted by $z \in \partial f(\beta)$, if we have

$$f(\beta + \Delta) \geq f(\beta) + \langle z, \Delta \rangle \qquad \text{for all } \Delta \in \mathbb{R}^p.$$

When $f(\beta) = \|\beta\|_1$, it can be seen that $z \in \partial\|\beta\|_1$ if and only if $z_j = \text{sign}(\beta_j)$ for all $j = 1, 2, \ldots, p$, where we allow sign(0) to be any number in the interval $[-1, 1]$. In application to the lasso program, we say that a pair $(\widehat{\beta}, \widehat{z}) \in \mathbb{R}^p \times \mathbb{R}^p$ is primal-dual optimal if $\widehat{\beta}$ is a minimizer and $\widehat{z} \in \partial\|\widehat{\beta}\|_1$. Any such pair must satisfy the zero-subgradient condition

$$-\frac{1}{N}\mathbf{X}^T(\mathbf{y} - \mathbf{X}\widehat{\beta}) + \lambda_N\widehat{z} = 0, \tag{11.32}$$

which is the analogue of a zero gradient condition in this nondifferentiable setting.

Our proof of Theorem 11.3 is based on a constructive procedure, known as a *primal-dual witness method* (PDW). When this procedure succeeds, it constructs a pair $(\widehat{\beta}, \widehat{z}) \in \mathbb{R}^p \times \mathbb{R}^p$ that are primal-dual optimal, and act as a witness for the fact that the lasso has a unique optimal solution with the

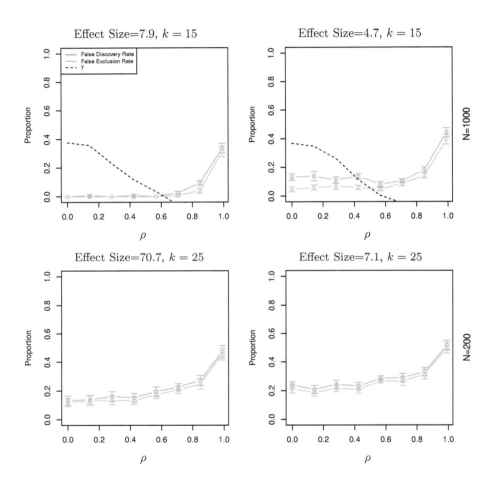

Figure 11.6 *Average false discovery and exclusion rates (with $\pm$ one standard error) from simulation experiments with $p = 500$ variables. In the top row $N = 1000$, and the size of S is $k = 15$. In the second row $N = 200$ and the subset size is $k = 25$. The effect size is the strength of real coefficients, as measured by an absolute Z statistic. Overall conclusion: when γ is favorable, and the signal is strong, recovery is good (top left). All other situations are problematic.*

correct signed support. Using $S = \text{supp}(\beta^*)$ to denote the support set of β^*, this procedure consists of the following steps:

Primal-dual witness (PDW) construction.

1. Set $\widehat{\beta}_{S^c} = 0$.
2. Determine $(\widehat{\beta}_S, \widehat{z}_S)$ by solving the k-dimensional oracle subproblem

$$\widehat{\beta}_S \in \arg\min_{\beta_S \in \mathbb{R}^k} \left\{ \frac{1}{2N} \|\mathbf{y} - \mathbf{X}_S \beta_S\|_2^2 + \lambda_N \|\beta_S\|_1 \right\}. \tag{11.33}$$

Thus $\widehat{z}_S$ is an element of subdifferential $\partial \|\widehat{\beta}_S\|_1$ satisfying the relation $\frac{1}{N} \mathbf{X}_S^T (\mathbf{y} - \mathbf{X}_S \widehat{\beta}_S) + \lambda_N \widehat{z}_S = 0$.
3. Solve for $\widehat{z}_{S^c}$ via the zero-subgradient Equation (11.32), and check whether or not the *strict dual feasibility* condition $\|\widehat{z}_{S^c}\|_\infty < 1$ holds.

To be clear, this procedure is *not* an implementable method for actually solving the lasso program (since it pre-supposes knowledge of the true support); rather, it is a proof technique for certifying variable-selection consistency of the lasso. Note that the subvector $\widehat{\beta}_{S^c}$ is determined in step 1, whereas the remaining three subvectors are determined in steps 2 and 3. By construction, the subvectors $\widehat{\beta}_S$, $\widehat{z}_S$ and $\widehat{z}_{S^c}$ satisfy the zero-subgradient condition (11.32). We say that the PDW construction succeeds if the vector $\widehat{z}_{S^c}$ constructed in step 3 satisfies the strict dual feasibility condition. The following result shows that this success acts as a witness for the lasso:

Lemma 11.2. If the PDW construction succeeds, then under the lower eigenvalue condition (11.29), the vector $(\widehat{\beta}_S, 0) \in \mathbb{R}^p$ is the unique optimal solution of the regularized lasso (11.3).

Proof: When the PDW construction succeeds, then $\widehat{\beta} = (\widehat{\beta}_S, 0)$ is an optimal solution with associated subgradient vector $\widehat{z} \in \mathbb{R}^p$ satisfying $\|\widehat{z}_{S^c}\|_\infty < 1$, and $\langle \widehat{z}, \widehat{\beta} \rangle = \|\widehat{\beta}\|_1$. Now let $\widetilde{\beta} \in \mathbb{R}^p$ be any other optimal solution of the lasso. If we introduce the shorthand notation $F(\beta) = \frac{1}{2N} \|\mathbf{y} - \mathbf{X}\beta\|_2^2$, then we are guaranteed that $F(\widehat{\beta}) + \lambda_N \langle \widehat{z}, \widehat{\beta} \rangle = F(\widetilde{\beta}) + \lambda_N \|\widetilde{\beta}\|_1$, and hence

$$F(\widehat{\beta}) - \lambda_N \langle \widehat{z}, \widetilde{\beta} - \widehat{\beta} \rangle = F(\widetilde{\beta}) + \lambda_N \big(\|\widetilde{\beta}\|_1 - \langle \widehat{z}, \widetilde{\beta} \rangle \big).$$

But by the zero-subgradient conditions for optimality (11.32), we have $\lambda_N \widehat{z} = -\nabla F(\widehat{\beta})$, which implies that

$$F(\widehat{\beta}) + \langle \nabla F(\widehat{\beta}), \widetilde{\beta} - \widehat{\beta} \rangle - F(\widetilde{\beta}) = \lambda_N \big(\|\widetilde{\beta}\|_1 - \langle \widehat{z}, \widetilde{\beta} \rangle \big).$$

By convexity of F, the left-hand side is negative, so that we must have $\|\widetilde{\beta}\|_1 \leq \langle \widehat{z}, \widetilde{\beta} \rangle$. Applying Hölder's inequality with the ℓ_1 and ℓ_∞ norms yields the upper bound $\langle \widehat{z}, \widetilde{\beta} \rangle \leq \|\widehat{z}\|_\infty \|\widetilde{\beta}\|_1$. These two inequalities together imply that $\|\widetilde{\beta}\|_1 = \langle \widetilde{z}, \widetilde{\beta} \rangle$. Since $\|\widetilde{z}_{S^c}\|_\infty < 1$, this equality can occur only if $\widetilde{\beta}_j = 0$ for all $j \in S^c$.

Thus, all optimal solutions are supported only on S, and hence can be obtained by solving the oracle subproblem (11.33). Given the lower eigenvalue bound (11.29), this subproblem is strictly convex, and so has a unique minimizer. $\qquad\square$

Based on Lemma 11.2, in order to prove parts (a) and (b) of Theorem 11.3, it suffices to show that the subvector $\widehat{z}_{S^c}$ constructed in step 3 satisfies the *strict dual feasibility* condition $\|\widehat{z}_{S^c}\|_\infty < 1$.

Establishing strict dual feasibility. Let us delve into the form of the subvector $\widehat{z}_{S^c}$ constructed in step 3. By using the fact that $\widehat{\beta}_{S^c} = \beta^*_{S^c} = 0$ and writing out the zero-subgradient condition (11.32) in block matrix form, we obtain

$$\frac{1}{N} \begin{bmatrix} \mathbf{X}_S^T \mathbf{X}_S & \mathbf{X}_S^T \mathbf{X}_{S^c} \\ \mathbf{X}_{S^c}^T \mathbf{X}_S & \mathbf{X}_{S^c}^T \mathbf{X}_{S^c} \end{bmatrix} \begin{bmatrix} \widehat{\beta}_S - \beta^*_S \\ 0 \end{bmatrix} - \frac{1}{N} \begin{bmatrix} \mathbf{X}_S^T \mathbf{w} \\ \mathbf{X}_{S^c}^T \mathbf{w} \end{bmatrix} + \lambda_N \begin{bmatrix} \widehat{z}_S \\ \widehat{z}_{S^c} \end{bmatrix} = \begin{bmatrix} 0 \\ 0 \end{bmatrix}.$$
(11.34)

Solving for the vector $\widehat{z}_{S^c} \in \mathbb{R}^{p-k}$ yields

$$\widehat{z}_{S^c} = \frac{1}{\lambda_N} \left\{ \frac{\mathbf{X}_{S^c}^T \mathbf{w}}{N} - \frac{\mathbf{X}_{S^c}^T \mathbf{X}_S}{N} (\widehat{\beta}_S - \beta^*_S) \right\}.$$
(11.35)

Similarly, using the assumed invertibility of $\mathbf{X}_S^T \mathbf{X}_S$ in order to solve for the difference $\widehat{\beta}_S - \beta^*_S$ yields

$$\widehat{\beta}_S - \beta^*_S = \underbrace{\left(\frac{\mathbf{X}_S^T \mathbf{X}_S}{N} \right)^{-1} \frac{\mathbf{X}_S^T \mathbf{w}}{N} - \lambda_N \left(\frac{\mathbf{X}_S^T \mathbf{X}_S}{N} \right)^{-1} \mathrm{sign}(\beta^*_S)}_{U_S}.$$
(11.36)

Substituting this expression back into Equation (11.35) and simplifying yields

$$\widehat{z}_{S^c} = \underbrace{\mathbf{X}_{S^c}^T \mathbf{X}_S (\mathbf{X}_S^T \mathbf{X}_S)^{-1} \mathrm{sign}(\beta^*_S)}_{\mu} + \underbrace{\mathbf{X}_{S^c}^T \left[\mathbf{I} - \mathbf{X}_S (\mathbf{X}_S^T \mathbf{X}_S)^{-1} \mathbf{X}_S^T \right] \left(\frac{\mathbf{w}}{\lambda_N N} \right)}_{V_{S^c}}.$$

By triangle inequality, we have

$$\|\widehat{z}_{S^c}\|_\infty \leq \|\mu\|_\infty + \|V_{S^c}\|_\infty.$$

Note that the vector $\mu \in \mathbb{R}^{p-k}$ is a deterministic quantity, and moreover, by the mutual incoherence condition (11.27), we have $\|\mu\|_\infty \leq 1 - \gamma$. The remaining quantity $V_{S^c} \in \mathbb{R}^{p-k}$ is a zero-mean Gaussian random vector, and we need to show that $\|V_{S^c}\|_\infty < \gamma$ with high probability.

For an arbitrary $j \in S^c$, consider the random variable

$$V_j := \mathbf{X}_j^T \underbrace{\left[\mathbf{I} - \mathbf{X}_S(\mathbf{X}_S^T\mathbf{X}_S)^{-1}\mathbf{X}_S^T\right]}_{\Pi_{S\perp}(\mathbf{X})} \left(\frac{\mathbf{w}}{\lambda_N N}\right).$$

Noting that the matrix $\Pi_{S\perp}(\mathbf{X})$ is an orthogonal projection matrix and using the column normalization condition (11.28), we see that each V_j is zero-mean with variance at most $\sigma^2 K_{\text{clm}}^2/(\lambda_N^2 N)$. Therefore, combining Gaussian tail bounds with the union bound, we find that

$$\mathbb{P}\left[\|V_{S^c}\|_\infty \geq \gamma/2\right] \leq 2\,(p-k)\,e^{-\frac{\lambda_N^2 N(\gamma/2)^2}{2\sigma^2 K_{\text{clm}}^2}}.$$

This probability vanishes at rate $2e^{-2\lambda_N^2 N}$ for the choice of λ_N given in the theorem statement.

Establishing ℓ_∞ bounds. Next we establish a bound on the ℓ_∞-norm of the difference vector $U_S = \widehat{\beta}_S - \beta_S^*$ from Equation (11.36). By the triangle inequality, we have

$$\|U_S\|_\infty \leq \left\|\left(\frac{\mathbf{X}_S^T\mathbf{X}_S}{N}\right)^{-1}\frac{\mathbf{X}_S^T\mathbf{w}}{N}\right\|_\infty + \left\|\left(\frac{\mathbf{X}_S^T\mathbf{X}_S}{N}\right)^{-1}\right\|_\infty \lambda_N, \qquad (11.37)$$

where we have multiplied and divided different terms by N for later convenience. The second term is a deterministic quantity, so that it remains to bound the first term. For each $i = 1, \ldots, k$, consider the random variable

$$Z_i := e_i^T \left(\frac{1}{N}\mathbf{X}_S^T\mathbf{X}_S\right)^{-1}\frac{1}{N}\mathbf{X}_S^T\mathbf{w}.$$

Since the elements of $\mathbf{w}$ are i.i.d. $N(0, \sigma^2)$ variables, the variable Z_i is zero-mean Gaussian with variance at most

$$\frac{\sigma^2}{N}\left\|\left(\frac{1}{N}\mathbf{X}_S^T\mathbf{X}_S\right)^{-1}\right\|_2 \leq \frac{\sigma^2}{C_{\min}N},$$

where we have used the eigenvalue condition (11.29). Therefore, again combining Gaussian tail bounds with the union bound, we find that

$$\mathbb{P}[\|U_S\|_\infty > t] \leq 2e^{-\frac{t^2 C_{\min}N}{2\sigma^2}+\log k}.$$

Let us set $t = 4\sigma\lambda_N/\sqrt{C_{\min}}$ and then observe that our choice of λ_N guarantees that $8N\lambda_N^2 > \log p \geq \log k$. Putting together these pieces, we conclude that $\|U_S\|_\infty \leq 4\sigma\lambda_N/\sqrt{C_{\min}}$ with probability at least $1 - 2e^{-c_2\lambda_N^2 N}$. Overall, we conclude that

$$\|\widehat{\beta}_S - \beta_S^*\|_\infty \leq \lambda_N\left[\frac{4\sigma}{\sqrt{C_{\min}}} + \|(\mathbf{X}_S^T\mathbf{X}_S/N)^{-1}\|_\infty\right],$$

with probability greater than $1 - 2e^{-c_2\lambda_N^2 N}$, as claimed.

11.5 Beyond the Basic Lasso

In this chapter, we have limited ourselves to discussion of the basic lasso, which involves the least-squares loss function combined with the ℓ_1-norm as a regularizer. However, many of the ingredients have straightforward extensions to more general cost functions, including logistic regression and other types of generalized linear models, as well as to more exotic forms of regularization, including the group lasso, nuclear norm, and other types of structured regularizers. Here we sketch out the basic picture, referring the reader to the bibliographic section for links to further details.

Consider an objective function of the form

$$F(\beta) = \frac{1}{N} \sum_{i=1}^{N} f(\beta; z_i), \qquad (11.38)$$

where the function $\beta \mapsto g(\beta; z_i)$ measures the fit of the parameter vector $\beta \in \mathbb{R}^p$ to the sample z_i. In the context of regression problems, each sample takes the form $z_i = (x_i, y_i) \in \mathbb{R}^p \times \mathbb{R}$, whereas in problems such as the graphical lasso, each sample corresponds to a vector $z_i = x_i \in \mathbb{R}^p$. Letting $\Phi : \mathbb{R}^p \to \mathbb{R}$ denote a regularizer, we then consider an estimator of the form

$$\widehat{\beta} \in \arg\min_{\beta \in \Omega} \left\{ F(\beta) + \lambda_N \Phi(\beta) \right\}. \qquad (11.39)$$

We can view $\widehat{\beta}$ as an estimate of the deterministic vector β^* that minimizes the population objective function $\bar{F}(\beta) := \mathbb{E}[f(\beta; Z)]$.

To put our previous discussion in context, the familiar lasso is a special case of this general M-estimator, based on the choices

$$f(\beta; x_i, y_i) = \frac{1}{2} \big(y_i - \langle x_i, \beta \rangle \big)^2, \quad \text{and} \quad \Phi(\beta) = \|\beta\|_1,$$

and with the optimization taking place over $\Omega = \mathbb{R}^p$. In the case of random design, say with covariates $x_i \sim N(0, \Sigma)$, the population objective function for linear regression takes the form $\bar{F}(\beta) = \frac{1}{2}(\beta - \beta^*)^T \Sigma (\beta - \beta^*) + \frac{1}{2}\sigma^2$.

Considering the general M-estimator (11.39), our goal here is to provide some intuition on how to analyze the error $\|\widehat{\beta} - \beta^*\|_2$. When $N < p$, then the objective function (11.38) can never be strongly convex: indeed, assuming that it is twice differentiable, the Hessian is a sum of N matrices in p dimensions, and so must be rank degenerate. As noted previously, the restricted eigenvalue condition is a special case of a more general property of cost functions and regularizers, known as restricted strong convexity. In particular, given a set $\mathcal{C} \subset \mathbb{R}^p$, a differentiable function F satisfies restricted strong convexity over $\mathcal{C}$ at β^* if there exists a parameter $\gamma > 0$ such that

$$F(\beta^* + \nu) - F(\beta^*) - \langle \nabla F(\beta^*), \nu \rangle \geq \gamma \|\nu\|_2^2 \qquad \text{for all } \nu \in \mathcal{C}. \qquad (11.40)$$

When F is twice differentiable, then this lower bound is equivalent to controlling the Hessian in a neighborhood of β^*, as in the definition (11.9)—see

Exercise 11.6 for details. Thus, in the special case of a least-squares problem, restricted strong convexity is equivalent to a restricted eigenvalue condition.

For what type of sets $\mathcal{C}$ can a condition of this form be expected to hold? Since our ultimate goal is to control the error vector $\widehat{\nu} = \widehat{\beta} - \beta^*$, we need only ensure that strong convexity hold over a subset $\mathcal{C}$ that is guaranteed—typically with high probability over the data—to contain the error vector. Such sets exist for regularizers that satisfy a property known as *decomposability*, which generalizes a basic property of the ℓ_1-norm to a broader family of regularizers. Decomposability is defined in terms of a subspace $\mathcal{M}$ of the parameter set Ω, meant to describe the structure expected in the optimum β^*, and its orthogonal complement $\mathcal{M}^\perp$, corresponding to undesirable perturbations away from the model structure. With this notation, a regularizer Φ is said to be decomposable with respect to $\mathcal{M}$ if

$$\Phi(\beta + \theta) = \Phi(\beta) + \Phi(\theta) \qquad \text{for all pairs } (\beta, \theta) \in \mathcal{M} \times \mathcal{M}^\perp. \tag{11.41}$$

In the case of the ℓ_1-norm, the model subspace is simply the set of all vectors with support on some fixed set S, whereas the orthogonal complement $\mathcal{M}^\perp$ consists of vectors supported on the complementary set S^c. The decomposability relation (11.41) follows from the coordinate-wise nature of the ℓ_1-norm. With appropriate choices of subspaces, many other regularizers are decomposable, including weighted forms of the lasso, the group lasso and overlap group lasso penalties, and (with a minor generalization) the nuclear norm for low-rank matrices. See the bibliographic section for further details.

Bibliographic Notes

Knight and Fu (2000) derive asymptotic theory for the lasso and related estimators when the dimension p is fixed; the irrepresentable condition (11.27) appears implicitly in their analysis. Greenshtein and Ritov (2004) were the first authors to provide a high-dimensional analysis of the lasso, in particular providing bounds on the prediction error allowing for the $p \gg N$ setting. The irrepresentable or mutual incoherence condition (11.27) was developed independently by Fuchs (2004) and Tropp (2006) in signal processing, and Meinshausen and Bühlmann (2006) as well as Zhao and Yu (2006) in statistics. The notion of restricted eigenvalues was introduced by Bickel, Ritov and Tsybakov (2009); it is a less restrictive condition than the restricted isometry property from Chapter 10. van de Geer and Bühlmann (2009) provide a comparison between these and other related conditions for proving estimation error bounds on the lasso. Candès and Tao (2007) defined and developed theory for the "Dantzig selector", a problem closely related to the lasso. Raskutti, Wainwright and Yu (2010) show that the RE condition holds with high probability for various types of random Gaussian design matrices; see Rudelson and Zhou (2013) for extensions to sub-Gaussian designs.

The proof of Theorem 11.1 is based on the work of Bickel et al. (2009), whereas Negahban et al. (2012) derive the lasso error bound (11.24) for ℓ_q-sparse vectors. The basic inequality technique used in these proofs is standard in the analysis of M-estimators (van de Geer 2000). Raskutti, Wainwright and Yu (2011) analyze the minimax rates of regression over ℓ_q-balls, obtaining rates for both ℓ_2-error and prediction error. Theorem 11.2(a) was proved by Bunea, Tsybakov and Wegkamp (2007), whereas part (b) is due to Bickel et al. (2009). The restricted eigenvalue condition is actually required by any polynomial-time method in order to achieve the "fast rates" given in Theorem 11.2(b), as follows from the results of Zhang, Wainwright and Jordan (2014). Under a standard assumption in complexity theory, they prove that no polynomial-time algorithm can achieve the fast rate without imposing an RE condition.

Theorem 11.3 and the primal-dual witness (PDW) proof is due to Wainwright (2009). In the same paper, sharp threshold results are established for Gaussian ensembles of design matrices, in particular concrete upper and lower bounds on the sample size that govern the transition from success to failure in support recovery. The proof of Lemma 11.2 was suggested by Caramanis (2010). The PDW method has been applied to a range of other problems, including analysis of group lasso (Obozinski et al. 2011, Wang, Liang and Xing 2013) and related relaxations (Jalali, Ravikumar, Sanghavi and Ruan 2010, Negahban and Wainwright 2011b), graphical lasso (Ravikumar et al. 2011), and methods for Gaussian graph selection with hidden variables (Chandrasekaran et al. 2012). Lee, Sun and Taylor (2013) provide a general formulation of the PDW method for a broader class of M-estimators.

As noted in Section 11.5, the analysis in this chapter can be extended to a much broader class of M-estimators, namely those based on decomposable regularizers. Negahban et al. (2012) provide a general framework for analyzing the estimation error $\|\widehat{\beta} - \beta^*\|_2$ for such M-estimators. As alluded to here, the two key ingredients are restricted strong convexity of the cost function, and decomposability of the regularizer.

Exercises

Ex. 11.1 For a given $q \in (0, 1]$, recall the set $\mathbb{B}_q(R_q)$ defined in Equation (11.7) as a model of soft sparsity.

(a) A related object is the weak ℓ_q-ball with parameters (C, α), given by

$$\mathbb{B}_{w(\alpha)}(C) := \left\{ \theta \in \mathbb{R}^p \mid |\theta|_{(j)} \leq C j^{-\alpha} \quad \text{for } j = 1, \ldots, p \right\}. \quad (11.42a)$$

Here $|\theta|_{(j)}$ denote the order statistics of θ in absolute value, ordered from largest to smallest (so that $|\theta|_{(1)} = \max_{j=1,2,\ldots,p} |\theta_j|$ and $|\theta|_{(p)} = \min_{j=1,2,\ldots,p} |\theta_j|$.) For any $\alpha > 1/q$, show that there is a radius R_q depending on (C, α) such that $\mathbb{B}_{w(\alpha)}(C) \subseteq \mathbb{B}_q(R_q)$.

(b) For a given integer $k \in \{1, 2, \ldots, p\}$, the best k-term approximation to a vector $\theta^* \in \mathbb{R}^p$ is given by

$$\Pi_k(\theta^*) := \arg\min_{\|\theta\|_0 \le k} \|\theta - \theta^*\|_2^2. \tag{11.42b}$$

Give a closed form expression for $\Pi_k(\theta^*)$.

(c) When $\theta^* \in \mathbb{B}_q(R_q)$ for some $q \in (0, 1]$, show that the best k-term approximation satisfies

$$\|\Pi_k(\theta^*) - \theta^*\|_2^2 \le (R_q)^{2/q} \left(\frac{1}{k}\right)^{\frac{2}{q}-1}. \tag{11.42c}$$

Ex. 11.2 In this exercise, we analyze an alternative version of the lasso, namely the estimator

$$\widehat{\beta} = \arg\min_{\beta \in \mathbb{R}^p} \|\beta\|_1 \quad \text{such that } \tfrac{1}{N}\|\mathbf{y} - \mathbf{X}\beta\|_2^2 \le C, \tag{11.43}$$

where the constant $C > 0$ is a parameter to be chosen by the user. (This form of the lasso is often referred to as *relaxed basis pursuit*.)

(a) Suppose that C is chosen such that β^* is feasible for the convex program. Show that the error vector $\widehat{\nu} = \widehat{\beta} - \beta^*$ must satisfy the cone constraint $\|\widehat{\nu}_{S^c}\|_1 \le \|\widehat{\nu}_S\|_1$.

(b) Assuming the linear observation model $y = \mathbf{X}\beta^* + w$, show that $\widehat{\nu}$ satisfies the basic inequality

$$\frac{\|\mathbf{X}\widehat{\nu}\|_2^2}{N} \le 2\frac{\|\mathbf{X}^T\mathbf{w}\|_\infty}{N} \|\widehat{\nu}\|_1 + \left\{C - \frac{\|w\|_2^2}{N}\right\}.$$

(c) Assuming a γ-RE condition on $\mathbf{X}$, use part (b) to establish a bound on the ℓ_2-error $\|\widehat{\beta} - \beta^*\|_2$.

Ex. 11.3 In this exercise, we sketch out the proof of the bound (11.24). In particular, we show that if $\lambda_N \ge \frac{\|\mathbf{X}^T\mathbf{w}\|_\infty}{N}$ and $\beta^* \in \mathbb{B}_q(R_q)$, then the Lagrangian lasso error satisfies a bound of the form

$$\|\widehat{\beta} - \beta^*\|_2^2 \le c \, R_q \, \lambda_N^{1-q/2}. \tag{11.44a}$$

(a) Generalize Lemma 11.1 by showing that the error vector $\widehat{\nu}$ satisfies the "cone-like" constraint

$$\|\widehat{\nu}_{S^c}\|_1 \le 3\|\widehat{\nu}_S\|_1 + \|\beta^*_{S^c}\|_1, \tag{11.44b}$$

valid for any subset $S \subseteq \{1, 2, \dots, p\}$ and its complement.

(b) Suppose that $\mathbf{X}$ satisfies a γ-RE condition over all vectors satisfying the cone-like condition (11.44b). Prove that

$$\|\widehat{\nu}\|_2^2 \le \lambda_N\{4\|\widehat{\nu}_S\|_1 + \|\beta^*_{S^c}\|_1\}.$$

again valid for any subset S of indices.

(c) Optimize the choice of S so as to obtain the claimed bound (11.44a).

Ex. 11.4 Consider a random design matrix $\mathbf{X} \in \mathbb{R}^{N \times p}$ with each row $x_i \in \mathbb{R}^p$ drawn i.i.d. from a $\mathcal{N}(0, \mathbf{\Sigma})$ distribution, where the covariance matrix $\mathbf{\Sigma}$ is strictly positive definite. Show that a γ-RE condition holds over the set $\mathcal{C}(S; \alpha)$ with high probability whenever the sample size is lower bounded as $N > c |S|^2 \log p$ for a sufficiently large constant c. (*Remark:* This scaling of the sample size is not optimal; a more refined argument can be used to reduce $|S|^2$ to $|S|$.)

Ex. 11.5 Consider a random design matrix $\mathbf{X} \in \mathbb{R}^{N \times p}$ with i.i.d. $\mathcal{N}(0, 1)$ entries. In this exercise, we show that the mutual incoherence condition (11.27) holds with high probability as long as $N > ck \log p$ for a sufficiently large numerical constant c. (*Hint:* For $N > 4k$, it is known that the event $\mathcal{E} = \{\lambda_{\min}(\frac{\mathbf{X}_S^T \mathbf{X}_S}{N}) \geq \frac{1}{4}\}$ holds with high probability.)

(a) Show that

$$\gamma = 1 - \max_{j \in S^c} \max_{z \in \{-1, +1\}^k} \underbrace{\mathbf{x}_j^T \mathbf{X}_S (\mathbf{X}_S^T \mathbf{X}_S)^{-1} z}_{V_{j,z}} .$$

(b) Recalling the event $\mathcal{E}$, show that there is a numerical constant $c_0 > 0$ such that

$$\mathbb{P}[V_{j,z} \geq t] \leq e^{-c_0 \frac{N t^2}{k}} + \mathbb{P}[\mathcal{E}^c] \qquad \text{for any } t > 0,$$

valid for each fixed index $j \in S^c$ and vector $z \in \{-1, +1\}^k$.

(c) Use part (b) to complete the proof.

Ex. 11.6 Consider a twice differentiable function $F : \mathbb{R}^p \to \mathbb{R}$ and a set $\mathcal{C} \subset \mathbb{R}^p$ such that

$$\frac{\nabla^2 F(\beta)}{\|\nu\|_2^2} \geq \gamma \|\nu\|_2^2 \qquad \text{for all } \nu \in \mathcal{C},$$

uniformly for all β in a neighborhood of some fixed parameter β^*. Show that the RSC condition (11.40) holds.

Bibliography

Agarwal, A., Anandkumar, A., Jain, P., Netrapalli, P. and Tandon, R. (2014), Learning sparsely used overcomplete dictionaries via alternating minimization, *Journal of Machine Learning Research Workshop* **35**, 123–137.

Agarwal, A., Negahban, S. and Wainwright, M. J. (2012a), Fast global convergence of gradient methods for high-dimensional statistical recovery, *Annals of Statistics* **40**(5), 2452–2482.

Agarwal, A., Negahban, S. and Wainwright, M. J. (2012b), Noisy matrix decomposition via convex relaxation: Optimal rates in high dimensions, *Annals of Statistics* **40**(2), 1171–1197.

Alizadeh, A., Eisen, M., Davis, R. E., Ma, C., Lossos, I., Rosenwal, A., Boldrick, J., Sabet, H., Tran, T., Yu, X., Pwellm, J., Marti, G., Moore, T., Hudsom, J., Lu, L., Lewis, D., Tibshirani, R., Sherlock, G., Chan, W., Greiner, T., Weisenburger, D., Armitage, K., Levy, R., Wilson, W., Greve, M., Byrd, J., Botstein, D., Brown, P. and Staudt, L. (2000), Identification of molecularly and clinically distinct subtypes of diffuse large b cell lymphoma by gene expression profiling, *Nature* **403**, 503–511.

Alliney, S. and Ruzinsky, S. (1994), An algorithm for the minimization of mixed L1 and L2 norms with application to Bayesian estimation, *Transactions on Signal Processing* **42**(3), 618–627.

Amini, A. A. and Wainwright, M. J. (2009), High-dimensional analysis of semdefinite relaxations for sparse principal component analysis, *Annals of Statistics* **5B**, 2877–2921.

Anderson, T. (2003), *An Introduction to Multivariate Statistical Analysis, 3rd ed.*, Wiley, New York.

Antoniadis, A. (2007), Wavelet methods in statistics: Some recent developments and their applications, *Statistics Surveys* **1**, 16–55.

Bach, F. (2008), Consistency of trace norm minimization, *Journal of Machine Learning Research* **9**, 1019–1048.

Bach, F., Jenatton, R., Mairal, J. and Obozinski, G. (2012), Optimization with sparsity-inducing penalties, *Foundations and Trends in Machine Learning* **4**(1), 1–106.

Banerjee, O., El Ghaoui, L. and d'Aspremont, A. (2008), Model selection through sparse maximum likelihood estimation for multivariate Gaussian or binary data, *Journal of Machine Learning Research* **9**, 485–516.

Baraniuk, R. G., Davenport, M. A., DeVore, R. A. and Wakin, M. B. (2008), A simple proof of the restricted isometry property for random matrices, *Constructive Approximation* **28**(3), 253–263.

Barlow, R. E., Bartholomew, D., Bremner, J. M. and Brunk, H. D. (1972), *Statistical Inference under Order Restrictions: The Theory and Application of Isotonic Regression*, Wiley, New York.

Beck, A. and Teboulle, M. (2009), A fast iterative shrinkage-thresholding algorithm for linear inverse problems, *SIAM Journal on Imaging Sciences* **2**, 183–202.

Benjamini, Y. and Hochberg, Y. (1995), Controlling the false discovery rate: a practical and powerful approach to multiple testing, *Journal of the Royal Statistical Society Series B.* **85**, 289–300.

Bennett, J. and Lanning, S. (2007), The netflix prize, in *Proceedings of KDD Cup and Workshop in conjunction with KDD*.

Bento, J. and Montanari, A. (2009), Which graphical models are difficulty to learn?, in *Advances in Neural Information Processing Systems (NIPS Conference Proceedings)*.

Berk, R., Brown, L., Buja, A., Zhang, K. and Zhao, L. (2013), Valid post-selection inference, *Annals of Statistics* **41**(2), 802–837.

Berthet, Q. and Rigollet, P. (2013), Computational lower bounds for sparse PCA, Technical report, Princeton University. arxiv1304.0828.

Bertsekas, D. (1999), *Nonlinear Programming*, Athena Scientific, Belmont MA.

Bertsekas, D. (2003), *Convex Analysis and Optimization*, Athena Scientific, Belmont MA.

Besag, J. (1974), Spatial interaction and the statistical analysis of lattice systems, *Journal of the Royal Statistical Society Series B* **36**, 192–236.

Besag, J. (1975), Statistical analysis of non-lattice data, *The Statistician* **24**(3), 179–195.

Bickel, P. J. and Levina, E. (2008), Covariance regularization by thresholding, *Annals of Statistics* **36**(6), 2577–2604.

Bickel, P. J., Ritov, Y. and Tsybakov, A. (2009), Simultaneous analysis of Lasso and Dantzig selector, *Annals of Statistics* **37**(4), 1705–1732.

Bien, J., Taylor, J. and Tibshirani, R. (2013), A Lasso for hierarchical interactions, *Annals of Statistics* **42**(3), 1111–1141.

Bien, J. and Tibshirani, R. (2011), Sparse estimation of a covariance matrix, *Biometrika* **98**(4), 807–820.

Birnbaum, A., Johnstone, I., Nadler, B. and Paul, D. (2013), Minimax bounds for sparse PCA with noisy high-dimensional data, *Annals of Statistics* **41**(3), 1055–1084.

Boser, B., Guyon, I. and Vapnik, V. (1992), A training algorithm for optimal margin classifiers, in *Proceedings of the Annual Conference on Learning Theory (COLT)*, Philadelphia, Pa.

Boyd, S., Parikh, N., Chu, E., Peleato, B. and Eckstein, J. (2011), Distributed optimization and statistical learning via the alternating direction method of multipliers, *Foundations and Trends in Machine Learning* **3**(1), 1–124.

Boyd, S. and Vandenberghe, L. (2004), *Convex Optimization*, Cambridge University Press, Cambridge, UK.

Breiman, L. (1995), Better subset selection using the nonnegative garrote, *Technometrics* **37**, 738–754.

Breiman, L. and Ihaka, R. (1984), Nonlinear discriminant analysis via scaling and ACE, Technical report, University of California, Berkeley.

Bühlmann, P. (2013), Statistical significance in high-dimensional linear models, *Bernoulli* **19**(4), 1212–1242.

Bühlmann, P. and van de Geer, S. (2011), *Statistics for High-Dimensional Data: Methods, Theory and Applications*, Springer, New York.

Bunea, F., She, Y. and Wegkamp, M. (2011), Optimal selection of reduced rank estimators of high-dimensional matrices, **39**(2), 1282–1309.

Bunea, F., Tsybakov, A. and Wegkamp, M. (2007), Sparsity oracle inequalities for the Lasso, *Electronic Journal of Statistics* pp. 169–194.

Burge, C. and Karlin, S. (1977), Prediction of complete gene structures in human genomic DNA, *Journal of Molecular Biology* **268**, 78–94.

Butte, A., Tamayo, P., Slonim, D., Golub, T. and Kohane, I. (2000), Discovering functional relationships between RNA expression and chemotherapeutic susceptibility using relevance networks, *Proceedings of the National Academy of Sciences* pp. 12182–12186.

Candès, E., Li, X., Ma, Y. and Wright, J. (2011), Robust Principal Component Analysis?, *Journal of the Association for Computing Machinery* **58**, 11:1–11:37.

Candès, E. and Plan, Y. (2010), Matrix completion with noise, *Proceedings of the IEEE* **98**(6), 925–936.

Candès, E. and Recht, B. (2009), Exact matrix completion via convex optimization, *Foundation for Computational Mathematics* **9**(6), 717–772.

Candès, E., Romberg, J. K. and Tao, T. (2006), Stable signal recovery from incomplete and inaccurate measurements, *Communications on Pure and Applied Mathematics* **59**(8), 1207–1223.

Candès, E. and Tao, T. (2005), Decoding by linear programming, *IEEE Transactions on Information Theory* **51**(12), 4203–4215.

Candès, E. and Tao, T. (2007), The Dantzig selector: Statistical estimation when p is much larger than n, *Annals of Statistics* **35**(6), 2313–2351.

Candès, E. and Wakin, M. (2008), An introduction to compressive sampling, *Signal Processing Magazine, IEEE* **25**(2), 21–30.

Caramanis, C. (2010), 'Personal communication'.

Chandrasekaran, V., Parrilo, P. A. and Willsky, A. S. (2012), Latent variable graphical model selection via convex optimization, *Annals of Statistics* **40**(4), 1935–1967.

Chandrasekaran, V., Sanghavi, S., Parrilo, P. A. and Willsky, A. S. (2011), Rank-sparsity incoherence for matrix decomposition, *SIAM Journal on Optimization* **21**, 572–596.

Chaudhuri, S., Drton, M. and Richardson, T. S. (2007), Estimation of a covariance matrix with zeros, *Biometrika* pp. 1–18.

Chen, S., Donoho, D. and Saunders, M. (1998), Atomic decomposition by basis pursuit, *SIAM Journal of Scientific Computing* **20**(1), 33–61.

Cheng, J., Levina, E. and Zhu, J. (2013), High-dimensional Mixed Graphical Models, *arXiv:1304.2810* .

Chi, E. C. and Lange, K. (2014), Splitting methods for convex clustering, *Journal of Computational and Graphical Statistics (online access)* .

Choi, Y., Taylor, J. and Tibshirani, R. (2014), Selecting the number of principal components: estimation of the true rank of a noisy matrix. arXiv:1410.8260.

Clemmensen, L. (2012), *sparseLDA: Sparse Discriminant Analysis.* R package version 0.1-6.
   URL: *http://CRAN.R-project.org/package=sparseLDA*

Clemmensen, L., Hastie, T., Witten, D. and Ersboll, B. (2011), Sparse discriminant analysis, *Technometrics* **53**, 406–413.

Clifford, P. (1990), Markov random fields in statistics, in G. Grimmett and D. J. A. Welsh, eds, *Disorder in physical systems*, Oxford Science Publications.

Cohen, A., Dahmen, W. and DeVore, R. A. (2009), Compressed sensing and best k-term approximation, *Journal of the American Mathematical Society* **22**(1), 211–231.

Cox, D. and Wermuth, N. (1996), *Multivariate Dependencies: Models, Analysis and Interpretation*, Chapman & Hall, London.

d'Aspremont, A., Banerjee, O. and El Ghaoui, L. (2008), First order methods for sparse covariance selection, *SIAM Journal on Matrix Analysis and its Applications* **30**(1), 55–66.

d'Aspremont, A., El Ghaoui, L., Jordan, M. I. and Lanckriet, G. R. G. (2007), A direct formulation for sparse PCA using semidefinite programming, *SIAM Review* **49**(3), 434–448.

Davidson, K. R. and Szarek, S. J. (2001), Local operator theory, random matrices, and Banach spaces, in *Handbook of Banach Spaces*, Vol. 1, Elsevier, Amsterdam, NL, pp. 317–336.

De Leeuw, J. (1994), Block-relaxation algorithms in statistics, in H. Bock,

W. Lenski and M. M. Richter, eds, *Information Systems and Data Analysis*, Springer-Verlag, Berlin.

Dobra, A., Hans, C., Jones, B., Nevins, J. R., Yao, G. and West, M. (2004), Sparse graphical models for exploring gene expression data, *Journal of Multivariate Analysis* **90**(1), 196 – 212.

Donoho, D. (2006), Compressed sensing, *IEEE Transactions on Information Theory* **52**(4), 1289–1306.

Donoho, D. and Huo, X. (2001), Uncertainty principles and ideal atomic decomposition, *IEEE Trans. Info Theory* **47**(7), 2845–2862.

Donoho, D. and Johnstone, I. (1994), Ideal spatial adaptation by wavelet shrinkage, *Biometrika* **81**, 425–455.

Donoho, D. and Stark, P. (1989), Uncertainty principles and signal recovery, *SIAM Journal of Applied Mathematics* **49**, 906–931.

Donoho, D. and Tanner, J. (2009), Counting faces of randomly-projected polytopes when the projection radically lowers dimension, *Journal of the American Mathematical Society* **22**(1), 1–53.

Dudoit, S., Fridlyand, J. and Speed, T. (2002), Comparison of discrimination methods for the classification of tumors using gene expression data, *Journal of the American Statistical Association* **97**(457), 77–87.

Edwards, D. (2000), *Introduction to Graphical Modelling, 2nd Edition*, Springer, New York.

Efron, B. (1979), Bootstrap methods: another look at the jackknife, *Annals of Statistics* **7**, 1–26.

Efron, B. (1982), *The Jackknife, the Bootstrap and Other Resampling plans*, Vol. 38, SIAM- CBMS-NSF Regional Conference Series in Applied Mathematics.

Efron, B. (2011), The bootstrap and Markov Chain Monte Carlo, *Journal of Biopharmaceutical Statistics* **21**(6), 1052–1062.

Efron, B. and Tibshirani, R. (1993), *An Introduction to the Bootstrap*, Chapman & Hall, London.

El Ghaoui, L., Viallon, V. and Rabbani, T. (2010), Safe feature elimination in sparse supervised learning, *Pacific journal of optimization* **6**(4), 667–698.

El Karoui, N. (2008), Operator norm consistent estimation of large-dimensional sparse covariance matrices, *Annals of Statistics* **36**(6), 2717–2756.

Elad, M. and Bruckstein, A. M. (2002), A generalized uncertainty principle and sparse representation in pairs of bases, *IEEE Transactions on Information Theory* **48**(9), 2558–2567.

Erdos, P. and Renyi, A. (1961), On a classical problem of probability theory, *Magyar Tud. Akad. Mat. Kutat Int. Kzl.* **6**, 215–220. (English and Russian summary).

Erhan, D., Bengio, Y., Courville, A., Manzagol, P.-A., Vincent, P. and Bengio, S. (2010), Why does unsupervised pre-training help deep learning?, *Journal of Machine Learning Research* **11**, 625–660.

Fan, J. and Li, R. (2001), Variable selection via nonconcave penalized likelihood and its oracle properties, *Journal of the American Statistical Association* **96**(456), 1348–1360.

Fazel, M. (2002), Matrix Rank Minimization with Applications, PhD thesis, Stanford. Available online: http://faculty.washington.edu/mfazel/thesis-final.pdf.

Feuer, A. and Nemirovski, A. (2003), On sparse representation in pairs of bases, *IEEE Transactions on Information Theory* **49**(6), 1579–1581.

Field, D. (1987), Relations between the statistics of natural images and the response properties of cortical cells, *Journal of the Optical Society of America* **A4**, 2379–2394.

Fisher, M. E. (1966), On the Dimer solution of planar Ising models, *Journal of Mathematical Physics* **7**, 1776–1781.

Fithian, W., Sun, D. and Taylor, J. (2014), Optimal inference after model selection, *ArXiv e-prints* .

Friedman, J., Hastie, T., Hoefling, H. and Tibshirani, R. (2007), Pathwise coordinate optimization, *Annals of Applied Statistics* **1**(2), 302–332.

Friedman, J., Hastie, T., Simon, N. and Tibshirani, R. (2015), *glmnet: Lasso and elastic-net regularized generalized linear models*. R package version 2.0.

Friedman, J., Hastie, T. and Tibshirani, R. (2008), Sparse inverse covariance estimation with the graphical Lasso, *Biostatistics* **9**, 432–441.

Friedman, J., Hastie, T. and Tibshirani, R. (2010*a*), Applications of the Lasso and grouped Lasso to the estimation of sparse graphical models, Technical report, Stanford University, Statistics Department.

Friedman, J., Hastie, T. and Tibshirani, R. (2010*b*), Regularization paths for generalized linear models via coordinate descent, *Journal of Statistical Software* **33**(1), 1–22.

Fu, W. (1998), Penalized regressions: the bridge versus the lasso, *Journal of Computational and Graphical Statistics* **7**(3), 397–416.

Fuchs, J. (2000), On the application of the global matched filter to doa estimation with uniform circular arrays, in *Proceedings of the Acoustics, Speech, and Signal Processing, 2000. on IEEE International Conference - Volume 05*, ICASSP '00, IEEE Computer Society, Washington, DC, USA, pp. 3089–3092.

Fuchs, J. (2004), Recovery of exact sparse representations in the presence of noise, in *International Conference on Acoustics, Speech, and Signal Processing*, Vol. 2, pp. 533–536.

Gannaz, I. (2007), Robust estimation and wavelet thresholding in partially linear models, *Statistics and Computing* **17**(4), 293–310.

Gao, H. and Bruce, A. (1997), Waveshrink with firm shrinkage, *Statistica Sinica* **7**, 855–874.

Geman, S. and Geman, D. (1984), Stochastic relaxation, Gibbs distributions, and the Bayesian restoration of images, *IEEE Transactions on Pattern Analysis and Machine Intelligence* **6**, 721–741.

Golub, G. and Loan, C. V. (1996), *Matrix Computations*, Johns Hopkins University Press, Baltimore.

Gorski, J., Pfeuffer, F. and Klamroth, K. (2007), Biconvex sets and optimization with biconvex functions: a survey and extensions, *Mathematical Methods of Operations Research* **66**(3), 373–407.

Gramacy, R. (2011), 'The monomvn package: Estimation for multivariate normal and student-t data with monotone missingness', CRAN. R package version 1.8.

Grazier G'Sell, M., Taylor, J. and Tibshirani, R. (2013), Adaptive testing for the graphical Lasso. arXiv: 1307.4765.

Grazier G'Sell, M., Wager, S., Chouldechova, A. and Tibshirani, R. (2015), Sequential selection procedures and false discovery rate control. arXiv: 1309.5352: To appear, Journal of the Royal Statistical Society Series B.

Greenshtein, E. and Ritov, Y. (2004), Persistency in high dimensional linear predictor-selection and the virtue of over-parametrization, *Bernoulli* **10**, 971–988.

Greig, D. M., Porteous, B. T. and Seheuly, A. H. (1989), Exact maximum a posteriori estimation for binary images, *Journal of the Royal Statistical Society Series B* **51**, 271–279.

Grimmett, G. R. (1973), A theorem about random fields, *Bulletin of the London Mathematical Society* **5**, 81–84.

Gross, D. (2011), Recovering low-rank matrices from few coefficients in any basis, *IEEE Transactions on Information Theory* **57**(3), 1548–1566.

Gu, C. (2002), *Smoothing Spline ANOVA Models*, Springer Series in Statistics, Springer, New York, NY.

Hammersley, J. M. and Clifford, P. (1971), Markov fields on finite graphs and lattices. Unpublished.

Hardoon, D. and Shawe-Taylor, J. (2011), Sparse canonical correlation analysis, *Machine Learning* **83**(3), 331–353.

Hastie, T., Buja, A. and Tibshirani, R. (1995), Penalized discriminant analysis, *Annals of Statistics* **23**, 73–102.

Hastie, T. and Mazumder, R. (2013), *softImpute: matrix completion via iterative soft-thresholded SVD*. R package version 1.0.
URL: *http://CRAN.R-project.org/package=softImpute*

Hastie, T. and Tibshirani, R. (1990), *Generalized Additive Models*, Chapman & Hall, London.

Hastie, T. and Tibshirani, R. (2004), Efficient quadratic regularization for expression arrays, *Biostatistics,* **5**, 329–340.

Hastie, T., Tibshirani, R. and Buja, A. (1994), Flexible discriminant analysis by optimal scoring, *Journal of the American Statistical Association* **89**, 1255–1270.

Hastie, T., Tibshirani, R. and Friedman, J. (2009), *The Elements of Statistical Learning: Data Mining, Inference and Prediction*, second edn, Springer Verlag, New York.

Hastie, T., Tibshirani, R., Narasimhan, B. and Chu, G. (2003), *pamr: Prediction Analysis for Microarrays in R*. R package version 1.54.1.
URL: *http://CRAN.R-project.org/package=pamr*

Hocking, T., Vert, J.-P., Bach, F. and Joulin, A. (2011), Clusterpath: an algorithm for clustering using convex fusion penalties., in L. Getoor and T. Scheffer, eds, *Proceedings of the Twenty-Eighth International Conference on Machine Learning (ICML)*, Omnipress, pp. 745–752.

Hoefling, H. (2010), A path algorithm for the fused Lasso signal approximator, *Journal of Computational and Graphical Statistics* **19**(4), 984–1006.

Hoefling, H. and Tibshirani, R. (2009), Estimation of sparse binary pairwise Markov networks using pseudo-likelihoods, *Journal of Machine Learning Research* **19**, 883–906.

Horn, R. A. and Johnson, C. R. (1985), *Matrix Analysis*, Cambridge University Press, Cambridge.

Hsu, D., Kakade, S. M. and Zhang, T. (2011), Robust matrix decomposition with sparse corruptions, *IEEE Transactions on Information Theory* **57**(11), 7221–7234.

Huang, J., Ma, S. and Zhang, C.-H. (2008), Adaptive Lasso for sparse high-dimensional regression models, *Statistica Sinica* **18**, 1603–1618.

Huang, J. and Zhang, T. (2010), The benefit of group sparsity, *The Annals of Statistics* **38**(4), 1978–2004.

Hunter, D. R. and Lange, K. (2004), A tutorial on MM algorithms, *The American Statistician* **58**(1), 30–37.

Ising, E. (1925), Beitrag zur theorie der ferromagnetismus, *Zeitschrift für Physik* **31**(1), 253–258.

Jacob, L., Obozinski, G. and Vert, J.-P. (2009), Group Lasso with overlap and graph Lasso, in *Proceeding of the 26th International Conference on Machine Learning, Montreal, Canada*.

Jalali, A., Ravikumar, P., Sanghavi, S. and Ruan, C. (2010), A dirty model for multi-task learning, in *Advances in Neural Information Processing Systems 23*, pp. 964–972.

Javanmard, A. and Montanari, A. (2013), Hypothesis testing in high-dimensional regression under the Gaussian random design model: Asymptotic theory. arXiv: 1301.4240.

Javanmard, A. and Montanari, A. (2014), Confidence intervals and hypothesis testing for high-dimensional regression, *Journal of Machine Learning Research* **15**, 2869–2909.

Jerrum, M. and Sinclair, A. (1993), Polynomial-time approximation algorithms for the Ising model, *SIAM Journal of Computing* **22**, 1087–1116.

Jerrum, M. and Sinclair, A. (1996), The Markov chain Monte Carlo method: An approach to approximate counting and integration, in D. Hochbaum, ed., *Approximation algorithms for NP-hard problems*, PWS Publishing, Boston.

Johnson, N. (2013), A dynamic programming algorithm for the fused Lasso and ℓ_0-segmentation, *Journal of Computational and Graphical Statistics* **22**(2), 246–260.

Johnson, W. B. and Lindenstrauss, J. (1984), Extensions of Lipschitz mappings into a Hilbert space, *Contemporary Mathematics* **26**, 189–206.

Johnstone, I. (2001), On the distribution of the largest eigenvalue in principal components analysis, *Annals of Statistics* **29**(2), 295–327.

Johnstone, I. and Lu, A. (2009), On consistency and sparsity for principal components analysis in high dimensions, *Journal of the American Statistical Association* **104**, 682–693.

Jolliffe, I. T., Trendafilov, N. T. and Uddin, M. (2003), A modified principal component technique based on the Lasso, *Journal of Computational and Graphical Statistics* **12**, 531–547.

Kaiser, H. (1958), The varimax criterion for analytic rotation in factor analysis, *Psychometrika* **23**, 187–200.

Kalisch, M. and Bühlmann, P. (2007), Estimating high-dimensional directed acyclic graphs with the PC algorithm, *Journal of Machine Learning Research* **8**, 613–636.

Kastelyn, P. W. (1963), Dimer statistics and phase transitions, *Journal of Mathematical Physics* **4**, 287–293.

Keshavan, R. H., Montanari, A. and Oh, S. (2010), Matrix completion from noisy entries, *Journal of Machine Learning Research* **11**, 2057–2078.

Keshavan, R. H., Oh, S. and Montanari, A. (2009), Matrix completion from a few entries, *IEEE Transactions on Information Theory* **56(6)**, 2980–2998.

Kim, S., Koh, K., Boyd, S. and Gorinevsky, D. (2009), L1 trend filtering, *SIAM Review, problems and techniques section* **51**(2), 339–360.

Knight, K. and Fu, W. J. (2000), Asymptotics for Lasso-type estimators, *Annals of Statistics* **28**, 1356–1378.

Koh, K., Kim, S. and Boyd, S. (2007), An interior-point method for large-scale ℓ_1-regularized logistic regression, *Journal of Machine Learning Research* **8**, 1519–1555.

Koller, D. and Friedman, N. (2009), *Probabilistic Graphical Models*, The MIT Press, Cambridge MA.

Koltchinskii, V., Lounici, K. and Tsybakov, A. (2011), Nuclear-norm penalization and optimal rates for noisy low-rank matrix completion, *Annals of Statistics* **39**, 2302–2329.

Koltchinskii, V. and Yuan, M. (2008), Sparse recovery in large ensembles of kernel machines, in *Proceedings of the Annual Conference on Learning Theory (COLT)*.

Koltchinskii, V. and Yuan, M. (2010), Sparsity in multiple kernel learning, *Annals of Statistics* **38**, 3660–3695.

Krahmer, F. and Ward, R. (2011), New and improved Johnson-Lindenstrauss embeddings via the restricted isometry property, *SIAM Journal on Mathematical Analysis* **43**(3), 1269–1281.

Lang, K. (1995), Newsweeder: Learning to filter netnews., in *Proceedings of the Twelfth International Conference on Machine Learning (ICML)*, pp. 331–339.

Lange, K. (2004), *Optimization*, Springer, New York.

Lange, K., Hunter, D. R. and Yang, I. (2000), Optimization transfer using surrogate objective functions (with discussion), *Computational and Graphical Statistics* **9**, 1–59.

Laurent, M. (2001), Matrix completion problems, in *The Encyclopedia of Optimization*, Kluwer Academic, pp. 221–229.

Lauritzen, S. L. (1996), *Graphical Models*, Oxford University Press.

Lauritzen, S. L. and Spiegelhalter, D. J. (1988), Local computations with probabilities on graphical structures and their application to expert systems (with discussion), *Journal of the Royal Statistical Society Series B* **50**, 155–224.

Le Cun, Y., Boser, B., Denker, J., Henderson, D., Howard, R., Hubbard, W. and Jackel, L. (1990), Handwritten digit recognition with a back-propogation network, in D. Touretzky, ed., *Advances in Neural Information Processing Systems*, Vol. 2, Morgan Kaufman, Denver, CO, pp. 386–404.

Le, Q., Ranzato, M., Monga, R., Devin, M., Chen, K., Corrado, G., Dean, J. and Ng, A. (2012), Building high-level features using large scale unsupervised learning, in *Proceedings of the 29th International Conference on Machine Learning*, Edinburgh, Scotland.

Lee, J. and Hastie, T. (2014), Learning the structure of mixed graphical models, *Journal of Computational and Graphical Statistics* . advanced online access.

Lee, J., Sun, D., Sun, Y. and Taylor, J. (2016), Exact post-selection inference, with application to the Lasso, *Annals of Statistics* **44**(4), 907–927.

Lee, J., Sun, Y. and Saunders, M. (2014), Proximal newton-type methods for minimizing composite functions, *SIAM Journal on Optimization* **24**(3), 1420–1443.

Lee, J., Sun, Y. and Taylor, J. (2013), On model selection consistency of m-estimators with geometrically decomposable penalties, Technical report, Stanford University. arxiv1305.7477v4.

Lee, M., Shen, H., Huang, J. and Marron, J. (2010), Biclustering via sparse singular value decomposition, *Biometrics* pp. 1086–1095.

Lee, S., Lee, H., Abneel, P. and Ng, A. (2006), Efficient L1 logistic regression, in *Proceedings of the Twenty-First National Conference on Artificial Intelligence (AAAI-06)*.

Lei, J. and Vu, V. Q. (2015), Sparsistency and Agnostic Inference in Sparse PCA, *Ann. Statist.* **43**(1), 299–322.

Leng, C. (2008), Sparse optimal scoring for multiclass cancer diagnosis and biomarker detection using microarray data, *Computational Biology and Chemistry* **32**, 417–425.

Li, L., Huang, W., Gu, I. Y. and Tian, Q. (2004), Statistical modeling of complex backgrounds for foreground object detection, *IEEE Transactions on Image Processing* **13**(11), 1459–1472.

Lim, M. and Hastie, T. (2014), Learning interactions via hierarchical group-Lasso regularization, *Journal of Computational and Graphical Statistics (online access)* .

Lin, Y. and Zhang, H. H. (2003), Component selection and smoothing in smoothing spline analysis of variance models, Technical report, Department of Statistics, University of Wisconsin, Madison.

Lockhart, R., Taylor, J., Tibshirani$_2$, R. and Tibshirani, R. (2014), A significance test for the Lasso, *Annals of Statistics (with discussion)* **42**(2), 413–468.

Loftus, J. and Taylor, J. (2014), A significance test for forward stepwise model selection. arXiv:1405.3920.

Lounici, K., Pontil, M., Tsybakov, A. and van de Geer, S. (2009), Taking advantage of sparsity in multi-task learning, Technical report, ETH Zurich.

Lustig, M., Donoho, D., Santos, J. and Pauly, J. (2008), Compressed sensing MRI, *IEEE Signal Processing Magazine* **27**, 72–82.

Lykou, A. and Whittaker, J. (2010), Sparse CCA using a Lasso with positivity constraints, *Computational Statistics & Data Analysis* **54**(12), 3144–3157.

Ma, S., Xue, L. and Zou, H. (2013), Alternating direction methods for latent variable Gaussian graphical model selection, *Neural Computation* **25**, 2172–2198.

Ma, Z. (2010), Contributions to high-dimensional principal component analysis, PhD thesis, Department of Statistics, Stanford University.

Ma, Z. (2013), Sparse principal component analysis and iterative thresholding, *Annals of Statistics* **41**(2), 772–801.

Mahoney, M. W. (2011), Randomized algorithms for matrices and data, *Foundations and Trends in Machine Learning in Machine Learning* **3**(2).

Mangasarian, O. (1999), Arbitrary-norm separating plane., *Operations Research Letters* **24**(1-2), 15–23.

Mazumder, R., Friedman, J. and Hastie, T. (2011), *Sparsenet*: Coordinate descent with non-convex penalties, *Journal of the American Statistical Association* **106**(495), 1125–1138.

Mazumder, R. and Hastie, T. (2012), The Graphical Lasso: New insights and alternatives, *Electronic Journal of Statistics* **6**, 2125–2149.

Mazumder, R., Hastie, T. and Friedman, J. (2012), *sparsenet: Fit sparse linear regression models via nonconvex optimization*. R package version 1.0. **URL:** *http://CRAN.R-project.org/package=sparsenet*

Mazumder, R., Hastie, T. and Tibshirani, R. (2010), Spectral regularization algorithms for learning large incomplete matrices, *Journal of Machine Learning Research* **11**, 2287–2322.

McCullagh, P. and Nelder, J. (1989), *Generalized Linear Models*, Chapman & Hall, London.

Meier, L., van de Geer, S. and Bühlmann, P. (2008), The group Lasso for logistic regression, *Journal of the Royal Statistical Society B* **70**(1), 53–71.

Meier, L., van de Geer, S. and Bühlmann, P. (2009), High-dimensional additive modeling, *Annals of Statistics* **37**, 3779–3821.

Meinshausen, N. (2007), Relaxed Lasso, *Computational Statistics and Data Analysis* pp. 374–393.

Meinshausen, N. and Bühlmann, P. (2006), High-dimensional graphs and variable selection with the Lasso, *Annals of Statistics* **34**, 1436–1462.

Meinshausen, N. and Bühlmann, P. (2010), Stability selection, *Journal of the Royal Statistical Society Series B* **72**(4), 417–473.

Mézard, M. and Montanari, A. (2008), *Information, Physics and Computation*, Oxford University Press, New York, NY.

Negahban, S., Ravikumar, P., Wainwright, M. J. and Yu, B. (2012), A unified framework for high-dimensional analysis of M-estimators with decomposable regularizers, *Statistical Science* **27**(4), 538–557.

Negahban, S. and Wainwright, M. J. (2011*a*), Estimation of (near) low-rank matrices with noise and high-dimensional scaling, *Annals of Statistics*

39(2), 1069–1097.

Negahban, S. and Wainwright, M. J. (2011*b*), Simultaneous support recovery in high-dimensional regression: Benefits and perils of $\ell_{1,\infty}$-regularization, *IEEE Transactions on Information Theory* **57**(6), 3481–3863.

Negahban, S. and Wainwright, M. J. (2012), Restricted strong convexity and (weighted) matrix completion: Optimal bounds with noise, *Journal of Machine Learning Research* **13**, 1665–1697.

Nelder, J. and Wedderburn, R. (1972), Generalized linear models, *J. Royal Statist. Soc. B.* **135**(3), 370–384.

Nemirovski, A. and Yudin, D. B. (1983), *Problem Complexity and Method Efficiency in Optimization*, John Wiley and Sons, New York.

Nesterov, Y. (2004), *Introductory Lectures on Convex Optimization*, Kluwer Academic Publishers, New York.

Nesterov, Y. (2007), Gradient methods for minimizing composite objective function, Technical Report 76, Center for Operations Research and Econometrics (CORE), Catholic University of Louvain (UCL).

Netrapalli, P., Jain, P. and Sanghavi, S. (2013), Phase retrieval using alternating minimization, in *Advances in Neural Information Processing Systems (NIPS Conference Proceedings)*, pp. 2796–2804.

Obozinski, G., Wainwright, M. J. and Jordan, M. I. (2011), Union support recovery in high-dimensional multivariate regression, *Annals of Statistics* **39**(1), 1–47.

Oldenburg, D. W., Scheuer, T. and Levy, S. (1983), Recovery of the acoustic impedance from reflection seismograms, *Geophysics* **48**(10), 1318–1337.

Olsen, S. (2002), 'Amazon blushes over sex link gaffe', CNET News. http://news.cnet.com/2100-1023-976435.html.

Olshausen, B. and Field, D. (1996), Emergence of simple-cell receptive field properties by learning a sparse code for natural images, *Nature* **381**.

Park, T. and Casella, G. (2008), The Bayesian Lasso, *Journal of the American Statistical Association* **103**(482), 681–686.

Parkhomenko, E., Tritchler, D. and Beyene, J. (2009), Sparse canonical correlation analysis with application to genomic data integration, *Statistical Applications in Genetics and Molecular Biology* **8**, 1–34.

Paul, D. and Johnstone, I. (2008), Augmented sparse principal component analysis for high-dimensional data, Technical report, UC Davis.

Pearl, J. (2000), *Causality: Models, Reasoning and Inference*, Cambridge University Press.

Pelckmans, K., De Moor, B. and Suykens, J. (2005), Convex clustering shrinkage, in *Workshop on Statistics and Optimization of Clustering (PASCAL), London, UK.*

Pilanci, M. and Wainwright, M. J. (2014), Randomized sketches of convex programs with sharp guarantees, Technical report, UC Berkeley. Full length version at arXiv:1404.7203; Presented in part at ISIT 2014.

Puig, A., Wiesel, A. and Hero, A. (2009), A multidimensional shrinkage thresholding operator, in *Proceedings of the 15th workshop on Statistical Signal Processing, SSP'09*, IEEE, pp. 113–116.

Raskutti, G., Wainwright, M. J. and Yu, B. (2009), Lower bounds on minimax rates for nonparametric regression with additive sparsity and smoothness, in *Advances in Neural Information Processing Systems 22*, MIT Press, Cambridge MA., pp. 1563–1570.

Raskutti, G., Wainwright, M. J. and Yu, B. (2010), Restricted eigenvalue conditions for correlated Gaussian designs, *Journal of Machine Learning Research* 11, 2241–2259.

Raskutti, G., Wainwright, M. J. and Yu, B. (2011), Minimax rates of estimation for high-dimensional linear regression over ℓ_q-balls, *IEEE Transactions on Information Theory* 57(10), 6976–6994.

Raskutti, G., Wainwright, M. J. and Yu, B. (2012), Minimax-optimal rates for sparse additive models over kernel classes via convex programming, *Journal of Machine Learning Research* 12, 389–427.

Ravikumar, P., Liu, H., Lafferty, J. and Wasserman, L. (2009), Sparse additive models, *Journal of the Royal Statistical Society Series B.* 71(5), 1009–1030.

Ravikumar, P., Wainwright, M. J. and Lafferty, J. (2010), High-dimensional ising model selection using ℓ_1-regularized logistic regression, *Annals of Statistics* 38(3), 1287–1319.

Ravikumar, P., Wainwright, M. J., Raskutti, G. and Yu, B. (2011), High-dimensional covariance estimation by minimizing ℓ_1-penalized log-determinant divergence, *Electronic Journal of Statistics* 5, 935–980.

Recht, B. (2011), A simpler approach to matrix completion, *Journal of Machine Learning Research* 12, 3413–3430.

Recht, B., Fazel, M. and Parrilo, P. A. (2010), Guaranteed minimum-rank solutions of linear matrix equations via nuclear norm minimization, *SIAM Review* 52(3), 471–501.

Rennie, J. and Srebro, N. (2005), Fast maximum margin matrix factorization for collaborative prediction, in *Proceedings of the 22nd International Conference on Machine Learning*, Association for Computing Machinery, pp. 713–719.

Rish, I. and Grabarnik, G. (2014), *Sparse Modeling: Theory, Algorithms, and Applications*, Chapman and Hall/CRC.

Rockafellar, R. T. (1996), *Convex Analysis*, Princeton University Press.

Rohde, A. and Tsybakov, A. (2011), Estimation of high-dimensional low-rank matrices, *Annals of Statistics* 39(2), 887–930.

Rosset, S. and Zhu, J. (2007), Adaptable, efficient and robust methods for regression and classification via piecewise linear regularized coefficient paths, *Annals of Statistics* **35**(3).

Rosset, S., Zhu, J. and Hastie, T. (2004), Boosting as a regularized path to a maximum margin classifier, *Journal of Machine Learning Research* **5**, 941–973.

Rothman, A. J., Bickel, P. J., Levina, E. and Zhu, J. (2008), Sparse permutation invariant covariance estimation, *Electronic Journal of Statistics* **2**, 494–515.

Rubin, D. (1981), The Bayesian Bootstrap, *Annals of Statistics* **9**, 130–134.

Rudelson, M. and Zhou, S. (2013), Reconstruction from anisotropic random measurements, *IEEE Transactions on Information Theory* **59**(6), 3434–3447.

Ruderman, D. (1994), The statistics of natural images, *Network: Computation in Neural Systems* **5**, 517–548.

Santhanam, N. P. and Wainwright, M. J. (2008), Information-theoretic limits of high-dimensional model selection, in *International Symposium on Information Theory*, Toronto, Canada.

Santosa, F. and Symes, W. W. (1986), Linear inversion of band-limited reflection seismograms, *SIAM Journal of Scientific and Statistical Computing* **7**(4), 1307–1330.

Scheffé, H. (1953), A method for judging all contrasts in the analysis of variance, *Biometrika* **40**, 87–104.

Schmidt, M., Niculescu-Mizil, A. and Murphy, K. (2007), Learning graphical model structure using l1-regularization paths, in *AAAI proceedings*.
URL: *http://www.cs.ubc.ca/ murphyk/Papers/aaai07.pdf*

Shalev-Shwartz, S., Singer, Y. and Srebro, N. (2007), Pegasos: Primal estimated sub-gradient solver for SVM, in *Proceedings of the 24th international conference on Machine learning*, pp. 807–814.

She, Y. and Owen, A. B. (2011), Outlier detection using nonconvex penalized regression, *Journal of the American Statistical Association* **106**(494), 626–639.

Simon, N., Friedman, J., Hastie, T. and Tibshirani, R. (2011), Regularization paths for Cox's proportional hazards model via coordinate descent, *Journal of Statistical Software* **39**(5), 1–13.

Simon, N., Friedman, J., Hastie, T. and Tibshirani, R. (2013), A sparse-group Lasso, *Journal of Computational and Graphical Statistics* **22**(2), 231–245.

Simon, N. and Tibshirani, R. (2012), Standardization and the group Lasso penalty, *Statistica Sinica* **22**, 983–1001.

Simoncelli, E. P. (2005), Statistical modeling of photographic images, in *Handbook of Video and Image Processing, 2nd Edition*, Academic Press,

Waltham MA, pp. 431–441.

Simoncelli, E. P. and Freeman, W. T. (1995), The steerable pyramid: A flexible architecture for multi-scale derivative computation, in *Int'l Conference on Image Processing*, Vol. III, IEEE Sig Proc Soc., Washington, DC, pp. 444–447.

Singer, Y. and Dubiner, M. (2011), Entire relaxation path for maximum entropy models, in *Proceedings of the Conference on Empirical Methods in Natural Language Processing (EMNPL 2011)*, pp. 941–948.

Srebro, N., Alon, N. and Jaakkola, T. (2005), Generalization error bounds for collaborative prediction with low-rank matrices, *Advances in Neural Information Processing Systems* .

Srebro, N. and Jaakkola, T. (2003), Weighted low-rank approximations, in *Twentieth International Conference on Machine Learning*, AAAI Press, pp. 720–727.

Srebro, N., Rennie, J. and Jaakkola, T. (2005), Maximum margin matrix factorization, *Advances in Neural Information Processing Systems* **17**, 1329–1336.

Stein, C. (1981), Estimation of the mean of a multivariate normal distribution, *Annals of Statistics* **9**, 1131–1151.

Stone, C. J. (1985), Additive regression and other non-parametric models, *Annals of Statistics* **13**(2), 689–705.

Taylor, J., Lockhart, R., Tibshirani2, R. and Tibshirani, R. (2014), Post-selection adaptive inference for least angle regression and the Lasso. arXiv: 1401.3889; submitted.

Taylor, J., Loftus, J. and Tibshirani2, R. (2016), Inference in adaptive regression via the Kac-Rice formula, *Annals of Statistics* **44**(2), 743–770.

Thodberg, H. H. and Olafsdottir, H. (2003), Adding curvature to minimum description length shape models, in *British Machine Vision Conference (BMVC)*, pp. 251–260.

Thomas, G. S. (1990), *The Rating Guide to Life in America's Small Cities*, Prometheus books. `http://college.cengage.com/mathematics/ brase/understandable_statistics/7e/students/datasets/mlr/ frames/frame.html`.

Tibshirani, R. (1996), Regression shrinkage and selection via the Lasso, *Journal of the Royal Statistical Society, Series B* **58**, 267–288.

Tibshirani, R., Bien, J., Friedman, J., Hastie, T., Simon, N., Taylor, J. and Tibshirani2, R. (2012), Strong rules for discarding predictors in Lasso-type problems, *Journal of the Royal Statistical Society Series B*. pp. 245–266.

Tibshirani, R. and Efron, B. (2002), Pre-validation and inference in microarrays, *Statistical Applications in Genetics and Molecular Biology* pp. 1–15.

Tibshirani, R., Hastie, T., Narasimhan, B. and Chu, G. (2001), Diagnosis of multiple cancer types by shrunken centroids of gene expression, *Proceedings of the National Academy of Sciences* **99**, 6567–6572.

Tibshirani, R., Hastie, T., Narasimhan, B. and Chu, G. (2003), Class prediction by nearest shrunken centroids, with applications to DNA microarrays, *Statistical Science* pp. 104–117.

Tibshirani, R., Saunders, M., Rosset, S., Zhu, J. and Knight, K. (2005), Sparsity and smoothness via the fused Lasso, *Journal of the Royal Statistical Society, Series B* **67**, 91–108.

Tibshirani$_2$, R. (2013), The Lasso problem and uniqueness, *Electronic Journal of Statistics* **7**, 1456–1490.

Tibshirani$_2$, R. (2014), Adaptive piecewise polynomial estimation via trend filtering, *Annals of Statistics* **42**(1), 285–323.

Tibshirani$_2$, R., Hoefling, H. and Tibshirani, R. (2011), Nearly-isotonic regression, *Technometrics* **53**(1), 54–61.

Tibshirani$_2$, R. and Taylor, J. (2011), The solution path of the generalized Lasso, *Annals of Statistics* **39**(3), 1335–1371.

Tibshirani$_2$, R. and Taylor, J. (2012), Degrees of freedom in Lasso problems, *Annals of Statistics* **40**(2), 1198–1232.

Trendafilov, N. T. and Jolliffe, I. T. (2007), DALASS: Variable selection in discriminant analysis via the LASSO, *Computational Statistics and Data Analysis* **51**, 3718–3736.

Tropp, J. A. (2006), Just relax: Convex programming methods for identifying sparse signals in noise, *IEEE Transactions on Information Theory* **52**(3), 1030–1051.

Tseng, P. (1988), Coordinate ascent for maximizing nondifferentiable concave functions, Technical Report LIDS-P ; 1840, Massachusetts Institute of Technology. Laboratory for Information and Decision Systems.

Tseng, P. (1993), Dual coordinate ascent methods for non-strictly convex minimization, *Mathematical Programming* **59**, 231–247.

Tseng, P. (2001), Convergence of block coordinate descent method for nondifferentiable maximization, *Journal of Optimization Theory and Applications* **109**(3), 474–494.

van de Geer, S. (2000), *Empirical Processes in M-Estimation*, Cambridge University Press.

van de Geer, S. and Bühlmann, P. (2009), On the conditions used to prove oracle results for the Lasso, *Electronic Journal of Statistics* **3**, 1360–1392.

van de Geer, S., Bühlmann, P., Ritov, Y. and Dezeure, R. (2013), On asymptotically optimal confidence regions and tests for high-dimensional models. arXiv: 1303.0518v2.

van Houwelingen, H. C., Bruinsma, T., Hart, A. A. M., van't Veer, L. J. and

Wessels, L. F. A. (2006), Cross-validated Cox regression on microarray gene expression data, *Statistics in Medicine* **45**, 3201–3216.

Vandenberghe, L., Boyd, S. and Wu, S. (1998), Determinant maximization with linear matrix inequality constraints, *SIAM Journal on Matrix Analysis and Applications* **19**, 499–533.

Vapnik, V. (1996), *The Nature of Statistical Learning Theory*, Springer, New York.

Vempala, S. (2004), *The Random Projection Method*, Discrete Mathematics and Theoretical Computer Science, American Mathematical Society, Providence, RI.

Vershynin, R. (2012), Introduction to the non-asymptotic analysis of random matrices, in *Compressed Sensing: Theory and Applications*, Cambridge University Press.

Vu, V. Q., Cho, J., Lei, J. and Rohe, K. (2013), Fantope projection and selection: A near-optimal convex relaxation of sparse PCA, in *Advances in Neural Information Processing Systems (NIPS Conference Proceedings)*, pp. 2670–2678.

Vu, V. Q. and Lei, J. (2012), Minimax rates of estimation for sparse PCA in high dimensions, in *15th Annual Conference on Artificial Intelligence and Statistics*, La Palma, Canary Islands.

Waaijenborg, S., Versélewel de Witt Hamer, P. and Zwinderman, A. (2008), Quantifying the association between gene expressions and DNA-markers by penalized canonical correlation analysis, *Statistical Applications in Genetics and Molecular Biology* **7**, Article 3.

Wahba, G. (1990), *Spline Models for Observational Data*, SIAM, Philadelphia, PA.

Wainwright, M. J. (2009), Sharp thresholds for noisy and high-dimensional recovery of sparsity using ℓ_1-constrained quadratic programming (Lasso), *IEEE Transactions on Information Theory* pp. 2183–2202.

Wainwright, M. J. and Jordan, M. I. (2008), Graphical models, exponential families and variational inference, *Foundations and Trends in Machine Learning* **1**(1–2), 1–305.

Wainwright, M. J., Simoncelli, E. P. and Willsky, A. S. (2001), Random cascades on wavelet trees and their use in modeling and analyzing natural images, *Applied Computational and Harmonic Analysis* **11**, 89–123.

Wang, H. (2014), Coordinate descent algorithm for covariance graphical Lasso, *Statistics and Computing* **24**(4), 521–529.

Wang, J., Lin, B., Gong, P., Wonka, P. and Ye, J. (2013), Lasso screening rules via dual polytope projection, in *Advances in Neural Information Processing Systems (NIPS Conference Proceedings)*, pp. 1070–1078.

Wang, L., Zhu, J. and Zou, H. (2006), The doubly regularized support vector machine, *Statistica Sinica* **16**(2), 589.

Wang, W., Liang, Y. and Xing, E. P. (2013), Block regularized Lasso for multivariate multiresponse linear regression, in *Proceedings of the 16th International Conference on Artifical Intelligence and Statistics*, Scottsdale, AZ.

Welsh, D. J. A. (1993), *Complexity: Knots, Colourings, and Counting*, LMS Lecture Note Series, Cambridge University Press, Cambridge.

Whittaker, J. (1990), *Graphical Models in Applied Multivariate Statistics*, Wiley, Chichester.

Winkler, G. (1995), *Image Analysis, Random Fields, and Dynamic Monte Carlo methods*, Springer-Verlag, New York, NY.

Witten, D. (2011), *penalizedLDA: Penalized classification using Fisher's linear discriminant.* R package version 1.0.
URL: *http://CRAN.R-project.org/package=penalizedLDA*

Witten, D., Friedman, J. and Simon, N. (2011), New insights and faster computations for the graphical Lasso, *Journal of Computational and Graphical Statistics* **20**, 892–200.

Witten, D. and Tibshirani, R. (2009), Extensions of sparse canonical correlation analysis, with application to genomic data, *Statistical Applications in Genetics and Molecular Biology* **8**(**1**), Article 28.

Witten, D. and Tibshirani, R. (2010), A framework for feature selection in clustering, *Journal of the American Statistical Association* **105**(**490**), 713–726.

Witten, D. and Tibshirani, R. (2011), Penalized classification using Fisher's linear discriminant, *Journal of the Royal Statistical Society Series B* **73**(5), 753–772.

Witten, D., Tibshirani, R. and Hastie, T. (2009), A penalized matrix decomposition, with applications to sparse principal components and canonical correlation analysis, *Biometrika* **10**, 515–534.

Wu, T., Chen, Y. F., Hastie, T., Sobel, E. and Lange, K. (2009), Genomewide association analysis by Lasso penalized logistic regression, *Bioinformatics* **25**(6), 714–721.

Wu, T. and Lange, K. (2008), Coordinate descent procedures for Lasso penalized regression, *Annals of Applied Statistics* **2**(1), 224–244.

Wu, T. and Lange, K. (2010), The MM alternative to EM, *Statistical Science* **25**(4), 492–505.

Xu, H., Caramanis, C. and Mannor, S. (2010), Robust regression and Lasso, *IEEE Transactions on Information Theory* **56**(7), 3561–3574.

Xu, H., Caramanis, C. and Sanghavi, S. (2012), Robust PCA via outlier pursuit, *IEEE Transactions on Information Theory* **58**(5), 3047–3064.

Yi, X., Caramanis, C. and Sanghavi, S. (2014), Alternating minimization for mixed linear regression, in *Proceedings of The 31st International Conference on Machine Learning*, pp. 613–621.

Yuan, M., Ekici, A., Lu, Z. and Monteiro, R. (2007), Dimension reduction and coefficient estimation in multivariate linear regression, *Journal of the Royal Statistical Society Series B* **69**(3), 329–346.

Yuan, M. and Lin, Y. (2006), Model selection and estimation in regression with grouped variables, *Journal of the Royal Statistical Society, Series B* **68**(1), 49–67.

Yuan, M. and Lin, Y. (2007a), Model selection and estimation in the Gaussian graphical model, *Biometrika* **94**(1), 19–35.

Yuan, M. and Lin, Y. (2007b), On the non-negative garrotte estimator, *Journal of the Royal Statistical Society, Series B* **69**(2), 143–161.

Yuan, X. T. and Zhang, T. (2013), Truncated power method for sparse eigenvalue problems, *Journal of Machine Learning Research* **14**, 899–925.

Zhang, C.-H. (2010), Nearly unbiased variable selection under minimax concave penalty, *Annals of Statistics* **38**(2), 894–942.

Zhang, C.-H. and Zhang, S. (2014), Confidence intervals for low-dimensional parameters with high-dimensional data, *Journal of the Royal Statistical Society Series B* **76**(1), 217–242.

Zhang, Y., Wainwright, M. J. and Jordan, M. I. (2014), Lower bounds on the performance of polynomial-time algorithms for sparse linear regression, in *Proceedings of the Annual Conference on Learning Theory (COLT)*, Barcelona, Spain. Full length version at http://arxiv.org/abs/1402.1918.

Zhao, P., Rocha, G. and Yu, B. (2009), Grouped and hierarchical model selection through composite absolute penalties, *Annals of Statistics* **37**(6A), 3468–3497.

Zhao, P. and Yu, B. (2006), On model selection consistency of Lasso, *Journal of Machine Learning Research* **7**, 2541–2567.

Zhao, Y., Levina, E. and Zhu, J. (2011), Community extraction for social networks, *Proceedings of the National Academy of Sciences* **108**(18), 7321–7326.

Zhou, S., Lafferty, J. and Wasserman, L. (2008), Time-varying undirected graphs, in *Proceedings of the Annual Conference on Learning Theory (COLT)*, Helsinki, Finland.

Zhu, J., Rosset, S., Hastie, T. and Tibshirani, R. (2004), 1-norm support vector machines, in *Advances in Neural Information Processing Systems*, Vol. 16, pp. 49–56.

Zou, H. (2006), The adaptive Lasso and its oracle properties, *Journal of the American Statistical Association* **101**, 1418–1429.

Zou, H. and Hastie, T. (2005), Regularization and variable selection via the elastic net, *Journal of the Royal Statistical Society Series B.* **67**(2), 301–

320.

Zou, H., Hastie, T. and Tibshirani, R. (2006), Sparse principal component analysis, *Journal of Computational and Graphical Statistics* **15**(2), 265–286.

Zou, H., Hastie, T. and Tibshirani, R. (2007), On the degrees of freedom of the Lasso, *Annals of Statistics* **35**(5), 2173–2192.

Zou, H. and Li, R. (2008), One-step sparse estimates in nonconcave penalized likelihood models, *The Annals of Statistics* **36**(4), 1509–1533.

Author Index

Index

ACS, *see* alternate convex search
Adaptive hypothesis test, 157
Adaptive lasso, 86
Additive
    matrix decomposition, 190–194
    model, 69–76
ADMM, 121
    applied to lasso, 122
Aliased, 60
Alternate convex search, 126
Alternating
    algorithm, 205
    direction method of multipliers
      see ADMM, 121
    minimization, 124
      partial optimum, 126
    regression, 237
    subspace algorithm, 126
Analysis of deviance, 33
ANOVA, 68
Applications
    20-newsgroups corpus, 32
    air pollution, 71
    arterial pressure, 271
    comparative genomic hybridiza-
      tion (CGH), 76
    crime, 10
    diabetes data, 140, 149, 159
    face silhouettes, 226
    handwritten digits, 37, 209
    helicopter data, 184, 193
    image processing, 271
    lymphoma, 42, 219
    mammal dentition, 232
    natural images, 271
    Netflix challenge, 170, 187, 215
    splice-site detection, 60

    video denoising, 184
    voting, 244, 257
Augmented SPCA algorithm, 213
Autoencoder, 236
    sparse, 210, 236
Auxiliary variables, 79
Average linkage, 227

Backfitting, 69–72
Base class, 36
Baseline hazard, 43
Basic inequality, 298, 313
Basis functions, 71
    Haar, 270
    multiscale, 271
    orthogonal, 269
    overcomplete, 274
Basis pursuit, 23, 276
Bayes
    decision boundary, 217
    rule, 217
Bayesian, 23
    lasso, 139, 144
    methods, 22, 139
Bellkor's Pragmatic Chaos, 172
Benjamini–Hochberg (BH) procedure,
    163
Best-additive approximation, 69
Best-subset selection, 22
Bet-on-sparsity principle, 24
Bias term (intercept), 7
Bias-variance tradeoff, 7
Biclustering , 190
Biconvex
    function, 124, 189, 207
    set, 125
Biconvexity, 124

For Product Safety Concerns and Information please contact our EU
representative GPSR@taylorandfrancis.com
Taylor & Francis Verlag GmbH, Kaufingerstraße 24, 80331 München, Germany